PARTICLES
AND FIELDS

PARTICLES AND FIELDS

Sixth Mexican Workshop

Morelia, Michoacán, México November 1997

EDITORS

Juan Carlos D'Olivo
ICN UNAM, México

Martí Ruiz-Altaba
IF-UNAM, México

Luis Villaseñor Cendejas
IFyM-UMSNH, México

American Institute of Physics

AIP CONFERENCE
PROCEEDINGS 445

Woodbury, New York

Editors:

Juan Carlos D'Olivo
Instituto de Ciencias Nucleares
Universidad Nacional Autonóma de México
Apdo. Postal 70-543
04510, México D.F.
MÉXICO
E-mail: dolivo@nuclecu.unam.mx

Martí Ruiz-Altaba
Instituto de Física
Universidad Nacional Autonóma de México
Apdo. Postal 20-364
01000, México D.F.
MÉXICO
E-mail: marti@ft.ifisicacu.unam.mx

Luis Villaseñor Cendejas
Instituto de Física y Matemáticas
Universidad Michoacana de San Nicolás de Hidalgo
Edificio C-3 Cd. Universitaria
Apdo. Postal 2-82
58040, Morelia, Michoacán
MÉXICO
E-mail: villasen@zeus.ccu.umich.mx

Authorization to photocopy items for internal or personal use, beyond the free copying permitted under the 1978 U.S. Copyright Law (see statement below), is granted by the American Institute of Physics for users registered with the Copyright Clearance Center (CCC) Transactional Reporting Service, provided that the base fee of $15.00 per copy is paid directly to CCC, 222 Rosewood Drive, Danvers, MA 01923. For those organizations that have been granted a photocopy license by CCC, a separate system of payment has been arranged. The fee code for users of the Transactional Reporting Service is: 1-56396-791-X/ 98 /$15.00.

© 1998 American Institute of Physics

Individual readers of this volume and nonprofit libraries, acting for them, are permitted to make fair use of the material in it, such as copying an article for use in teaching or research. Permission is granted to quote from this volume in scientific work with the customary acknowledgment of the source. To reprint a figure, table, or other excerpt requires the consent of one of the original authors and notification to AIP. Republication or systematic or multiple reproduction of any material in this volume is permitted only under license from AIP. Address inquiries to Office of Rights and Permissions, 500 Sunnyside Boulevard, Woodbury, NY 11797-2999; phone: 516-576-2268; fax: 516-576-2499; e-mail: rights@aip.org.

L.C. Catalog Card No. 98-87299
ISBN 1-56396-791-X
ISSN 0094-243X
DOE CONF- 9711168

Printed in the United States of America

CONTENTS

Preface vii
Acknowledgments ix
Organizing Committee ix

COURSES AND LECTURES

I. Experiment

Heavy Ion Physics at CERN: Present and Future 5
    B. Alessandro and A. Marzari Chiesa
Past, Present, and Future Experiments at LEP 42
    M. Pepe Altarelli
Detection Techniques of Ultra High Energy Cosmic Rays 64
    H. Salazar
Precision Measurement of the Σ^0 Hyperon Mass 80
    M. H. L. S. Wang, E. P. Hartouni, M. N. Kreisler, J. Uribe,
    M. D. Church, E. E. Gottschalk, B. C. Knapp, B. J. Stern, L. R. Wiencke,
    D. C. Christian, G. Gutierrez, A. Wehmann, C. Avilez, J. Félix, G. Moreno,
    M. Forbush, F. R. Huson, and J. T. White

II. Phenomenology

The Standard Model and Beyond 93
    G. Altarelli
Cosmological Phase Transitions and Baryogenesis 143
    M. Quirós
The Highest Energy Cosmics Rays, Photons, and Neutrinos 185
    E. Zas
The Width and Height of Interferometry 228
    A. Ayala

III. Theory

Brane Engineering 243
    C. Gómez and R. Hernández
A Brief Introduction to Duality and D-branes 279
    N. Quiroz and B. Zwiebach

SEMINARS

Dirac Confinement of a Heavy Quark-Light Quark System $(Q,\bar{q})$ in
High Orbital Angular Momentum States 297
    M. A. Avila

Pre-Big-Bang in String Cosmology 301
    M. Borunda and M. Ruiz-Altaba
Braided Spin Groups 305
    A. Criscuolo, M. Rosenbaum, and J. D. Vergara
Quark Mass Ratios from Mixing Angles and Flavor Symmetry Considerations. 309
    M. de Coss and R. Huerta
A Model for a Dynamically Induced Large *CP*-Violating Phase. 312
    D. Delépine
Photon Dispersion Relations in a Nuclear Medium. 316
    J. C. D'Olivo and J. F. Nieves
Study of Λ_0 Polarization in Exclusive *pp* Reactions at 27.5 GeV/c 320
    J. Félix, C. Avilez, D. C. Christian, M. D. Church, M. Forbush,
    E. E. Gottschalk, G. Gutierrez, E. P. Hartouni, S. D. Holmes, F. R. Huson,
    D. A. Jensen, B. C. Knapp, M. N. Kreisler, G. Moreno, J. Uribe, B. J. Stern,
    M. H. L. S. Wang, A. Wehmann, L. R. Wiencke, and J. T. White
String Effective Potential with Massive Ends. 325
    G. Germán and Y. Jiang
τ Weak Magnetic Dipole Moment 332
    G. A. González-Sprinberg
$W^{\pm}$ and Z^0 Production at LEP/LHC Energies 336
    A. Gutiérrez and A. Rosado
Finite Soft SUSY Breaking Terms. 344
    T. Kobayashi, J. Kubo, M. Mondragón, and G. Zoupanos
Non-Noether Charges in Classical Mechanics 348
    J. L. Lucio M., A. Cabo, and V. M. Villanueva
Concerning *CP* Violation in 331 Models 353
    J. C. Montero, V. Pleitez, and O. Ravínez
The Dirac Equation as Spontaneously Broken Supersymmetry 357
    C. Ramírez
The CKM Matrix from a Scheme of Flavour Symmetry Breaking 361
    E. Rodríguez-Jáuregui and A. Mondragón
Regge Trajectories and the Renormalization Group. 365
    C. R. Stephens, A. Weber, J. C. López Vieyra, S. Dilcher, and P. O. Hess
Current Status of Finite Unified Models. 369
    L. E. Velasco-Sevilla
Classical and Path Integral Analysis of a Solvable Model with Gribov Problems. 373
    V. M. Villanueva, J. Govaerts, and J. L. Lucio M.
Calibration of Water Čerenkov Detectors for Ultraenergetic Cosmic Rays ... 378
    L. Villaseñor

List of Participants. 383
Author Index. 387

Preface

The Sixth Mexican Workshop on Particles and Fields (VI-TMPC) was the continuation of an enduring tradition in Mexico designed to promote the development of High Energy Physics within the scientific community, especially among students and young researchers. In alternate years, the Mexican School and the Mexican Workshop take place in different places of the Republic. On this occasion, the beautiful city of Morelia, Michoacán, hosted our event. A large number of students learned about the current status of High Energy Physics, from the interesting new results found at LEP to the latest developments in string theory. Besides courses and lectures, we had three round table discussions on Experiment, Phenomenology and Theory, respectively, which provoked a lively debate on the future of particle physics.

The VI-TPC was organized by the Division of Particles and Fields of the Mexican Physical Society. Most of the event was held in Spanish, but the Proceedings have been written in English. We have classified their contents in three largely intersecting areas: Experimental, Phenomenological, and Theoretical High Energy Physics. In addition, we have appended the written version of some of the seminars presented by the participants on their research work.

Financial support for the VI-TMPC was generously provided by various institutions (listed separately). We are grateful to our colleagues in the Organizing Committee (Francisco Astorga, Gabriel López-Castro, Fernando Quevedo, and Alfonso Rosado) for their tireless efforts before and during the workshop. We wish to extend our thanks also to Augusto García, who chaired the round table on phenomenology, to Matías Moreno, whose concern for funding for the students remains unabated, and most particularly to Trinidad Ramírez, whose organizing efficiency assured the success of the meeting. The valuable help given by Dina Zamudio is also acknowledged. Last, but not least, we would like to thank the speakers for their excellent lectures and prompt delivery of their written contributions.

Juan Carlos D'Olivo
ICN-UNAM

Martí Ruiz-Altaba
IF-UNAM

Luis Villaseñor
IFyM-UMSNH

Acknowledgments

The VI-TMPC was sponsored by the following institutions:

- CENTRO DE INVESTIGACIÓN Y DE ESTUDIOS AVANZADOS
    - Dirección General
    - Departamento de Física (Zacatenco)
- CENTRO LATIONAMERICANO DE FÍSICA, MÉXICO
- CONSEJO NACIONAL DE CIENCIA Y TECNOLOGÍA, MÉXICO
- UNIVERSIDAD DE GUANAJUATO
- UNIVERSIDAD MICHOACANA DE SAN NICOLÁS HIDALGO
- UNIVERSIDAD NACIONAL AUTÓNOMA DE MÉXICO
    - Coordinación de la Investigación Científica
    - Coordinación General de Estudios de Posgrado
    - Dirección General de Asuntos del Personal Académico
    - Dirección General de Intercambio Académico
    - Instituto de Ciencias Nucleares
    - Instituto de Física
    - Facultad de Ciencias

Organizing Committee

ASTORGA, Francisco
D'OLIVO, Juan Carlos
LÓPEZ-CASTRO, Gabriel
QUEVEDO, Fernando
ROSADO, Alfonso
RUIZ-ALTABA, Martí
VILLASEÑOR, Luis

COURSES AND LECTURES

I. Experiment

Heavy Ion Physics at CERN: present and future

Bruno Alessandro and Alberta Marzari Chiesa

I.N.F.N. and Dipartimento di Fisica Sperimentale dell'Università di Torino, Italy

Abstract. After a general introduction on the very high energy heavy ion interactions, the CERN heavy ion program is presented. Three CERN experiments are described in details: NA38/50 (J/Ψ suppression), NA45/CERES (e^+e^- production) and WA85/97 (multi-strange particle production). The ALICE experiment, to be built in the next years and foreseen at the CERN Large Hadron Collider (LHC) is also extensively described.

INTRODUCTION

According to Quantum Chromodynamics (QCD), a transition from hadronic matter to a plasma of quarks and gluons should occur when nuclear matter is compressed to a sufficiently high density and temperature [1], [2]. The QCD potential, which in vacuum increases linearly with distance giving rise to a strong attractive force that confines quarks and gluons into hadrons, in dense matter goes to zero at large distances and consequently quarks and gluons are free. The condition for the transition from hadronic matter (HM) to the quark-gluon plasma (QGP) is that the density of the constituents is sufficiently high. Naively speaking, it is impossible to define $q\bar{q}$ or qqq as a specific hadron, since in any hadronic volume there are many other possible partners [3].

QCD lattice calculations predict this transition and provide estimates for critical values ranging from 140 to 180 MeV for the temperature and $2-3$ GeV/fm^3 for the energy density [4].

The QGP was the state of the matter in the first instants of the Universe (for time $t < 10^{-6}$s after the "big-bang"); it is possible that QGP is present in the core of neutron stars. In the laboratory, QGP can be obtained, as a transient state, by means of very high energy ion-ion collisions.

We are confident to have high energy densities in nucleus-nucleus collisions at high energy because in these interactions the degree of slowing-down (i.e. the stopping power) is quite large. In fact, in a central nucleus-nucleus collision there are many nucleon-nucleon interactions, and it was measured that in a p-p interaction at 100 GeV/c the fraction of the incident momentum carried out by the outgoing proton has a mean value $<x> \approx 0.6$, and therefore 40% of the incident momentum is lost [1].

Moreover, relativistic heavy ion collisions are the only mean to create a strongly interacting system which can be studied in thermodynamical terms. To do so, the system under study has to consist of many particles, so that macroscopic variables can apply; it must have a size much larger than the mean free path of the constituents ($\approx 0.5 fm$ for quarks at densities of ≈ 2 GeV/fm^3), since several collisions per particle must occur, and finally have large energy density. Collisions of nuclei at ultrarelativistic energies have proven to fulfill these conditions: the system created in a Pb-Pb collision has a volume of the order of 1000 fm^3, consists of ≈ 1000 particles, shows clear evidence of rescattering (more than one collision/particle) and, at SpS energies, has an energy density ≈ 20 times larger than in a nucleus and ≈ 4 times larger than in a hadron.

High energy ion beams have been available for experiments since 1986 both at CERN and Brookhaven: ^{16}O and ^{32}S nuclei were accelerated up to a momentum per nucleon of 200 GeV/c at CERN till 1992, while a beam of ^{28}Si at 14.5 GeV/c was available at AGS. Accelerators were upgraded in order to produce heavy ion beams, leading up in 1993 to the gold beam of 12 A·GeV at the AGS and finally in 1995 to the Lead beam of 158 A·GeV at the SpS.

GLOBAL FEATURES

Geometry and kinematics

A schematic picture of a collision between two relativistic nuclei is shown in fig. 1. The incoming nuclei are Lorentz-contracted: their transverse size is equal to the nuclear section, while their thickness is ≈ 1 fm. Given an impact parameter b, the nucleons can be separated into *participants*, which undergo primary nucleon-nucleon collisions, and *spectators*, which continue along their original direction with modest perturbation. Collisions with $b \approx 0$ are defined *central*. Collisions with $b \approx r_1 + r_2$ are defined *peripheral*.

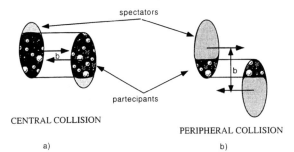

FIGURE 1. *Schematic representation of nucleus-nucleus central and pheripheral collisions*

In the laboratory system most secondary particles are emitted in the forward direction, but spectators and participants behave differently; spectators are emitted at very forward angles: $\theta_{spect} \approx p_{Fermi}/p_{beam} \approx 100$ MeV/200 GeV ≈ 0.5 mrad, while participants interact producing secondary particles which are emitted at larger angles: $\theta_{partic} \approx p_x/p_z \approx 500$ MeV/$\beta\gamma m \approx 500$ MeV/10 GeV ≈ 50 mrad. They can therefore be separated with a detector covering only the very forward direction (ZDC=Zero Degree Calorimeter).

To describe the final state, since secondary particles can be classified as target fragments, beam fragments or particles produced in collisions of the participant nucleons, it is important to define a variable which transforms additively under a Lorentz transformation, because for such a variable the shape of the distribution is invariant going from a reference system to another. This variable is the *rapidity* of the particles, defined as:

$$y = 0.5 \, ln[(E + p_z)/(E - p_z)] \tag{1}$$

or its approximation for relativistic momenta, the *pseudorapidity*:

$$\eta = -ln \, tan(\theta/2) \tag{2}$$

The beam and target fragments will be found close to the beam and target rapidities, y_P and y_T respectively, while the particles produced in the collision at large angles will populate the central rapidity region. The greater the incident energy, the greater the separation between y_P and y_T.

Transverse Energy distribution and estimate of Energy Density

The energy which is lost by the incident nucleus reappears mainly in the form of many "soft" mesons. It is convenient to describe the many particles produced by a global variable which is directly measurable and closely related to the energy density produced. This is the transverse energy defined by:

$$E_T = \sum E_i sin\theta_i$$

where i is summed over all the particles and θ_i is the emission angle of particle i in the laboratory reference frame.

So far, many experiments have measured the E_T distributions for several systems, using different beams, energies, projectiles and targets (fig. 2). The shape of the distributions is always the same, even if the maximum E_T reached is quite different.

The shape in fact reflects the geometry of the interaction: it can be reproduced calculating the probability distribution of the number of possible nucleon-nucleon interactions, calculated as a function of the impact parameter from the overlap

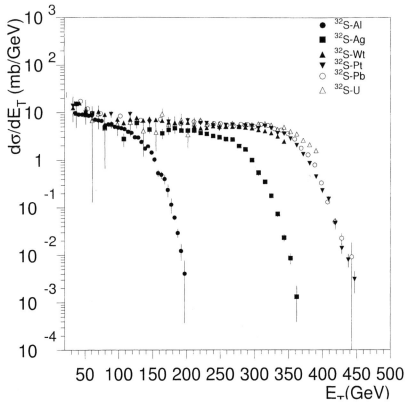

FIGURE 2. *Transverse energy differential cross-section d σ/dE_T for different targets*

integral of the two interacting nuclei : $\Omega = \int \rho_1 \rho_2 dS$ (ρ_i=nuclear density distributions). Simple geometrical considerations allow thus to evaluate cross sections, number of participants and related kinematical quantities as shown in fig. 3.

Since only the stopped energy is significant, and this is re-emitted isotropically, the energy density ε can be estimated through the measurement of the transverse energy, making geometrical and dynamical assumptions. Two approaches can be taken, based on opposite hypothesis. The first one assumes that, as a consequence of the interaction between many nucleons, a "fireball" has been formed, which then explodes isotropically; the second one is based on the Bjorken hydrodynamical model [6], valid for an ultra-relativistic regime. At SpS energies, both models are inadequate to fit the situation: the rapidity distribution is not flat like it should be in the ultrarelativistic regime, nor bell-shaped like it should be in the low energy one. We are therefore aware of the fact that we can estimate only roughly the energy density.

- **Low energy regime: the Fireball model**

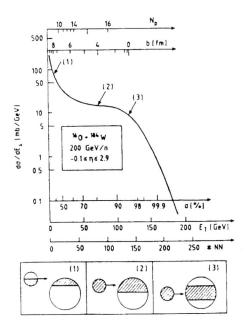

FIGURE 3. Fit to the E_T cross section from NA34 data with a geometrical parametrization based on indipendent nucleon-nucleon collisions. For a given E_T the model gives the number NN of nucleon-nucleon collisions, the number N_p of projectile participants, the impact parameter b and the fraction of the total geometrical cross section σ below a given E_T

In this model the transverse energy attained in the interaction is compared with the maximum observable E_T. The maximum energy in the center of mass system (c.m.s.) for an interaction between n_P projectile and n_T target nucleons is

$$E^{max} = \sqrt{s} - m(n_P + n_T)$$

If this energy is emitted isotropically then:

$$E_T^{max} = \frac{E^{max} \int sin\theta d\Omega}{\int d\Omega} = \frac{\pi}{4} \cdot E^{max}$$

The ratio between the maximum experimentally observed E_T and E_T^{max} gives the so-called "stopping power". The maximum energy density will be

$$\varepsilon_{F.B.}^{max} = E_T^{max}/V$$

The volume V ($V = \pi R_o^2 n_P^{2/3} \cdot 2R_o n_T^{1/3}$) is the cilinder (see fig. 4) cut by the projectile in the target nucleus. It must be divided by the γ factor to take into account the Lorentz contraction: $\varepsilon_{F.B.}^{max} = \gamma E_T^{max}/V$. Assuming a linear dependence of ε on E_T:

$$\varepsilon_{F.B.} = \varepsilon_{F.B.}^{max} E_T^{exp}/E_T^{max} = E_T^{exp}/V \qquad (3)$$

The energy density calculated in this way is in general overestimated at SpS energies (where the situation is far from the hypothesis of the model): in $O - W$ at 200 GeV per nucleon $\varepsilon_{F.B.} \approx 4.8 GEv/fm^3$!

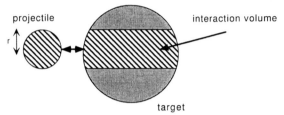

FIGURE 4. *Sketch of interaction volume in ion-ion collision*

- High energy regime: the Bjorken hydrodynamical model

The two interacting nuclei, in their c.m.s., appear strongly contracted because of their relativistic speed: after the interaction these disks will move in opposite directions with speed $\approx c$, and the region between them, free of baryons, will expand cylindrically. The energy will be deposited in the "interaction volume", defined, in the hypothesis of a central collision between two nuclei of radius R separated from a distance d, as the cylinder $\pi R^2 \cdot d$.

Particles (mainly pions) are not produced immediately: the production time is estimated to be $\tau_o \approx 1 fm/c$. In a slice Δz around $z = 0$, at $t = \tau_o$ there are particles with $v_z \leq \Delta z/\tau_o = \Delta v_z$. The energy carried by these particles will be $\Delta E \approx \Delta E_T$. The energy density can therefore be calcuated as follows:

$$\varepsilon_{Bj} = \Delta E/V = \frac{\Delta E_T}{s \cdot \Delta z} = \frac{\Delta E_T}{s \tau_o \Delta v_z}$$

where s is the intersecting area of the two nuclei. Since $\lim_{p \to 0} y = \frac{v}{c}$ we get

$$\varepsilon_{Bj} = \frac{1}{sc\tau_o} \frac{\Delta E_T}{\Delta y} \qquad (4)$$

The energy density, calculated in this way for S-S and S-Au collisions at 200 AGeV is 1.3 and 2.6 GeV/fm^3 while for Pb-Pb collisions at 158 AGeV a value of 3.2 GeV/fm^3 has been obtained. Pb-Pb collisions at 158 AGeV provide up to now the highest initial energy density in the largest volume.

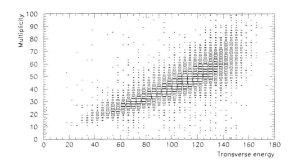

FIGURE 5. *Correlation between multiplicity and transverse energy.*

Multiplicity distribution

Transverse energy and (charged) multiplicity distributions are strongly correlated, as can be seen in fig. 5 [7]. These measurements, if performed at the same time, allow to estimate the transverse energy per charged particle, and hence the mean p_t as a function of the energy density. A behaviour typical of the phase transition, was looked for also in the first generation SpS experiments. All the results obtained till now are in favour of a $<p_t>$ nearly independent of ε (and E_t). The multiplicity gives therefore the same information as the transverse energy, and it can be used as a measurement of the centrality.

Baryon content

The baryon content of the interaction volume depends on the degree of stopping, therefore on the energy and on the mass number A of the interacting nuclei. In case of full-stopping we will have a baryon rich hadron gas, since all the participant nucleons are stopped, while at very high energy ($E \sim$ TeV) the slowed down baryons after the collision can still have enough momentum to proceed forward, and move away from the region of collision, so that the central rapidity region is populated only by mesons. At AGS energies the degree of stopping is very high: in Au-Au collisions at 14.5 AGeV the number of nucleons in the interaction volume can be as high as 400, while the number of pions is ~ 700 and $\frac{dN_{ch}^\pi}{dy} \sim 150$. At SpS energies, in Pb-Pb central collisions, the number of nucleons in the interaction volume is less than a hundreds and $\frac{dN_{ch}}{dy} \sim 500$.

Temperature

The temperature can be extracted from the transverse momentum spectra, which, for 0.2 GeV/c < p_T <1 GeV/c, exibit a typical thermal spectrum

$$f(p_t) \propto exp(-m_t/T)$$

where m_T is the transverse mass, defined as $m_T = \sqrt{P_t^2 + m^2}$. Lower p_T are polluted by resonance decays, and larger ones by initial state scattering. Measuring the inverse slope of the m_T distribution, one can obtain T.

Global features: summary

Summarizing what has been said until now, we look for QGP in relativistic heavy ion interactions because they consist of many nucleon-nucleon collisions, and the energy lost in each of these collisions is high. We are interested in *central* collisions because these are the interactions in which the energy density is the highest, and therefore we need a good measurement of the impact parameter. This can be done using the very forward energy (through a Zero Degree Calorimeter), the transverse energy and the charged multiplicity.

The transverse energy (or the multiplicity) per rapidity unit allows to estimate the energy density reached in the interaction. Unfortunately one always needs a model: in heavy ion interactions two models are usually used, based on opposite hypothesis and therefore valid in different situations: the Fireball model and the Bjorken model, valid in a ultrarelativistic regime. At SpS energies both models are inadequate, but traditionally ε is calculated with the Bjorken model. This is useful at least when different situations are compared. More sophisticated models [8] can be used when a more realistic evaluation is needed.

From the measurements done until now, at SpS energies, on systems going up to Pb-Pb, it seems that the energy densities predicted for the deconfinement have already been reached.

CERN SPS HEAVY ION EXPERIMENTS

Space-time evolution and Signatures of QGP

The space-time evolution of a central collision at high energy is shown in fig. 6. In the first instants, nucleon-nucleon collisions redistribute a fraction of the original energy into other degrees of freedom, materializing into quarks and gluons after a time $\tau_o \approx 1$ fm/c (formation time). Parton-parton interactions might then lead to the formation of a QGP. If created, the QGP would rapidly cool down via expansion and evaporation, going through a "mixed phase" in which hadrons and "blobs" of plasma would coexist, and finally condensate into a state of ordinary hadrons, first interacting and finally decoupling (freezout). The final state reflects the complex evolution of the system, and different observables carry information on its different aspects and stages.

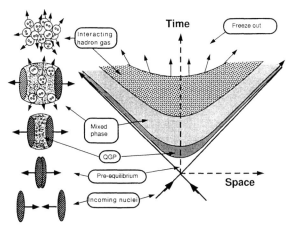

FIGURE 6. *Space-time evolution of a collision between two ultra-relativistic heavy ions*

Initial conditions. The measurement of global event features is necessary to specify the initial conditions. As explained before, from E_{ZDC}, E_T and multiplicity one can derive, in a more or less model-dependent way, the impact parameter and the energy density.

Thermodynamic variables. The *phase diagram* of hot hadronic matter can be derived from the measurement of energy density and temperature. A phase transition involving a large latent heat would manifest itself in a characteristic shape of the T vs ε dependence: the T would first grow with ε, then remain constant while the additional energy goes into the latent heat, and finally grow again. The problem in this case is that the secondary particles are in general strong interacting and hence the final state does not keep the information about the initial one. Dileptons (originated by real or virtual photons) which decouple at early times, represent the black body thermal radiation of the system. They are the most direct measure of the system temperature, but are extremely difficult to measure. The space-time evolution of the system, and in particular the freeze-out volume, can be measured via identical particle interferometry (HBT). Finally, multiplicity fluctuations could signal the critical phenomena linked to a phase transition.

Chiral symmetry restoration. In QGP, the quarks should lose the effective mass which they carry when confined in a hadron, and recover their "bare" mass. In other words, with the quark masses small and almost equal, the QGP would be chirally symmetric. This should manifest itself in a change in strangeness production (strangeness enhancement), and in changes of the hadron masses. Mass, width and decays of particles such as ρ, ω and Φ should undergo sharp modifications.

Deconfinement. As a sufficiently dense ordinary plasma can prevent the formation of atoms by electromagnetic screening of nuclei from electrons (the "Mott transition"), so would a QGP screen the colour force between quarks. Ordinary mesons, being bound states of (abundantly produced) light quarks, would not suffer such suppression. But the heavy charm-anticharm pairs necessary to make J/Ψ or Ψ' mesons are produced only rarely, in very hard quark collisions that can happen only in the initial instants of a nuclear collision. If such a pair is eventually produced in a state of deconfined quarks and gluons, ideally a QCD plasma, the evolution of an initial $c\bar{c}$ pair towards its final J/Ψ or Ψ' state could be blocked. A suppression of J/Ψ and Ψ' in central nuclear collisions, in comparison with the yield derived from the first generation nucleonic collisions, was thus expected in a dynamics proceeding via a deconfinement phase [9]. Of course, several competing effects exist, e.g. absorption, and a very accurate analysis would be required to isolate them from the melting.

Other observables have been proposed to look for the QGP: anomalous multiplicity fluctuations, plasma droplets, massive photons (thermal mass: in QGP $m_\gamma \approx 20 MeV$), strangelets (stable matter with equal u, d, s content).

Brief review of CERN SpS Heavy Ion Experiments

At CERN, several experiments took data at the SpS, with O and S beams until 1992, and with a Pb beam after 1994. All the Pb experiments can be seen as second generation experiments since they are either upgrades of experimental apparata used with the lighter nuclei, or are new projects, like NA49, based on the experience acquired in the first years of experimental physics with ultrarelativistic nuclear collisions.

Two strategies have been used to cope with the enormous particle density implied by collisions of Pb ions on a fixed target: the first one has been used by NA49 apparatus, the only general purpose experiment active with the Pb beam. It consists of a very high track resolution maintained over large volumes ($40 m^3$ TPC, 180000 channels) in order to achieve the goal of tracking ~ 1400 charged particles per event, resulting in a raw data volume of 8 Mbyte after zero suppression. The NA49 experiment is a charged hadron exclusive 4π spectrometer, capable of identifying both charged hadrons as $\pi^+, \pi^-, K^+, K^-, p, \bar{p}$ and neutral strange particles as K^o, Λ and $\bar{\Lambda}$. Moreover, it is the only experiment being sensitive to any observable that may show non statistical event-by-event fluctuations. The second strategy has been used by the other Collaborations, that have built very selective apparata which are sensitive to a particular signal. For instance, WA85/97 studies hyperon production at the *Omega* facility, NA45/CERES studies electron pair production at low mass, NA44 is a focussing spectrometer devoted to the study of hadronic correlations, NA50 studies muon pair production with high acceptance for J/Ψ and Ψ' vector mesons, seen in their $\mu^+\mu^-$ decay channel, WA98 studies direct photon production and NA52 looks for exotics charge to mass ratios in the strangelets search.

New data provided by these experiments are copious: in this paper we will report on three of them, namely NA38/NA50, NA45/CERES and WA85/97.

QGP Signatures: charmonium suppression

The original idea of J/Ψ suppression was proposed by Matsui and Satz in 1986 [9] considering quark deconfinement. The potential binding $c\bar{c}$ to J/Ψ in vacuum is given by:

$$V(r) = \sigma r - \frac{\alpha}{r} \qquad (5)$$

where σ is the string tension, r the separation between quark and antiquark, and the term α/r describes the Coulomb part. In a medium, the presence of other colour charges leads to a screening of the colour force and the potential becomes:

$$V(r) = \sigma r \left(\frac{1-e^{-\mu r}}{\mu r}\right) - \frac{\alpha}{r} e^{-\mu r} \qquad (6)$$

where $\mu = 1/r_D$ the inverse of the colour screening radius is related to the temperature of the system ($\mu \propto \sqrt{T}$).

In the case in which the state is deconfined, the first term of the potential disappears because $\sigma = 0$ while the second term has a finite range of interaction of the order r_D. The potential binding the $c\bar{c}$ in a QGP state becomes

$$V(r)_{QGP} = -\frac{\alpha}{r} e^{-\mu r} \qquad (7)$$

When the temperature of the system increases and μ is so large that $r_D < r_{J/\Psi}$ the range of the interaction is so small that prevents the formation of the $c\bar{c}$ bound states. The temperature predicted for this effect is $T \approx 200$ MeV.

After the plasma phase, at the hadronisation time, there are a few $c\bar{c}$ pairs and the c and $\bar{c}$ are usually too far to bind together. Since the quarks u, d and s are present in large number, the c and $\bar{c}$ quarks tend to form with them open charm hadrons. Therefore a deconfining medium suppresses the J/Ψ and somewhat enhances the D and $\bar{D}$ (open charm mesons) production.

In fig. 7 the r_D value as a function of T is compared with the J/Ψ and Ψ' binding radius (0.30 fm and $\approx$ 0.6 fm respectively). It is evident that the Ψ' is expected to be suppressed at energy densities lower than those necessary for J/Ψ suppression.

The J/Ψ production in heavy ion collisions was studied since 1986 by the NA38 collaboration, with Oxigen and Sulphur beams [10]. When the Lead beam was available at the CERN SpS, the NA38 apparatus was improved, and the collaboration was enlarged: the J/Ψ (and Ψ') production in Pb-Pb interactions is studied by the NA50 collaboration. To have a comparison with a system in which QGP is not present, these experiments took also data with proton beams.

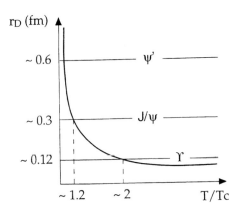

FIGURE 7. *Colour screening radius r_D as a function of the temperature compared with the J/Ψ and Ψ' binding radius*

The J/Ψ production is studied by detecting its decay in two muons: the apparatus is optimized to detect and measure the muon pairs produced in the collision when the $\mu^+\mu^-$ invariant mass is around 3 Gev/c^2. These experiments must take data at very high intensity ($\approx 10^7$ Pb/s) because the J/Ψ production cross section is around 4500 μb and the $J/\Psi(\Psi')$ branching ratio in two muons is very little ($J/\Psi \to \mu^+\mu^- \approx 6 \cdot 10^{-2}$; $\Psi' \to \mu^+\mu^- \approx 7 \cdot 10^{-3}$). This imposes severe conditions of counting rate and radiation damage.

In fig.8 the NA50 apparatus is shown. It consists of a multiple active target, an electromagnetic calorimeter, a ZDC calorimeter, a multiplicity detector, a dimuon spectrometer and several beam counters. The target system is composed of several subtargets surrounded by quarz blades, that allow to identify the subtarget in which the first interaction occurred, and to tag the re-interactions. In Na38 the subtargets were surrounded by cylindrical scintillators, with the same purpose. The electromagnetic calorimeter is made of scintillating fibers imbedded in lead and divided in 30 cells. It measures the released transverse neutral energy which is connected, as explained before, with the centrality of the collision and with the energy density reached in the interaction. The centrality can also be measured with a Zero Degree Calorimeter (ZDC) and with a multiplicity detector (MD), which measures also the angular distribution of the charged secondaries. ZDC and MD were not present in NA38. The spectrometer is composed of a beam dump to absorb hadrons, four multiwire proportional chambers (MWPC) before a toroidal magnet and four MWPC after the magnet. This system allows to measure the momenta and the sign of the crossing particles (muons). Four scintillator hodoscopes are used to trigger the data aquisition system.

In fig. 9 a $\mu^+\mu^-$ invariant mass spectrum (including background) for Pb-Pb collisions is presented [11]. The spectrum is the superposition of contributions

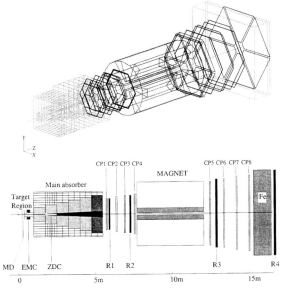

FIGURE 8. *NA50 setup*

coming from different origins: J/Ψ and Ψ' resonances, Drell-Yan and background pairs, and a small contribution from the semi-leptonic decay of D and $\bar{D}$ mesons. From this spectrum the relative yield for J/Ψ production can be obtained. To avoid systematic errors, the J/Ψ yield is sometimes presented normalized to the underlying Drell-Yan production cross section, which was experimentally proved to be independent of the collision system.

In fig. 10 we show the results for different projectiles and targets: all the cross sections measured at 450 GeV/c are rescaled to 200 GeV/c. All the points, from p-Be to S-U fit well with a power low $(AB)^\alpha$, giving $\alpha_{J/\Psi} = 0.92 \pm 0.015$. Since $\alpha-1$ is foreseen for a hard process like the J/Ψ production, this means that there is a "J/Ψ suppression".

Explanations of this "suppression", in term of normal hadronic matter, were suggested. The most successfull one is the following, proposed in 1991 by Gerschel and Hufner [12]. The cross section for J/Ψ production in an interaction between a projectile nucleus A and a target nucleus B can be written according to the Glauber approach [1] as:

$$\sigma_{AB \to \Psi} = A_{proj} A_{target} \cdot \sigma_{pN \to \Psi} e^{-L\rho_o \sigma_{abs}}$$

where the exponential represents the absorption, ρ_o is the nuclear matter density, σ_{abs} is the J/Ψ-nucleon absorption cross-section and L is the average length crossed by the J/Ψ inside the nuclear matter. L depends on the impact parameter and can be calculated with a simple geometrical model.

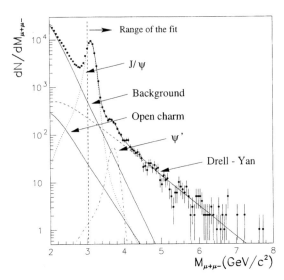

FIGURE 9. $\mu^+\mu^-$ *invariant mass spectrum (including background) in Pb-Pb collisions measured by NA50 experiment*

The ratio $J/\Psi/Drell - Yan$ as a function of L, obtained by NA38 for different systems, is shown in fig. 11: all the data, from p-Be to S-U, are very well described by the previous formula, with a value of σ_{abs} of 6.2 ± 0.6 mb. Now, since in $p - A$ collisions the plasma state is not formed, and since the exponential describing the absorption is the same in proton-nucleus and in nucleus-nucleus interactions, the conclusion was that no QGP is necessary to explain the J/Ψ suppression in heavy ion collisions.

The problem is that the absorption cross section found in this fit is too far from the theoretical prediction (2-3 mb): the absorbed object cannot be the J/Ψ, but some "pre-hadronic" state.

The explanation was proposed by Kharzeev and Satz [15], following a result of Braaten [13], who studied the quarkonium production to explain the "anomalous" charmonium production cross section found by CDF and D0 at Fermilab [14].

The main mechanism for J/Ψ production in hadronic collisions is gluon fusion, (see fig. 12) but the whole process can be divided in three phases:

- Two gluons interact in a hard process to form a $c\bar{c}$ pair that is a colour octet

- this state neutralizes its colour radiating or absorbing a gluon and hence becoming a colour singlet

- the c and $\bar{c}$, being at a distance $\approx r_{J/\Psi}$ bind together forming a J/Ψ.

Each phase is characterized by a specific time and it can be shown that the object crossing the nuclear matter is the colour octet. In other words, the produced $c\bar{c}$

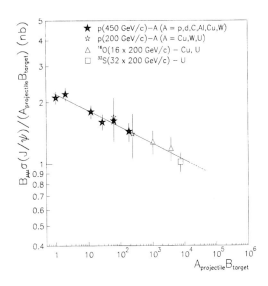

FIGURE 10. *J/Ψ suppression in proton-nucleus and nucleus-nucleus collisions*

pair is not immediately a colour singlet state, but it consists of several terms:

$$| \Psi > = a \,|\, (c\bar{c})_1 > + b \,|\, (c\bar{c})_8 g > + ... \qquad (8)$$

in which every state is a colour singlet, also if $c\bar{c}$ is not neutral. The second term becomes important in an extended strong interacting medium.

In this model, called "Colour Octet Model", the $(c\bar{c})_8$ crosses the nuclear matter together with a collinear gluon that neutralizes its colour. This pre-resonant state $c\bar{c}g$ can either be absorbed by the surrounding matter or it can happen than after a time $\tau_8 = 1/\sqrt{2 m_c E_g}$ (E_g is the energy of emitted or absorbed gluon), the $(c\bar{c})_8$ absorbs the gluon thus becoming a colour singlet $(c\bar{c})_1$ and subsequently the J/Ψ meson (see fig. 13). Kharzeev and Šatz [15] have estimated that the effective cross section of this state in ground state hadronic matter is $\sigma_{abs} \approx$ 6-7 mb, in good agreement with the result of the phenomenological fit to the experimental data.

In conclusion, all the data obtained by NA38 on J/Ψ suppression can be explained by this absorption model: up to S-U no evidence of the existence of QGP has been found by this experiment.

For Ψ' the situation is different, but again the results can be explained in terms of hadronic matter. In p-nucleus interactions the ratio Ψ'/Ψ is constant (fig. 14): this can be explained with the Colour Octet Model because what is absorbed is a state $c\bar{c}g$ before the formation of the resonance. In S-U the Ψ' is more suppressed than J/Ψ. For this it is necessary to add a further mechanism, in a later stage of the collision, when the final Ψ' is formed and interacts with the "comovers". These

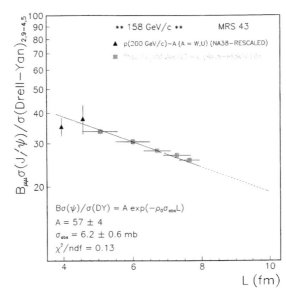

FIGURE 11. J/Ψ suppression in proton-nucleus and nucleus-nucleus collisions as a function of L

"comovers" are the particles, like pions, created in the collision. Some of them can have enough energy to destroy the loosly bound Ψ', but they never have enough energy to destroy the J/Ψ.

Now we come to the NA50 results, that is to Pb-Pb interactions. In these interactions the mean energy density is not much higher than in S-U; however, in a localized but rather large region of space, the Pb-Pb system has an energy density which is higher than the maximum value of the S-U system ($\approx 40\%$ higher) [8].

This change may be very important as it can help to bring the system above the critical energy density needed to observe the phase transition. The results obtained by NA50 seem indeed to indicate [11] that in the core of the interaction volume the energy density was sufficient to create a partonic, rather than a hadronic state, co-travelling with the emerging $J/\Psi(\Psi')$. For the J/Ψ in fact, only the bin corresponding to the "peripheral" events falls onto the 6.2 mb absorption line, while the other Pb-Pb data fall well below this line, with increasing centrality (see fig. 15). An additional break-up mechanism thus appear to set in with central $Pb-Pb$ collisions. The fact that something new happens to the J/Ψ while Ψ' is suppressed as in $S-U$ was explained in different ways. Capella [16] justified this effect through more efficient co-movers, present in this more dense medium, which can destroy also the J/Ψ. It is however difficult to explain this different behaviour between $S-U$ and $Pb-Pb$: to justify all the results, from $p-A$ to $Pb-Pb$, new parameters must be introduced. Kharzeev and Satz [17], Blaizot et al. [8] propose that in the central region of $Pb-Pb$ interactions a partonic state, co-travelling with the emerging J/Ψ

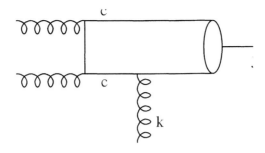

FIGURE 12. J/Ψ production mechanism through the colour singlet $c\bar{c}$

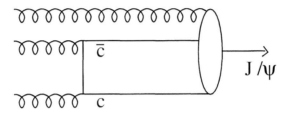

FIGURE 13. J/Ψ production mechanism through the colour singlet $c\bar{c}g$

has been formed, producing the Debye screening which can explain this degree of suppression, far beyond the 6 mb of pre-hadronization breakup.

Electron pair production : NA45/CERES experiment

A baryon free Quark Gluon Plasma has an high (and equal) density of quarks and antiquarks: it is therefore a "natural" environment for high rate production of lepton pairs (e^+e^- or $\mu^+\mu^-$) via $q-\bar{q}$ annihilation. The mean free path of the leptons are expected to be quite large and the leptons are not likely to suffer further collisions after they are produced. On the other hand, the production rate and the momentum distribution of the l^+l^- pair depend on the momentum distribution of the q and $\bar{q}$ in the plasma, which are governed by the temperature. They carry therefore information on the thermodynamic state of the medium at the moment of their production. In fact, the invariant mass distribution of the dileptons from $q\bar{q}$ annihilation approximately behaves as:

$$dN_{l+l-}/dM \simeq M^{1/2}e^{-M/T}$$

where $T \simeq$ temperature of the QGP.

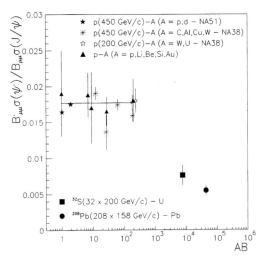

FIGURE 14. *Ratio Ψ'/Ψ as a function of AB in p-nucleus and nucleus-nucleus interactions*

The dilepton yield is sensitive to QGP formation and to its temperature, provided that dilepton from QGP can be identified from other dilepton sources [1].

Several non-QGP dilepton sources are present, mainly in the invariant mass region below 1.5 GeV. Many resonances, a part from J/Ψ and Ψ', can decay in two leptons: ρ, ω and Φ are the most important. Another (less important) source is the semi-leptonic decay of the charmed mesons: if a couple $D\bar{D}$ is created, decays like $D^+ \to \mu^+ \bar{K}^o \nu_\mu$ and $D^- \to \mu^- K^o \bar{\nu}_\mu$ lead in fact to simultaneous emission of two leptons. At low invariant mass, the hadron gas contributes to the production of dileptons via reactions of the type: $\pi\pi \to \gamma^*(\rho^*) \to \mu^+\mu^-$. On the contrary, the Drell-yan process, already mentioned, is specially important at high mass.

The experimental challenge is to disentangle direct electromagnetic radiation from the contribution of these "conventional" hadronic sources. The contribution to the mass spectrum of these processes is well known in normal conditions: the experimental mass spectrum is therefore compared with the mass spectrum calculated with a Montecarlo, in which the ingredients are all the known sources. Any difference can be ascribed to "new physics".

Electron pair production in proton-nucleus and nucleus-nucleus collisions is measured at CERN by the NA45/CERES experiment. It consists [18] of two azimuthally symmetric Ring Imaging Cherenkov (RICH) and a superconducting magnet between the two RICH (see fig. 16). The Cherenkov threshold is very high ($\gamma > 32$), in order to suppress the signals coming from the hadrons produced in the collision, and to detect only the electrons (and positrons). The Cherenkov photons from the RICH are registered in two UV-detectors which are placed upstream the

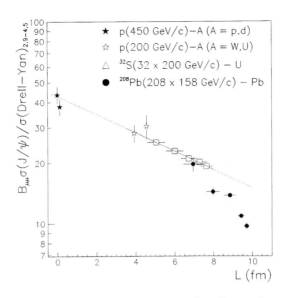

FIGURE 15. *J/Ψ suppression in proton-nucleus and nucleus-nucleus interactions and the anomalous suppression obtained by NA50 with Pb-Pb collisions*

target and are therefore not subject to the large flux of forward going charged particles. The information of the detectors is read out via two-dimensional arrays of ≈ 50000 pads each, allowing the unambiguous reconstruction of single photon hits. Two silicon drift detectors (SDD1 and SDD2) located downstream and close the target are used for a good vertex reconstruction. They improve also the recognition in the first RICH. This apparatus allows identification, tracking and momentum measurement of the electrons and it is therefore able to reconstruct the invariant mass of the electron pairs.

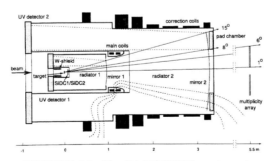

FIGURE 16. *The NA45/CERES spectrometer*

This experiment took data with Sulphur and Lead beams [19], and with protons.

In fig. 17 the e^+e^- mass spectrum obtained in p-Be interactions is shown. The experimental data are compared with the Montecarlo calculation, in which the hadron decays contributions (from π^0 to Φ), extrapolated from p-p collisions, are taken into account. The agreement is very good: the "conventional" hadronic sources fully exhaust the observed yield.

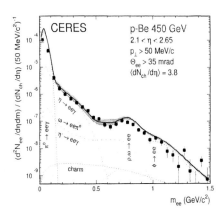

FIGURE 17. *Mass spectra of inclusive e^+e^- pairs in 450 GeV p-Be collisions. Full circles are data, straight lines are the various contributions from hadron decays*

The situation is different for S-Au electron pairs. As can be seen in fig. 18 in the region (0.2-1.5) GeV/c^2, the hadronic sources are not able to reproduce the shape of the spectrum, which shows a significant excess in the dilepton yield. The enhancement factor, defined as the ratio of the integral of the data over the integral of the predicted sources, is found to be $R = 5.0 \pm 0.7(stat) \pm 2.0(syst.)$.

This result is confirmed by the Pb-Au collisions data at 158 $AGeV$ [19] (fig. 19), where the enhancement factor is $R = 4.6 \pm 1.8 \pm 2.7$.

From these results we have a clear indication that other mechanisms must be added to explain the data in nucleus-nucleus collisions. $\pi - \pi$ annihilation to dileptons is usually considered to be the most important source of dilepton production [24] in a dense medium, but also the relative yields of hadronic species may change in central nucleus-nucleus collisions with respect to p-p and p-A interactions. Measurements of NA38/NA50 [20] actually indicate that the ratio $\Phi/(\rho+\omega)$ increases by a factor four from d+U to central Pb-Pb collisions. Also the open charm production may be enhanced in heavy ion collisions: NA38 observes an enhancement in the mass region around $2GeV/c^2$, but a similar effect could be present also at lower masses.

Many calculations were performed [21] [22]: they treat the collision dynamics in different ways and include different additional sources of dileptons (π annihilation, Φ enhancement, etc..) but keep the meson masses fixed at their vacuum values.

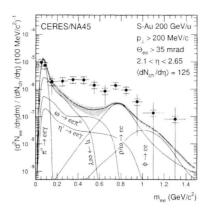

FIGURE 18. *Mass spectra of inclusive e^+e^- pairs in 200 AGeV S-Au collisions. Full circles are data, straight lines are the contributions from hadron decays*

All these calculations agree fairly well with each other (see fig. 20), but they fail to reproduce the data below the ρ peak.

On the contrary, if it is assumed that, in hot and dense matter, the masses of the ρ and ω mesons may change, then the agreement with the data is quite good [22] (see fig. 20). Such a mass shift was suggested as a signature for the restoration of chiral symmetry [23], when the quarks and hence the hadrons loose their mass. An alternative approach assumes that there is an interaction broadening in the medium which leads to a shift in the decay branching ratios of the resonances [25], [26]. The shape of the experimental distribution is also well reproduced. The foreseen improvements of CERES mass resolution will be essential to distinguish between these two hypothesis. In any case there is an indication that in these collisions a dense (partonic or hadronic) interacting matter is formed.

Strangeness enhancement : WA85/WA97 experiment

In hadron interaction, strange particle production occurs via associated production. The reaction which requires the minimum energy is:

$$p + p \rightarrow p + K + \Lambda, Q = M_\Lambda + M_K - M_p = 670 MeV$$

Strange anti-baryon production requires more energy:

$$p + p \rightarrow p + p + \Lambda + \bar{\Lambda}, Q = 2m_\Lambda = 2.2 GeV$$

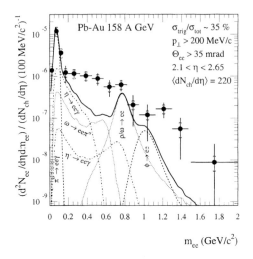

FIGURE 19. *Mass spectra of inclusive e^+e^- pairs in 158 AGeV Pb-Au collisions. Full circles are data, straight lines are the contributions from hadron decays*

$$p + p \rightarrow p + p + K + \bar{K}, Q = 2M_K = 986 MeV$$

As mentioned before, in QGP the quarks have their bare mass ($m_s = 150 MeV/c^2$) and hence the energy threshold is lower, being the energy to produce a couple $s\bar{s}$: $2m_s = \approx 300 MeV$. Moreover, in a volume rich of u and d quarks, the relative production of s and $\bar{s}$ is increased because of the Pauli exclusion principle ("Pauli blocking").

The strangeness content in hadron matter and in a QGP are therefore different. The strange particle production seems to be favoured if this new state of matter is formed [27].

An enhancement of the production of strange particles with respect to non strange ones has been reported with light ion beams comparing the production ratios between strange and non strange particles in $A - B$ collisions versus $p - A$ collisions. As a matter of fact, such increase has been found in K/π, in $\bar{\Lambda}/\bar{p}$, in $\Phi/(\rho+\omega)$, etc..This observed enhancement is surely compatible with the onset of a deconfined medium [28], [29], but it could also appear in a dense hadronic fireball with a sufficiently long lifetime [30], [31]. In the framework of a purely hadronic scenario, however, a strong enhancement of multistrange baryons or multistrange antibaryons is not foreseen since their production cross-section is too small. Ω^- (S=3) for instance, can be produced via $\pi + \pi \rightarrow \Omega + \bar{\Omega}$, but the threshold is very high, since $m_\Omega \approx 1670 MeV$. It can be produced also through a long chain of reactions, but this takes a long time:$\approx 100 fm/c$ while the time of a N-N collision is $\approx 5 - 10 fm/c$. Therefore the study of (multi)strange baryon production is of

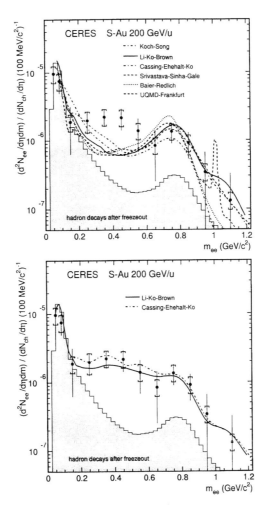

FIGURE 20. *Mass spectra of inclusive e^+e^- pairs in 200 AGeV S-Au collisions with different theoretical predictions: upper keeping mesons masses at their vacuum levels; lower ρ and ω masses shifted by medium*

great importance in the QGP search.

The WA85 experiment was designed to measure the production of strange particles in central S-W collisions at 200 $AGeV$, in the central rapidity region with medium to high p_T [32]. The heart of the detector is the tracking system, formed by seven MWPC placed inside the 1.8 Tesla magnetic field (in which also the target is placed) and four MWPC outside the field. A silicon microstrips multiplicity detector, placed above and below the beam, 15 cm downstream from the target, measures the charged multiplicity in the pseudorapidity range $2.1 \leq \eta \leq 3.4$. A zero-degree hadron calorimeter located along the beam line 25 m downstream from the target measures the centrality of the collision.

The main results obtained by this experiment in S-W interactions are the following [33]:

- the ratio $R_{S-W} = \bar{\Xi}^+/\bar{\Lambda}$ in the range $1.2 < p_T < 3.0$ GeV/c and $2.3 < y_{lab} < 3.0$ is

$$R_{S-W} = 0.21 \pm 0.02$$

while the corresponding ratio R_{p-p} in p-p interactions measured by the AFS Collaboration [34], at the CERN ISR, is

$$R_{p-p} = 0.06 \pm 0.02$$

- the ratio $(\Omega^- + \bar{\Omega}^+)/(\Xi^- + \bar{\Xi}^+)$ is significantly higher than the corresponding ratio in p-p interactions.

WA97, which is an improved version of WA85, took data with the CERN lead beam. A schematic view of the WA97 apparatus is is drawn in fig. 21 [35]. The main innovation is a sylicon telescope, used as a tracking chamber. It consists of six planes of pixel detectors (pixel size: $75 \times 500 \mu m^2$) and five planes of silicon microstrip detectors, placed in the magnetic field at 90 cm from the target. This system allows a precise mesurement of the momenta and angles of the particles produced in the decays.

The results [36] obtained analyzing the data collected in 1995 are presented in fig. 22, where the Ω and Ξ baryon production is reported for $p - Pb$ and $Pb - Pb$ collisions. The ratio $(\Omega^- + \bar{\Omega}^+)/(\Xi^- + \bar{\Xi}^+)$ is enhanced by a factor 3.2 ± 1.1 when going from $p - p$ to $Pb - Pb$ collisions. An enhancement of about a factor 2 of the Ξ^- hyperon production relative to Λ is also observed when comparing $p - S$ to $Pb - Pb$ interactions. Finally, the WA97 collaboration measured the transverse mass distributions for Λ and Ξ^- hyperons: their inverse slopes have been found systematically larger in $Pb - Pb$ than in $S - S$ interactions by about $\approx 40 - 50 MeV/c^2$. The results obtained so far by this experiment are very interesting, in particular the Ω/Ξ ratio seems to be hardly compatible with a purely hadronic colliding system.

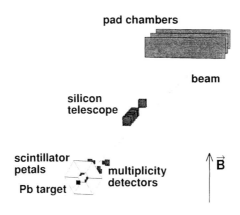

FIGURE 21. *Schematic view of the WA97 apparatus*

Conclusions

From the CERN SpS heavy ion program, in which the boundary conditions between hadron gas and QGP were explored, we can *at least* conclude that

- high energy heavy ion experiments are feasible
- the energy density is high ($\gg \rho_{nucleus}$)
- new collective phenomena are seen
- it is not at present excluded that the boundary has been crossed.

THE ALICE EXPERIMENT AT LHC

Introduction

The heavy ion experiments performed so far proved that relativistic heavy ions are the best mean to create a strongly interacting system which can be studied in thermodynamical terms. Their results suggest that heavy-ion collisions at higher energy will provide an ideal tool for the study of nuclear matter in a regime of thermodynamic behaviour, and will very likely provide conclusive evidence of quark deconfinement. Two very high energy colliders will accelerate heavy ions in the next years: RHIC at Brookhaven (100+100 GeV per nucleon), and LHC at CERN ($\approx 2.7 + 2.7$ TeV per nucleon). At these energies, all parameters relevant to the formation of the QGP will be more favourable: the energy density, the volume and lifetime of the system, the relaxation times are all expected, from montecarlo

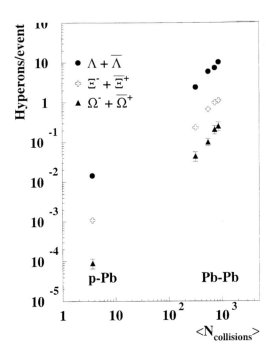

FIGURE 22. *Hyperon yields as a function of the average number of inelastic nucleon-nucleon collisions*

simulations, to improve by a large factor, typically 2-3 at RHIC and ≈ 10 at LHC, compared to Pb-Pb collisions at the SpS. The nuclear matter will reach a regime of thermodinamical equilibrium and the central rapidity region should have a very low baryon density, a situation similar to the early state of universe and more near to the hypothesis of the QCD calculations. At LHC, the very large number of secondaries produced in a $Pb-Pb$ collision (up to ≈ 8000 per unit rapidity), will provide both a formidable experimental challenge and a unique experimental opportunity. It will be possible to measure a large number of observables on an *event-by-event* basis: impact parameter, dN/dy, particle ratios and p_t spectra (π, γ, K, p) and, of particular importance, size and lifetime from interferometry. It will be possible in this way to study each event as a thermodynamical system, and to access correlations and non-statistical fluctuations which would be washed out when averaging over many events.

We will describe here the only experiment foreseen for LHC: ALICE (A Large Ion Collider Experiment). Contrary to RHIC, which is a dedicated accelerator, LHC will have in fact first of all proton beams, and two experimental zones (of the three available) are foreseen for p-p experiments.

To explore this physical potential, ALICE has been designed to search in a comprehensive way for qualitative and quantitative changes in composition and structure of the final states as a function of energy density. Since one of the most striking signatures expected for the QGP is an anomalously small cross section for the production of heavy vector mesons [39], the ALICE design incorporates a forward muon spectrometer, dedicated to the study of heavy meson resonances via their decay in two muons. Being the only heavy ion experiment at LHC, the ALICE Collaboration (Alessandria, Aligarh, Athens, Attikis, Bari, Beijing, Bergen, Birmingham, Bombay, Bratislava, Budapest, Cagliari, Calcutta, Catania, CERN, Chandigarh, Clermont-Ferrand, Copenhagen, Cracow, Darmstadt, Dubna, Frankfurt, Gatchina, Geneva, Heidelberg, Heidelberg MPI, Ioannina, Jaipur, Jammu, IHEP, ITEP, Kharkov, Kiev, Kosice, Kurchatov, Legnaro, Lund, Marburg, Mexico, Minsk, Moscow INR, Munster, Nantes, Novosibirsk, Oak Ridge, Orsay, Oslo, Padua, Prague, Rehovot, Rez, Rome, Salerno, Sarov, St. Petersburg, Strasbourg, Trieste, Turin, Utrecht, Warsaw, Wuhan, Yerevan, Zagreb) has therefore proposed [45,42,43] to build a dedicated, general-purpose detector, which will address most sensitive observables, detecting hadrons, di-leptons, and photons. In addition to the main running with Pb ions, collisions of ions of lower mass (Ca) are foreseen, to vary the energy density, while running with proton beams will provide reference data. The possibility of asymmetric collisions, i.e. of collisions of nuclei of different A, is currently under study.

In order to establish and analyse the existence of the QGP, a number of observables have to be studied in a systematic and comprehensive way. The ALICE strategy is to study a number of *specific signals* in the same experiment together with *global information* about the events. Therefore the experiment will measure the flavour content and phase-space distribution event by event for a large number of particles whose momenta and masses are of the order of the typical energy scale involved (temperature ≈ Λ_{QCD} ≈ 200 MeV). A part from the *global event features* (impact parameter, overlap volume, number of constituents participating in the interaction), this apparatus will detect and measure *prompt photons, Strange particles* and *J/Ψ and Y* families. *Multiplicity fluctuations*, which are a signature for the critical phenomena at the onset of a phase transition will be searched for; *particle interferometry*, which measures the expansion time in the mixed phase, expected to be long in the case of a first-order phase transition will be also studied, event by event.

Design Considerations [37]

The average event rate for Pb–Pb collisions at the LHC, given the maximum luminosity of 1.8 x 10^{27} $cm^{-2}s^{-1}$, a luminosity half–life of ≈ 10 hours and an inelastic cross– section of 5.5 b, will only be about 2000–4000 minimum-bias collisions per second, depending on whether one or two experiments are running at the same time. Of these events, approximately 2–3% (< 100 events/s) correspond to the most interesting central collisions. This low interaction rate has a crucial role in the design of the experiment, leading to an approach which combines large acceptance with simple central collision triggers. At the same time, it allows the use of slow but high–granularity detectors, like the time projection chamber (TPC) and the silicon drift detectors (SDD's). It is expected to collect a few 10^7 central events/year. The experiment must be designed to cope with the highest anticipated multiplicities, which are estimated for Pb–Pb central collisions at 8000 charged particles per unit of rapidity. The rapidity acceptance has to be large enough to allow the study of particle ratios, p_t spectra and HBT (Hanbury-Brown-Twiss [40]) radii on an event-by-event basis, meaning several thousand reconstructed particles per event. Detecting the decay of particles at $p_t < m$ requires about 2 units in rapidity (for masses above 1–2 GeV) and corresponding coverage in azimuth. In particular, efficient rejection of low-mass Dalitz decays, which is needed for the lepton-pair measurements, can only be approached with full azimuthal coverage. The rapidity coverage of the central detector ($|\eta| < 0.9$) has been chosen as a compromise between acceptance and cost. To be sensitive to the global event structure, $dN_{ch}/d\eta$ will be measured with multiplicity detectors in a large rapidity window ($|\eta| < 4.5$).

The design of the ALICE tracking system has therefore primarily been driven by the requirement for safe and robust track finding. It uses mostly three-dimensional hit information and dense tracking with many points in a weak magnetic field. The field strength, ≈ 0.2 T, is a compromise between momentum resolution, low momentum acceptance, and tracking efficiency. The momentum cut-off should be as low as possible (< 100 MeV/c), in order to study collective effects associated with large length scales. A low-p_t cut-off is also mandatory to reject the soft conversion and Dalitz background in the lepton-pair spectrum. The most stringent requirement on momentum resolution in the low-p_t region is posed by identical particle interferometry owing to the large source radii and the corresponding narrow correlation enhancement. In the intermediate energy regime, the mass resolution should be of the order of the natural width of the ω and ϕ in order to maximize the signal-to-background ratio and, more importantly, to study mass and width of these mesons in the dense medium. At high momenta, the resolution has to be sufficient to measure the spectrum of jets via leading particles. The detection of hyperons, and even more of charmed mesons, requires in addition a high–resolution vertex detector close to the beam pipe.

The momentum range for particle identification can be restricted for the bulk of the hadronic signals to a few times the average p_t (>97% of all charged particles are below $p_t = 2$ GeV/c). Good π/K/p separation (better than 3-4σ) is needed on a track-by-track basis for the abundant soft hadrons in order to study HBT with identified particles, decays (hyperons, $\phi \to$ K K), and event-by-event particle ratios. A statistical analysis (separation better than 2–3σ) will be sufficient to measure inclusive particle ratios and p_t spectra in the mini-jet region. The e/π rejection has to be sufficient to reduce the additional combinatorial background due to misidentification to below the level remaining from unrejected Dalitz pairs.

The accuracy of the single inclusive photon spectra will be determined in the range of interest ($p_t < 10$ GeV/c) by systematic errors on photon-reconstruction efficiency and the knowledge of the decay background. A 5% sensitivity requires a knowledge of the reconstruction efficiency as well as the π^0 cross-section to ≤ 3% and the η cross-section to ≤ 10%. An acceptable systematic error can be obtained only at low channel occupancy and therefore requires a calorimeter with small Molière radius at a large distance (≈ 5 m) from the vertex.

DETECTOR LAYOUT AND PERFORMANCE

A longitudinal view of the ALICE experiment is shown in fig. 23, in which the central horizontal axis in the figure is the line of flight of the two counter-rotating beams. The central part, which covers $\pm$ 45° ($|\eta| < 0.9$) over the full azimuth, is embedded in a large magnet with a weak solenoidal field. It consists of the Inner Tracking System with six layers of high-resolution silicon detectors, the cylindrical Time Projection Chamber, a barrel particle identification array (Time of Flight - TOF), a small–area detector at large distance for the identification of high momentum particles (Ring Imaging Cherenkov - RICH - detectors), and a single-arm electromagnetic calorimeter of high density crystals. There is a muon detector covering the very forward region ($2.4 \leq \eta \leq 4$). This is constructed of a low-Z absorber very close to the vertex followed by a spectrometer with a dipole magnet and, finally, an iron wall to select the muons. Not visible are the array of Multiplicity Counters, located near the beam pipe, and the Zero–Degree Calorimeters (ZDC), which are located $\approx$ 90 m downstream. The open geometry of the detector mantains the possibility of future modifications or upgrades.

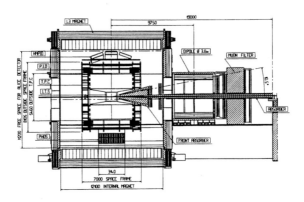

FIGURE 23. *Longitudinal view of the ALICE detector*

Inner Tracking System The inner tracker provides secondary vertex reconstruction for hyperon decays, tracking and identification by species of low-p_t particles, and improved momentum resolution for the higher-p_t particles which also traverse the time projection chamber. The six cylindrical layers are located at r=3.9, 7.6, 14, 24, 40 and 45 cm. Four layers will have analog readout to provide particle identification via dE/dx in the $1/\beta^2$ region, which will give the ITS a stand-alone capability as low-p_t particle spectrometer. Because of the particle density, the innermost four layers need to be truly two-dimensional devices. The use of silicon pixel detectors for the two innermost plane is therefore compulsory, while silicon drift detectors will be used for the following two. SDD's are ideally suited to this experiment, in which very high particle multiplicities are coupled with relatively low event rates. Like gaseous drift detectors, they exploit the measurement of the transport time of the charge deposited by a traversing particle to the collection electrodes to localize the impact point in one of the dimensions. This improves both position resolution ($\sigma_z = 25$ μm) and multitrack capability at the expense of speed. The outer layers, will be equipped with silicon micro-strip detectors. Since the minimization of the material thickness is an absolute priority, the support and cooling system is the object of a specific, extensive development program: we aim at an average thickness, including detectors, below 0.6% of X_0 per layer and at a temperature stability of 0.1 °C.

Time Projection Chamber The TPC has an inner radius of 90 cm, given by the maximum acceptable hit density (0.1 cm^{-2}), and an outer radius of 250 cm, given by the length required for a dE/dx resolution of $< 7\%$. This resolution allows electron identification up to ≈ 3 GeV/c. The design is optimized for good double-track resolution; in particular, the use of Ne/CH$_4$/CO$_2$ (88/10/2) minimises electron diffusion and reduces the space charge.

Performance of the Tracking System For the highest particle multiplicity considered a reconstruction efficiency in the TPC of $\approx 95\%$ was found, practically independent of p_t down to 100 MeV/c and with a negligible number of ghosts. The efficiency of connecting tracks from the TPC to the ITS is also better than 90%, resulting in an overall tracking efficiency for the TPC and the ITS combined of $\approx$ 90% for transverse momenta above 100 MeV/c (for pions). Low p_t electrons (and pions) are reconstructed in the ITS used as a stand-alone tracker after the hits of the higher-momentum tracks are removed. The efficiency is shown in fig. 24. The momentum resolution $\Delta p/p$ is shown for pions in fig. 25. It is generally better than 1.5 % for the relevant momentum range. The momentum resolution for electrons below 100 MeV/c, obtained with the ITS only, is better than 8%.

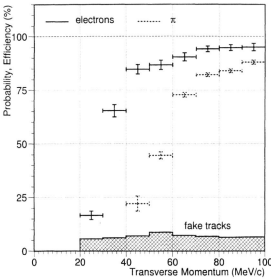

FIGURE 24. *Track-finding efficiency for low-momentum pions and electrons in the ITS used as stand-alone tracker*

The angular resolution for both polar and azimuthal angles is well below 1 mrad, except for very low momenta. the impact parameter resolution is better than 100 μm for p_t larger than 600 MeV (for kaons). The effective mass resolution of a particle decaying into an e^+e^- pair is $\Delta m/m \approx 1\%$, i.e. 8 MeV/c^2 for ρ and ω, 10 MeV/c^2 for ϕ, and 30 MeV/c^2 in the J/ψ region. Particle interferometry depends on two-track resolution, momentum resolution and acceptance; single-event pion interferometry should be feasible with a relative error of $\approx$ 20% up to effective sizes of 15 fm. In fig. 26 is shown an example of a single-event correlation function for R=7 fm at $dN_{ch}/dy = 8000$. For the hyperons, after reconstruction and identification, we can accept 50 $\pm$ 10% of the Λ's with $p_t > 600$ MeV/c, with very little background.

Particle Identification System Two separate systems will provide hadron identification. The barrel TOF at r=3.5 m will provide the identification of the bulk of

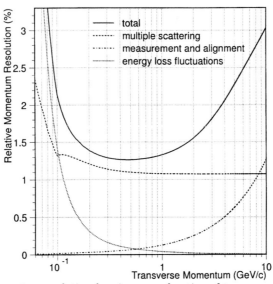

FIGURE 25. *Momentum resolution for pions as a function of transverse momentum, showing separately its main components*

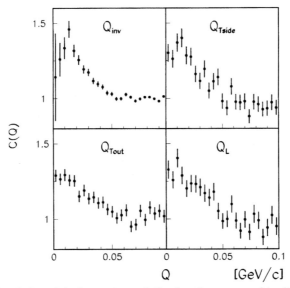

FIGURE 26. *Simulation of single event correlation function measured in ALICE with effective radius set to 7 fm*

the particles for event-by-event analysis and the e/π rejection needed in order to keep the backgrounds small in the electron pairs mass plots. The small-area RICH will provide the particle identification needed to measure inclusive particle ratios up to higher momenta. For the TOF, two different options are currently under test, of which the Pestov counter, with a timing resolution better than 100 ps, would be the preferred choice if it is proven that large area systems can be reliably manufactured and operated. The Pestov Spark Counter is a single-gap, high–pressure gaseous detector working in the streamer mode; the use of a semiconductive anode keeps the discharge local and enables a high count-rate capability. Prototypes and simulations show that at the expected particle density the $3\,\sigma$ limit of π/K identification is ≈ 2.1 GeV/c. Parallel plate chambers (PPC), which feature a more modest but still sufficient time resolution of ≈ 160 ps, are considered as a fall–back solution.

For the high-momentum particle identification the preferred solution is a RICH detector of the proximity focusing type, placed at a distance of ≈ 4.5m from the beam axis. It will use a liquid-Freon radiator and a UV detector based on the use of a low-gain MWPC with pad readout. It will use as photocathode a thin layer of CsI evaporated onto the pad plane. The RICH will extend the $3\,\sigma$ limit of π/K identification to ≈ 3.4 GeV/c.

Particle identification is achieved combining the information of the TOF with the dE/dx measurements in the TPC and in the ITS. The combined separation power, shown in fig. 27, is more than 3σ for π and K identification up to 2 GeV/c and for protons up to 3 GeV/c. This includes the bulk of the produced hadrons, thus allowing the study of particle ratios on an event–by–event basis, the HBT analysis, and the reconstruction of decays.

Thanks to the combination of particle identification and vertexing capabilities, it will be possible to measure in ALICE the production of both charged and neutral charmed mesons with a significance (signal over square root of the background) of about 30.

Electron Pairs As mentioned before, the ability to measure di-electrons depends essentially on the capability to recognize and remove from the sample the electrons from Dalitz decays, whose combinations form most of the background in the invariant mass spectrum. It must be kept in mind that there will be several electrons from Dalitz decays per event, which enter the plot in all possible combinations. The invariant-mass spectrum of electrons passing all our cuts for 5×10^7 recorded events is shown in fig. 28 a) for the mass region of the ω and ϕ mesons and in fig. 29 b) for the J/Ψ region The importance of excellent mass resolution ($\Delta m/m \approx 1\%$) is evident.

Photon Spectrometer (PHOS) Prompt photons, π^0's and η's will be measured in a single-arm, high-resolution electromagnetic calorimeter. Prompt photons are a small fraction of the meson decay photons, which must be accurately known before the former can be determined. The goal is to measure all photons over a solid angle large enough so that both π^0's and η's can be reconstructed (from 2γ decays) and their cross-section determined above some minimal p_t. The prompt photon cross-

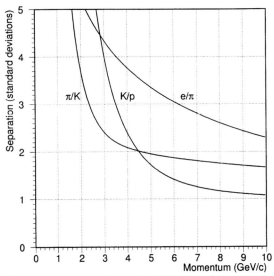

FIGURE 27. *Combined particle separation with TPC ($\sigma = 7\%$) and TOF ($\sigma = 100$ ps) at $r = 3.5$ m*

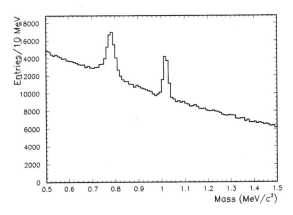

FIGURE 28. *Invariant-mass distributions, after cuts, for $5 \cdot 10^7$ central events, including tracking efficiency and momentum resolution in the ω/ϕ region*

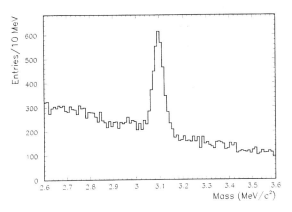

FIGURE 29. *Invariant-mass distributions, after cuts, for $5 \cdot 10^7$ central events, including tracking efficiency and momentum resolution in the J/Ψ region*

section is then obtained by subtraction of the decay photons from the measured γ spectrum. This requires little overlap of photon showers, i.e. high granularity, and good energy and spatial resolution. The reliable measurement of single photons [42] requires cell sizes not exceeding one Molière radius R_M and a pixel occupancy not exceeding 3%. At this occupancy, the gamma reconstruction efficiency should be measurable to an accuracy of $\approx 4\%$. The PHOS will be made of 36,000 elements, and located 5 m vertically beneath the interaction region; it has to be built from a material with small Molière radius and high light-output. PbWO$_4$, lead tungstate, seems best suited, since it is very dense, it scintillates and it is rather easy to grow, to machine and to handle.

Large Rapidity Detectors A multiplicity counter array close to the interaction region will measure the pseudorapidity distribution of charged particles over most of the phase space ($|\eta| < 4.5$). Several options are under study, including micro-channel plates and silicon pad detectors. A set of small calorimeters will be used to measure and trigger on the impact parameter of the collisions. Owing to their different Z/A values, it is possible to separate in space the neutron and proton spectators and the beam particles ($Z/A \simeq 0.4$) by means of the LHC dipole D1. Therefore, the neutron and proton spectators will be detected in two distinct calorimeters, made of tungsten with embedded quartz fibres, located on both sides of the interaction region ≈ 90 m downstream in the machine tunnel.

The Forward Muon Spectrometer The forward muon spectrometer will allow the study of vector meson resonances like $J/\Psi, \psi', Y, Y'$ and Y'' via their $\mu^+\mu^-$ decay. The signals will appear on a continuum due to B and D meson decays and Drell-Yan processes. Enhancement of Φ production and suppression of heavy quarkonia will be signatures of QGP production. In ALICE these phenomena would be studied as a function of the centrality, p_T and A. To achieve this, the spectrometer must have an efficiency for dimuons better than 90% and a mass resolution better than 100 MeV in the Y region and better than 70 MeV in the J/Ψ region. The momentum precision must be about 1%. A very good separation between the various states of the Y family is essential to understand the onset of deconfinement in this region.

The muon spectrometer consists of a composite absorber ($\approx 10\lambda_{INT}$) starting close to the interaction point(one meter) to reduce the μ background due to π and K decays. The absorber is carefully designed with layers of both high and low Z materials to reduce multiple scattering and particle leakage. It is followed by a large dipole magnet with a nominal field of 0.7 T, giving a 3 Tm field integral. The dipole will accept μ at angles smaller than 9 degrees. A small angle absorber with a central hole will shield the angles from 0 to 2 degrees and allow the non interacting Pb ions to traverse the spectrometer. Ten planes of thin multiwire proportional chambers with cathode pad readout placed in front, inside and following the dipole will provide the tracking information for the muons. Finally the muons will be identified as the particles crossing a second absorber ($\approx 10\lambda_{INT}$ of iron) at the end of the spectrometer. Two resistive plate chamber planes will detect these particles and trigger the spectrometer.

Acknowledgment We would like to thank the organizing commettee of the XX Taller de Particulas y campos for giving us the opportunity to present the heavy ion CERN SpS program and the ALICE experiment to the Mexican High Energy Physics community.

REFERENCES

1. C.Y. Wong, Introduction to High-Energy Heavy-Ion Collisions, World Scientific, 1994, and references therein
2. J. Cleymans et al., Phys. Rep. 130, 217-292 (1986)
3. H. Satz, CERN TH/96-172 (1996)
4. E. Laermann, Nucl. Phys. A 610, 1c-12c (1996)
5. H.R. Schmidt and J. Schukraft, J.Phys. G19 (1993)
6. J.D. Bjorken, Phys. Rev. D 27, 140, (1983)
7. M.Idzik, talk presented at "Frontier Detectors for frontier Physics, Elba (1997)
8. J.P.Blaizot and J.Y.Ollitrault, Phys.Rev.Lett. 77, 1703 (1996)
9. T. Satz and T. Matsui, Phys. Rev. Lett. B 178, 416 (1986)
10. C. Baglin et al., Phys. Lett. B 220, 471 (1989)
11. M.C. Abreu et al., Phys. Lett. B 410, 327 (1997)
12. C. Gerschel and J. Hufner, Nucl.Phys. A544, 513 (1992)
13. E. Braaten et al., Ann.Rev.Nucl.Part.Sci. 46, 197, (1996)
14. A.Sansoni, CDF Collaboration, NPA 610, 373c (1996)
15. D. Kharzeev and H. Satz, Phys.Lett. B 366, 316 (1996)
16. A.Capella et al., Phys. Lett. B 393, 431 (1997)
17. D.Kharzeev et al., Z.Phys. C 74, 307 (1997)
18. R. Baur et al., Nucl.Inst.Meth. A343, 87 (1994)

19. T.Ullrich et al. (CERES Coll), Nucl.Phys. A610, 317 (1996)
20. D.Jouan for the NA50 Collaboration, QM'97
21. D.K.Srivastava et al., Phys.Rev.C 53, R567 (1996); J.Sollfrank et al., nucl-th/960729, (1996); V.Koch and C.Song, nucl-th 9606028, (1996);
22. G.K.Li et al., Phys.Rev.Lett. 75, 4007 (1995); W.Cassing et al., Phys. Lett. B 363, 35 (1995); W.Cassing et al., Phys.Lett. B 377, 5 (1996)
23. R.Pisarski, Phys.Lett. B 110, 155 (1982); G.E.Brown and M.Rho, Phys.Rev.Lett. 66, 2720 (1991); T.Hatsuda and S.H.Lee, Phys.Rev. C 46, 34 (1992)
24. K.Kajantie, Phys.Rev. D 34, 2746 (1986); J.Cleymans et al., Z.Phys. C 52, 517 (1991), P.Koch, Z.Phys. C 48, 283 (1993); B.Kaempfer et al., Phys.Rev. C 49, 1132 (1994)
25. B.Friman and H.J.Pirner, nucl-th/9701016 (1997)
26. R.Rapp et al., hep-ph/9702210 (1997)
27. J.Rafelski and B.Mueller, Phys.Rev.Lett. 48, 1066 (1982)
28. B. Müller and J. Rafelski, Phys.Rev.Lett. 56, 2324 (1986)
29. J. Rafelski, Phys.Lett. B262, 333 (1991)
30. T. Matsui et al., Phys. Rev. D 34, 2047 (1986)
31. K.S. Lee et al., Phys. Rev. C 37, 1452 (1988)
32. WA85 Coll., CERN/SPSC/84-76 P206 (1984), CERN/SPSC/87-18 P206 (1987), CERN/SPSC/88-20 P206 (1988)
33. Di Bari et al. (WA85 Coll.), Nucl.Phys. A590, 307c (1995)
34. T. Åkesson et al., Nucl.Phys. B246, 1 (1984)
35. G. Alexeev et al., Nucl.Phys. A590, 139c (1995)
36. H. Helstrup et al. (WA97 Coll.), Nucl.Phys. A610, 165c (1996)
37. E.Chiavassa and P.Giubellino, Rev.Mex.Fis. 43, Sup.1, 144 (1997)
38. Xin-Nian Wang, Nuclear Physics A590 (1995), 47, and references therein.
39. H. Satz, Nuclear Physics A590 (1995), 63, and references therein.
40. H. Gutbrod and J. Rafelski, eds., Particle Production in Highly Excited Matter, Plenum, 1993. See the article by W.A. Zajc for an introduction to HBT in hadronic interactions.
41. H.R. Schmidt and J. Schukraft, J. Phys. **G19** (1993)
42. ALICE Technical Proposal N. Ahmad et al., CERN/LHCC/95-71 (1995).
43. ALICE TP addendum, S. Beolé et al., CERN/LHCC/96-32
44. Proc. of the Large Hadron Collider Workshop, Aachen, Germany, October 1990, ECFA 90-133, CERN 90-10, Vol. I, p. 188-217, and Vol. III, p. 1057-1235.
45. N. Antoniou et al., Letter of Intent for A Large Ion Collider Experiment, CERN/LHCC/93-16 (1993).

Past, present and future experiments at LEP

Monica Pepe Altarelli

INFN - Laboratori Nazionali di Frascati

Abstract. Recent data on precision tests of the standard model at LEP are presented and compared with the theoretical expectations. These results are obtained by a preliminary analysis of all the data collected at LEP between 1990 and 1995. The most recent LEP2 measurements of the W mass, of fermion pair production cross sections and results on searches for the Higgs at energies well above the Z resonance are also included.

INTRODUCTION

The results presented here are based on the analysis of ~ 16 million Z decays, collected by the four LEP experiments during the years 1990-1995. The data consist of the hadronic and leptonic cross sections, the leptonic forward-backward asymmetries, the τ polarization asymmetries, the $b\bar{b}$ and $c\bar{c}$ partial widths and forward-backward asymmetries and the $q\bar{q}$ charge asymmetry [1]. Information on the individual results and a detailed list of references can be found in [2]. Some of the electroweak results from SLD and the TEVATRON are also included, as well as the LEP2 results on the direct determination of the W mass and on searches for the Higgs boson.

Z LINE-SHAPE AND LEP ENERGY CALIBRATION

The parameters m_Z and Γ_Z are extracted by a scan of the Z resonance, *i.e.*, by measuring the cross sections $e^+e^- \to f\bar{f}$ for hadronic ($q\bar{q}$) and leptonic ($\ell^+\ell^-$) final states as a function of $\sqrt{s} \sim m_Z$. The number of selected events and the systematic errors on the event selections are shown in Table 1. In total, approximately 16.4 million events are used in the analysis of the Z line-shape.

The luminosity is measured by counting small angle Bhabha scattering events and by dividing by the predicted theoretical cross section σ_{ref} : $\mathcal{L} = \frac{N_{\text{Bhabha}}}{\sigma_{\text{ref}}}$. The precision of the theoretical prediction is therefore of the same importance as the accuracy of the measurements. Small angles are chosen because in this region the cross section is dominated by t-channel photon exchange which can be calculated

		ALEPH	DELPHI	L3	OPAL
Number of events	$q\bar{q}$	4164K	3556K	3358K	3701K
	$\ell^+\ell^-$	485K	382K	317K	454K
Syst. error	$q\bar{q}$	0.07%	0.10%	0.05%	0.08%
	e^+e^-	0.30%	0.60%	0.25%	0.25%
	$\mu^+\mu^-$	0.12%	0.30%	0.30%	0.15%
	$\tau^+\tau^-$	0.20%	0.60%	0.65%	0.45%
Experimental syst. error on luminosity		0.07%	0.09%	0.09%	0.05%

TABLE 1. Number of selected events and systematic errors of the event selection used for the analysis of the Z line-shape. The data sample corresponds to an integrated luminosity of ~ 140 pb^{-1} collected between 1990 and 1995 by each LEP experiment.

from QED with the required precision, while contributions from Z-exchange interference effects are small ($\sim 0.2\%$ to $\sim 1\%$ of the Born cross section depending on the angular acceptance of the luminosity monitor). The Bhabha reference cross section σ_{ref} is calculated using the Monte Carlo program BHLUMI [3]. The theoretical uncertainty associated to this Monte Carlo generator is 0.11% and is common to all four LEP experiments. This uncertainty has recently been decreased from 0.16% to 0.11% [4] with respect to previous calculations, reflecting into a more accurate determination of the hadronic cross section.

The LEP energy uncertainty [5,6] has an important impact on the determination of m_Z and Γ_Z: the error on the mass is in fact dominated by the calibration error, while the error on the width due to relative point to point calibration uncertainties is almost as large as the statistical one.

The measurements of m_Z and Γ_Z are completely dominated by the high statistics scans of the resonance performed in 1993 and 1995, when data were collected at about 2 GeV above and below the peak (denoted by p±2 in the following). The errors on m_Z and Γ_Z associated to the beam energy calibration are approximately given by:

$$\Delta m_Z \simeq 0.5 \Delta(E_{+2} + E_{-2}) \tag{1}$$

$$\Delta \Gamma_Z \simeq \frac{\Gamma_Z}{(E_{+2} - E_{-2})} \Delta(E_{+2} - E_{-2}), \tag{2}$$

i.e., m_Z is affected by the uncertainty on the absolute energy scale, while Γ_Z is only influenced by the error on the difference in energy between the scan points.

The very accurate determination of the average energy of the beams in LEP is based on the resonant spin depolarization technique [7]. This technique relies on the fact that, under favourable circumstances, a finite transverse polarization can build up naturally in e$^+$e$^-$ storage rings due to the interactions of the electrons or positrons with the guiding magnetic field (through the so-called Solokov-Ternov mechanism). The degree of polarization can be measured by the angular distribution of scattered polarized laser light. By exciting the beam with a transverse RF field, the transverse polarization can be destroyed if the RF frequency is in phase with the spin precession frequency (*spin tune*). The measurement of the depolarization frequency determines the spin tune ν_s which is proportional to the beam energy E_b via the electron magnetic moment anomaly

$$\nu_s = \frac{g_e - 2}{2}\frac{E_b}{m_e} = \frac{E_b[\text{GeV}]}{0.44065}. \quad (3)$$

The intrinsic resolution of the method is of the order of 0.2 MeV. However such measurements were only possible outside normal data taking periods with separated beams, typically at the end of fills. In order to interpolate possible energy changes between two resonant depolarization measurements it was therefore necessary to develop a detailed model of the accelerator behaviour which tracks the time evolution of the properties of the magnets, current, field, temperature, as well as the geometrical properties of the ring. The current measurement of the LEP energy scale is almost finalised [8]; its uncertainty produces errors of $\Delta m_Z \sim 1.5$ MeV and $\Delta\Gamma_Z \sim 1.5$ MeV.

The description of the information contained in the hadronic and leptonic cross sections and lepton forward-backward asymmetries is based on nine independent parameters: m_Z, Γ_Z, σ_h^0, R_e, R_μ, R_τ, $A_{FB}^{0,e}$, $A_{FB}^{0,\mu}$, $A_{FB}^{0,\tau}$. These parameters are chosen because they are most directly related to the experimental quantities and are weakly correlated.

The parameter σ_h^0 is the hadronic cross section after deconvolution of initial state radiation, which, at the peak, takes the form:

$$\sigma_h^0 \equiv \frac{12\pi}{m_Z^2}\frac{\Gamma_{ee}\Gamma_{had}}{\Gamma_Z^2}. \quad (4)$$

The pole asymmetry $A_{FB}^{0,f}$ can be expressed directly in terms of the ratio of the vector (g_{Vf}) and axial vector (g_{Af}) coupling constants of the neutral current to fermion f:

$$A_{FB}^{0,f} \equiv \frac{3}{4}\mathcal{A}_e\mathcal{A}_f \quad (5)$$

with:

$$\mathcal{A}_f \equiv \frac{2g_{Vf}g_{Af}}{g_{Vf}^2 + g_{Af}^2} = \frac{2g_{Vf}/g_{Af}}{1 + (g_{Vf}/g_{Af})^2}. \quad (6)$$

At tree level g_{Vf} and g_{Af} can be written as

$$g_{Vf} = I_{3_f} - 2Q_f\sin^2\theta_{\text{eff}}^{\text{lept}} \quad (7)$$
$$g_{Af} = I_{3_f}, \quad (8)$$

i.e., the asymmetries determine the ratio g_{Vf}/g_{Af}, allowing a measurement of $\sin^2\theta_W^{eff}$ defined as

$$\sin^2\theta_{\text{eff}}^{\text{lept}} \equiv \frac{1}{4}(1 - g_{V\ell}/g_{A\ell}). \qquad (9)$$

On the other hand the Z partial width are proportional to the sum of the coupling squared: $\Gamma(Z \to f\bar{f}) \propto g_{Vf}^2 + g_{Af}^2$. Therefore, the measurement of asymmetries and Z partial widths allow the determination of the Z couplings to fermions.

The parameter R_ℓ gives, for each lepton species, the ratio of the hadronic and the leptonic partial widths.

The number of fitted quantities is reduced to five when lepton universality is assumed. Tables 2 and 3 show the results obtained when combining the data of the four collaborations for the nine and five parameter fit, respectively. The average correlation coefficients for the five parameter fit are given in Table 4.

The Z mass is now known with a relative precision of $2 \cdot 10^{-5}$, *i.e.*, as accurately as G_f and much better than the QED coupling constant at the Z mass ($\Delta\alpha/\alpha = 70 \cdot 10^{-5}$).

Parameter	Average Value
m_Z (GeV)	91.1867±0.0020
Γ_Z (GeV)	2.4948±0.0025
σ_h^0 (nb)	41.486±0.053
R_e	20.757±0.056
R_μ	20.783±0.037
R_τ	20.823±0.050
$A_{FB}^{0,e}$	0.0160±0.0024
$A_{FB}^{0,\mu}$	0.0163±0.0014
$A_{FB}^{0,\tau}$	0.0192±0.0018

TABLE 2. Average line-shape and forward-backward asymmetry parameters from the data of the four LEP experiments, without the assumption of lepton universality.

Starting from these primary measurements one can derive important additional quantities such as, for example, $\Gamma_{\ell\ell}$, Γ_{had} and Γ_{inv}, which are shown in Table 5. Using the results of Table 5 on the ratio $\Gamma_{\text{inv}}/\Gamma_{\ell\ell}$ and taking the standard model prediction for $\Gamma_{\nu\nu}/\Gamma_{\ell\ell}$ (1.991 ± 0.001), the number of light neutrino species can be derived:

$$N_\nu = 2.993 \pm 0.011, \qquad (10)$$

showing that only three fermion generations exist with $m_\nu < 45$ GeV.

Parameter	Average Value
m_Z (GeV)	91.1867±0.0020
Γ_Z (GeV)	2.4948±0.0025
σ_h^0 (nb)	41.486±0.053
R_ℓ	20.775±0.027
$A_{FB}^{0,\ell}$	0.0171±0.0010

TABLE 3. Average line-shape and forward-backward asymmetry parameters from the results of the four LEP experiments, assuming lepton universality.

	m_Z	Γ_Z	σ_h^0	R_ℓ	$A_{FB}^{0,\ell}$
m_Z	1.00	0.05	−0.01	−0.02	0.06
Γ_Z	0.05	1.00	−0.16	−0.00	0.00
σ_h^0	−0.01	−0.16	1.00	0.14	0.00
R_ℓ	−0.02	0.00	0.14	1.00	0.01
$A_{FB}^{0,\ell}$	0.06	0.00	0.00	0.01	1.00

TABLE 4. The correlation matrix for the set of parameters given in Table 3.

Without Lepton Universality:	
Γ_{ee} (MeV)	83.94±0.14
$\Gamma_{\mu\mu}$ (MeV)	83.84±0.20
$\Gamma_{\tau\tau}$ (MeV)	83.68±0.24

With Lepton Universality:	
$\Gamma_{\ell\ell}$ (MeV)	83.91±0.10
Γ_{had} (MeV)	1743.2±2.3
Γ_{inv} (MeV)	500.1±1.8

TABLE 5. Partial decay widths of the Z boson, derived from the results of the nine and five-parameter fit.

The line-shape results presented here are still preliminary. At present, only ALEPH and DELPHI have used the most recent determination of the LEP energy errors; L3 and OPAL have evaluated the effect of the new energies on their results. In addition no preliminary results are quoted yet for the 1995 OPAL lepton cross sections. Final results on the LEP energy calibrations for the years 1993-1995 should be available soon.

THE HADRONIC γZ INTERFERENCE TERM

In the fitting procedure described in the previous section, the interference between the continuum and the Z resonance amplitude was fixed to the value predicted by the standard model. However, this assumption can be tested by measuring the interference term directly from the data by fitting the hadronic cross sections in the framework of the so-called S-matrix approach [9].

Measurements of the hadronic cross section at centre-of-mass energies far away from the Z pole are especially sensitive to the parameters describing the interference between photon and Z-boson exchange $j_{\text{had}}^{\text{tot}}$. Starting in 1995 LEP has been taking data above the Z peak (LEP2) collecting an integrated luminosity per experiment of approximately 6 pb^{-1} at $\sqrt{s}$ of 130-136 GeV, 11 pb^{-1} just above the WW threshold, 11 pb^{-1} at $\sqrt{s} = 172$ GeV and 60 pb^{-1} at 183 GeV. The results presented here include the data collected at $\sqrt{s}$ from 130 GeV to 172 GeV [10].

As one moves above the Z peak, the fast decrease of the cross section favours the radiation of hard photons from the initial state boosting the effective $f\bar{f}$ centre-of-mass energy back to the Z mass (the so-called "radiative Z return"). QED initial state radiation was also important at LEP1, leading to a 74 % reduction of the cross section due to soft photon emission. At LEP2, however, the effect becomes dramatic: for example, at $\sqrt{s} = 180$ GeV, in 70% of the hadronic events hard photon emission from the initial state leads to events with a final state invariant mass $m_{q\bar{q}} = \sqrt{s'} \simeq m_Z$. Such "LEP1 type of events" are useless for the measurement of the high energy cross sections. Therefore, the analyses impose a cut on the final state invariant mass, typically demanding $\sqrt{\frac{s'}{s}} > 0.85$.

The combination of the high energy hadronic cross sections from the four LEP experiments leads to the following result for the hadronic γZ interference:

$$j_{\text{had}}^{\text{tot}} = 0.14 \pm 0.14, \qquad (11)$$

to be compared with the standard model prediction of 0.22.

The hadronic γZ interference term is strongly correlated with m_Z (-75% correlation). The release of the assumption that the hadronic interference term be fixed to the standard model prediction allows a model independent determination of the Z mass, leading to the following result: $m_Z = 91.1882 \pm 0.0031$ GeV, in excellent agreement with the value presented in the previous Section.

By now, the precision on $j_{\text{had}}^{\text{tot}}$ obtained by the four LEP experiments can only be marginally improved by including low energy data from TRISTAN. The fit to the LEP and TOPAZ [11] cross sections in fact gives $j_{\text{had}}^{\text{tot}} = 0.14 \pm 0.12$.

ASYMMETRIES AND $\sin^2\theta_{\text{eff}}^{\text{lept}}$

In the second Section, results were shown for the measurement of the lepton forward-backward asymmetries. Various asymmetries can be determined at the Z, from which the effective electroweak mixing parameters $\sin^2\theta_{\text{eff}}^{\text{lept}}$ is extracted allowing independent tests of the standard model.

The τ polarization is determined by measuring the longitudinal polarization of τ pairs produced in Z decays. It is defined as

$$\mathcal{P}_\tau \equiv \frac{\sigma_R - \sigma_L}{\sigma_R + \sigma_L}, \qquad (12)$$

where σ_R and σ_L are the τ-pair cross sections for the production of a right-handed and left-handed τ^-, respectively. The angular dependence of P_τ as a function of the angle θ between the e^- and the τ^- is given by:

$$\mathcal{P}_\tau(\cos\theta) = -\frac{\mathcal{A}_\tau(1 + \cos^2\theta) + 2\mathcal{A}_e\cos\theta}{1 + \cos^2\theta + 2\mathcal{A}_\tau\mathcal{A}_e\cos\theta}, \qquad (13)$$

with $\mathcal{A}_e$ and $\mathcal{A}_\tau$ as defined in Equation (6). When averaged on all production angles P_τ is a measurement of $\mathcal{A}_\tau$, while as a function of $\cos\theta$, $\mathcal{P}_\tau$ provides nearly independent determinations of $\mathcal{A}_\tau$ and $\mathcal{A}_e$, allowing thus to test universality of the couplings of the Z to e and τ. When combining the results from the four LEP experiments, the average values for $\mathcal{A}_\tau$ and $\mathcal{A}_e$ are:

$$\mathcal{A}_\tau = 0.1411 \pm 0.0064 \qquad (14)$$
$$\mathcal{A}_e = 0.1399 \pm 0.0073. \qquad (15)$$

The measurements included in the above averages do not yet make use of the full LEP1 statistics. The ALEPH analysis, in particular, includes data only up to 1992. Some improvements in the results can therefore be expected, especially for $\mathcal{A}_e$, a quantity still dominated by statistical errors. The new LEP average results for the b and c forward-backward asymmetries are:

$$A_{\text{FB}}^{0,\text{b}} = 0.0984 \pm 0.0024 \qquad (16)$$
$$A_{\text{FB}}^{0,\text{c}} = 0.0741 \pm 0.0048, \qquad (17)$$

where all corrections due to the energy shift to $\sqrt{s} = m_Z$, initial state radiation and QCD effects are already taken into account. The QCD corrections combine theoretical calculations with experimental effects evaluated by each experiment from a Monte Carlo simulation following the procedure detailed in Ref. [12].

One can take advantage of the large hadron statistics and measure the average quark charge asymmetry for all hadronic events. To infer the original quark charge, one relies on the fact that the leading particles in a jet carry information on their primary charge. The present value of $\sin^2\theta_{\text{eff}}^{\text{lept}}$ from inclusive hadronic charge asymmetries at LEP is:

$$\sin^2\theta_{\text{eff}}^{\text{lept}} = 0.2322 \pm 0.0010.$$

The effective electroweak mixing parameter $\sin^2\theta_{\text{eff}}^{\text{lept}}$ defined as in Equation 9 is derived from the measurements of the various asymmetries. The results on $\sin^2\theta_{\text{eff}}^{\text{lept}}$ are displayed in Figure 1 together with the standard model prediction as a function of m_H. Also shown is the result obtained from the measurement of the left-right asymmetry A_LR at SLD [13]. A longitudinal polarization P for the e^- beam is needed in this case. The left-right asymmetry is defined as

$$A_\text{LR}(e^+e^- \to X) = \frac{\sigma_\text{L} - \sigma_\text{R}}{\sigma_\text{L} + \sigma_\text{R}} \frac{1}{\text{P}}, \quad (18)$$

σ_L and σ_R being the cross sections for $e^-_{\text{LR}} + e^+ \to X$ and X stands for any channel. This measurement does not require any final state analysis beyond event selection: one needs to count the number of Z produced with left- and right electron beams, and measure the average electron beam polarization.

The difference in the dependence of the asymmetries on the parameter $\mathcal{A}_\text{f}$ of Equation 6 ($A_\text{FB}^{0,\text{f}} = \frac{3}{4}\mathcal{A}_\text{e}\mathcal{A}_\text{f}$, $\mathcal{P}_\tau = \mathcal{A}_\tau$, $A_\text{LR} = \text{P}\mathcal{A}_\text{e}$) has a strong impact on the sensitivity to $\sin^2\theta_{\text{eff}}^{\text{lept}}$. It can easily be shown that:

$$\delta\sin^2\theta_{\text{eff}}^{\text{lept}} \approx \frac{1}{8}\delta\mathcal{P}_\tau \approx \frac{1}{8\text{P}}\delta A_\text{LR} \approx \frac{1}{2}\delta A_\text{FB}^{0,\ell} \approx \frac{1}{6}\delta A_\text{FB}^{0,\text{b}}. \quad (19)$$

As shown in Figure 1, the two most precise determinations of $\sin^2\theta_{\text{eff}}^{\text{lept}}$ derived from the measurements of $A_\text{FB}^{0,\text{b}}$ and A_LR are three standard deviations apart. Both measurements have small systematic uncertainties. The result for $A_\text{FB}^{0,\text{b}}$ is still largely based on a preliminary set of results. In the case of A_LR, which also includes a preliminary result for the 1996 subset of the data, the systematic error is one third of the statistical one and derives essentially from the uncertainty on the electron beam polarization. SLD is approved to run into 1998, adding approximately another 300,000 events to their present sample of 200,000 hadronic events. This should lead to a reduction of the error on A_LR by a factor of approximately 1.6.

The uncertainty associated to the value of the fine structure constant $\alpha(m_Z^2)$ induces an error on the standard model prediction of $\sin^2\theta_{\text{eff}}^{\text{lept}}$ as large as the present experimental uncertainty. If the value $\alpha(m_Z^2) = 1/(128.896 \pm 0.090)$ [14] is used, this translates into an uncertainty on the standard model prediction of 0.00023, to be compared to an experimental one of the same size.

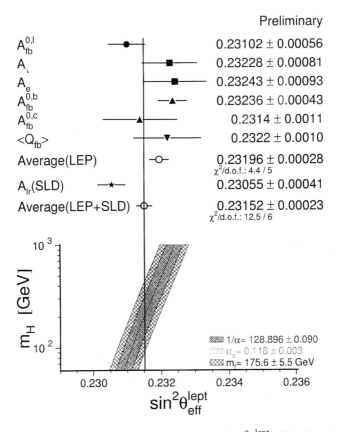

FIGURE 1. Comparison among different determinations of $\sin^2\theta_{\text{eff}}^{\text{lept}}$. Also shown is the standard model prediction as a function of m_{H}. The width of the band reflects the present uncertainties on $\alpha(m_Z^2)$, $\alpha_s(m_Z^2)$ and m_t.

EFFECTIVE VECTOR AND AXIAL-VECTOR COUPLING CONSTANTS AND HEAVY QUARK COUPLINGS

The effective vector and axial-vector couplings can be extracted from the measurement of the purely leptonic observables, *i.e.*, the partial widths of the Z into leptons, the lepton forward-backward asymmetries, the tau polarization and tau polarization asymmetries. The results of the fit to the LEP+SLD data are given in Table 6 showing that lepton universality is well verified. The neutrino coupling

	g_{Vf}	g_{Af}
e	-0.03844 ± 0.00071	-0.50111 ± 0.00043
μ	-0.0358 ± 0.0032	-0.50098 ± 0.00065
τ	-0.0365 ± 0.0015	-0.50103 ± 0.00074
ℓ	-0.03793 ± 0.00058	-0.50103 ± 0.00031
ν	0.50125 ± 0.00092	0.50125 ± 0.00092

TABLE 6. Lepton couplings g_{Vf} and g_{Af} extracted from lepton asymmetries and tau polarization for LEP and SLD data combined. The neutrino coupling derived from from the invisible width assuming three light neutrino species is also given.

to the Z are extracted from from the invisible width assuming three light neutrino species and taking $g_{V\nu} = g_{A\nu} = g_\nu$.

The measurements of the b and c forward-backward asymmetries determine the products $A_{\rm FB}^{0,\rm f} = \frac{3}{4}\mathcal{A}_e\mathcal{A}_f$ (Equation 5). One can therefore extract $\mathcal{A}_f$, once $\mathcal{A}_e$ is known. By combining the values of $\mathcal{A}_e$ obtained from the lepton forward-backward asymmetry $A_{\rm FB}^{0,\ell}$ ($\mathcal{A}_e = 0.1510 \pm 0.0044$) and from the $\mathcal{P}_\tau(\cos\theta)$ measurements of Equations 14 and 15 ($\mathcal{A}_e = 0.1406 \pm 0.0048$), one obtains $\mathcal{A}_e = 0.1461 \pm 0.0033$ and hence

$$\mathcal{A}_b = 0.897 \pm 0.029 \, (\text{LEP}) \qquad (20)$$
$$\mathcal{A}_c = 0.674 \pm 0.047$$

This result for $\mathcal{A}_b$ is approximately -1.3 σ lower than the standard model prediction $\mathcal{A}_b^{\rm SM} = 0.935$. The measurements of the left-right forward-backward asymmetries for b and c quarks at SLD [15] lead to the following direct determinations of $\mathcal{A}_b$ and $\mathcal{A}_c$:

$$\mathcal{A}_b = 0.900 \pm 0.050 \, (\text{SLD}) \qquad (21)$$
$$\mathcal{A}_c = 0.650 \pm 0.058$$

The LEP and SLD results can be combined by assuming the standard model prediction for $\mathcal{A}_e$ ($\mathcal{A}_e^{SM} = 0.1470$, taken from the fit in Table 8). This choice allows in fact to decouple from the well known two standard deviations discrepancy between A_{LR} from SLD and $\mathcal{A}_e$ from LEP, which is not related to heavy flavour couplings. Under this hypothesis, the LEP-SLD combination gives:

$$\mathcal{A}_b = 0.893 \pm 0.020 \; (\text{LEP} + \text{SLD with} \, \mathcal{A}_e = \mathcal{A}_e^{SM}) \qquad (22)$$
$$\mathcal{A}_c = 0.663 \pm 0.035,$$

leading to a 2.1 standard deviations for $\mathcal{A}_b$ with respect to the standard model prediction, while $\mathcal{A}_c$ agrees well with expectations.

R_b

The measurement of the b partial width is particularly important due to the additional quadratic m_t dependence present in the Z→ $b\bar{b}$ vertex.

The lifetime/mass (and lepton) double tag measurements are the most precise ones [16–19]. If a high purity b tag is applied to each event hemisphere, by measuring the number of single and double tagged hemispheres, one can extract R_b and the b efficiency ε_b. The charm and light quark efficiencies ε_c, ε_{uds} and the correlation in tagging efficiency between b hemispheres ρ_b must be taken from a Monte Carlo simulation and represent a major source of systematic errors. It is therefore crucial to be able to keep the correlation and the light quark background as small as possible.

There are several identified sources of hemisphere-hemisphere correlations:

- Geometrical effects induce a positive correlation (if a b-hadron is on the edge of the vertex detector angular acceptance, so is the other, since they tend to be back to back).

- Gluon emission induces a positive correlation by lowering the momenta of both b-hadrons.

- Correlations are also possible through the sharing of a common primary vertex. For example if one b hadron has a long decay length, it will probably be tagged. The resolution on the primary vertex will however degrade due to the lower track multiplicity, making the tag of the second b hadron less likely.

This last source of correlation is the dominant one for the LEP experiments. Its impact was drastically reduced in the ALEPH and DELPHI analyses by reconstructing a primary vertex for each hemisphere separately, using tracks from that hemisphere only. For the SLD result, such an effect is much less important due to their precise knowledge of the e^+e^- interaction point (7μ perpendicular to the beam direction and 35 μ along the beam direction).

The introduction of tags based not only on lifetime but also on information from secondary vertices allows to considerably reduce the charm background and the corresponding systematic uncertainties. SLD and ALEPH exploit the difference in mass between b and charm hadrons, whereas DELPHI also includes information from the energy of all particles originating from the secondary vertex and their rapidity. Table 7 shows the purity and efficiency of the b-tags for the most recent LEP and SLD results. SLD achieves the highest efficiency and purity because

	ALEPH	DELPHI	L3	OPAL	SLD
b Purity %	97.8	98.0	86.4	90.5	97.6
b Efficiency %	22.7	32.4	23.7	23.1	47.9

TABLE 7. Purity and efficiency of the b-tags from the most recent LEP and SLD results.

of the small interaction region which allows to reduce the radius of the internal layer of their vertex detector. However, their measurement suffers from the low statistics available. ALEPH and DELPHI improve the statistical accuracy of their measurements by making use of several different tags based, for example, on event shapes or on leptons from semileptonic b decays. The average combined preliminary LEP+SLD value

$$R_b = 0.2170 \pm 0.0009 \tag{23}$$

is now consistent with the standard model prediction of 0.2158 ± 0.0003.

THE W MASS

Two methods are used at LEP to measure the W mass:

- One procedure requires a measurement of the total W-pair cross section close to threshold where the size of the cross section is most sensitive to m_W.

- The second procedure is based on the direct mass reconstruction from the W decay products at energies above threshold, exploiting in particular the two- and four-jet topologies.

Since the two methods in principle have comparable statistical sensitivity but largely independent systematic uncertainties, they can be regarded as complementary. However, since most of the data at LEP2 are collected above threshold, the direct reconstruction method provides the most accurate determination of m_W.

In 1996 the LEP centre of mass energy was raised just above the W pair threshold. The energy that provides optimal sensitivity to m_W is $(\sqrt{s})^{opt} \simeq 2m_W + 0.5$ GeV, i.e., $(\sqrt{s})^{opt} \simeq 161$ GeV. Based on a luminosity of approximately 10 pb^{-1} per experiment the average W-pair cross section at $\sqrt{s} = 161$ GeV is measured to be

$$\sigma_{W^+W^-} = 3.69 \pm 0.45 \, \text{pb} \tag{24}$$

From a comparison of the cross section measurement with a theoretical calculation which has m_W as a free parameter, the W mass is inferred to be

$$m_W = 80.40^{+0.22}_{-0.21} \pm 0.03 \, \text{GeV}, \tag{25}$$

where the first error is experimental and the second is due to the LEP energy calibration.

Each experiment also accumulated 10 pb^{-1} at 172 GeV, where the W-pair cross section is approximately 10 times larger than at $\sqrt{s} = 161$ GeV, yielding 100 W pair events per experiment observed at this energy. By directly reconstructing the jet-jet invariant masses from the channels $W^+W^- \to q\bar{q}q\bar{q}$, $q\bar{q}l\nu$, the W mass is measured to be

$$m_W = 80.53 \pm 0.18 \, \text{GeV}. \tag{26}$$

The error includes a LEP energy uncertainty of 30 MeV and, for the four-quark channel, the potential source of systematics associated to interconnection phenomena which may obscure the separate identities of the two W bosons (so-called colour reconnection and Bose-Einstein correlations). This effect is estimated to contribute 50 MeV to the result of Equation 26. Attempts have been made to evaluate with the data the importance of such phenomena but no conclusive tests are available yet due to the limited statistics.

The combination of the two results of Equations 25 and 26 gives

$$m_W = 80.48 \pm 0.14 \, \text{GeV}. \tag{27}$$

In 1997, another ~ 60pb^{-1} per experiment were collected at $\sqrt{s} = 183$ GeV, so that approximately 800 additional WW pairs were observed per experiment. However no results on m_W from this sample are available yet[1].

If the result of Equation 27 is combined with the measurement from hadron colliders ($m_W = 80.41 \pm 0.09$ [20]) one obtains the following average:

$$m_W = 80.43 \pm 0.08 \, \text{GeV}. \tag{28}$$

---

[1] At the 1998 Rencontres de Moriond, the LEP Collaborations have presented the following new preliminary W mass measurement based on all the available statistics: $m_W = 80.35 \pm 0.090$ GeV.

STANDARD MODEL FITS

All the results described in the previous Sections can be compared with the standard model predictions. These results, with other precision electroweak measurements obtained outside LEP, are summarised in Table 8.

		Measurement with total error	Standard model	Pull
a)	LEP			
	m_Z [GeV]	91.1867 ± 0.0020	91.1866	0.0
	Γ_Z [GeV]	2.4948 ± 0.0025	2.4966	-0.7
	σ_h^0 [nb]	41.486 ± 0.053	41.467	0.4
	R_ℓ	20.775 ± 0.027	20.756	0.7
	$A_{FB}^{0,\ell}$	0.0171 ± 0.0010	0.0162	0.9
	$\mathcal{A}_\tau$	0.1411 ± 0.0064	0.1470	-0.9
	$\mathcal{A}_e$	0.1399 ± 0.0073	0.1470	-1.0
	$\sin^2\theta_{\text{eff}}^{\text{lept}}$ ($\langle Q_{FB}\rangle$)	0.2322 ± 0.0010	0.23152	0.7
	m_W [GeV]	80.48 ± 0.14	80.375	0.8
b)	SLD			
	$\sin^2\theta_{\text{eff}}^{\text{lept}}$ (A_{LR} [13])	0.23055 ± 0.00041	0.23152	-2.4
	$\mathcal{A}_b$	0.900 ± 0.050	0.935	-0.7
	$\mathcal{A}_c$	0.650 ± 0.058	0.668	-0.3
c)	LEP and SLD Heavy Flavour			
	R_b	0.2170 ± 0.0009	0.2158	1.3
	R_c	0.1734 ± 0.0038	0.1723	0.2
	$A_{FB}^{0,b}$	0.0984 ± 0.0024	0.1031	-2.0
	$A_{FB}^{0,c}$	0.0741 ± 0.0048	0.0736	0.1
c)	$p\bar{p}$			
	m_W [GeV] ($p\bar{p}$ [21])	80.41 ± 0.09	80.375	0.4
	m_t [GeV] ($p\bar{p}$ [22–24])	175.6 ± 5.5	173.1	0.4

TABLE 8. Electroweak measurements for the 1997 summer conferences. The results shown in columns 3 and 4 derive from a fit to the standard model parameters including all data with the Higgs mass treated as a free parameter.

The main goal is to extract information on the unknown parameters of the theory, in particular on m_H.

Table 9 shows the result of the standard model fit to the LEP data alone and to all data of Table 8, with and without inclusion of the direct measurement of the top mass [22–24] and of the W mass. The second column of results in Table 9 shows what is obtained when the direct determinations of m_t and m_W are left out of the fit. These fits make use of the electroweak libraries described in [25].

As one can see, the LEP data favour a light top and a light Higgs. The precision

	LEP	all data except m_t and m_W	all data
m_t [GeV]	155^{+14}_{-11}	157^{+10}_{-9}	173.1 ± 5.4
m_H [GeV]	83^{+168}_{-49}	41^{+64}_{-21}	115^{+116}_{-66}
$\log(m_H/\text{GeV})$	$1.92^{+0.48}_{-0.39}$	$1.62^{+0.41}_{-0.31}$	$2.06^{+0.30}_{-0.37}$
$\alpha_s(m_Z^2)$	0.121 ± 0.003	0.120 ± 0.003	0.120 ± 0.003
χ^2/d.o.f.	8/9	14/12	17/15
$\sin^2\theta^{\text{lept}}_{\text{eff}}$	0.23188 ± 0.00026	0.23153 ± 0.00023	0.23152 ± 0.00022
m_W [GeV]	80.298 ± 0.043	80.329 ± 0.041	80.375 ± 0.030

TABLE 9. Results of the fits to LEP data alone and to all data of Table 8 with and without inclusion of the direct measurements of m_W and m_t. The bottom part of the Table gives results for m_W and $\sin^2\theta^{\text{lept}}_{\text{eff}}$.

on m_W of 41 MeV obtained from radiative correction sets a goal for the direct measurements of the W mass at LEP and the TEVATRON. In fact, when the precision on m_W from direct measurements will match that obtained from the radiative corrections, this will provide an additional powerful test of the theory, in complete analogy with the top case.

The radiative corrections in the standard model depend logarithmically on the Higgs mass. When all the data are included, the fit gives the following result: $\log(m_H/\text{GeV}) = 2.06^{+0.30}_{-0.37}$, i.e, $\log(m_H/\text{GeV})$ falls right into the interval determined by the lower limit from direct searches at LEP2 ($m_H \gtrsim 88$ GeV, preliminary ALEPH result which is going to be discussed in the next Section) and by the theoretical upper limit derived from the consistency of the perturbative approach ($m_H \lesssim 800$ GeV).

The errors on the results shown in Table 9 do not include theoretical uncertainties in the standard model predictions such as those due to missing higher order corrections, as studied in the workshop on 'Precision calculations for the Z resonance' [26]. If these uncertainties are also taken into account, one obtains a 95% confidence level upper limit on m_H of approximately 420 GeV.

HIGGS BOSON SEARCHES

At LEP2, the dominant mechanism for producing the standard model Higgs boson is the so-called Higgs-strahlung process, $e^+e^- \to HZ$. The corresponding diagram is shown in Figure 2a), together with the corresponding cross sections as a function of m_H at LEP1 energies, at 161 GeV and at 172 GeV (Figure 2b)).

a)   b)

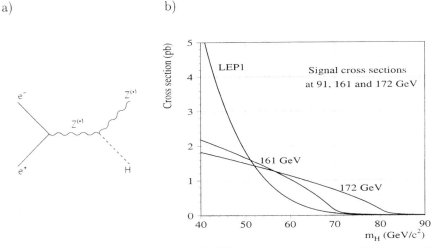

FIGURE 2. a) Feynman diagram for the Higgs-strahlung process; and b) corresponding production cross sections as a function of the Higgs boson mass at LEP1 energies, at 161 GeV and at 172 GeV.

A sizeable cross section is obtained up to $m_H \sim \sqrt{s} - m_Z$, so that an energy larger than 190 GeV is needed to extend the search above $m_H \simeq m_Z$. The present combined limit from the four LEP collaborations as presented at EPS-HEP-97, Jerusalem, [27] gives $m_H \gtrsim 77$ GeV at 95% CL. However the preliminary results based on the analysis of the data at $\sqrt{s} = 183$ GeV raise this limit to approximately 88 GeV [28].

The following event topologies are studied:

i) the four jet channel ($Z \to q\bar{q}$, $H \to b\bar{b}$)

ii) the missing energy channel ($Z \to \nu\bar{\nu}$, $H \to b\bar{b}$)

iii) the leptonic channel ($Z \to e^+e^-, \mu^+\mu^-$, $H \to b\bar{b}$)

iv) the $\tau^+\tau^- q\bar{q}$ channel ($Z \to \tau^+\tau^-$, $H \to q\bar{q}$ and vice-versa).

While at LEP1 energies the signal to noise ratio was as small as 10^{-6} due to the very high $q\bar{q}$ cross section, at LEP2 the signal to noise ratio is much more favourable, increasing to $\sim 1\%$. In order to reduce the residual background, use is made of b-tagging techniques which exploit the large branching ratio of the Higgs into $b\bar{b}$.

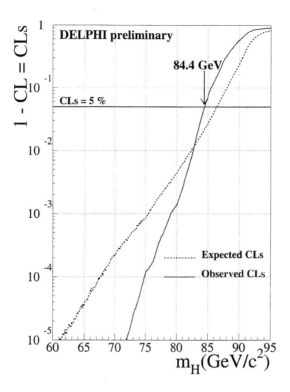

FIGURE 3. Preliminary DELPHI limit on the standard model Higgs boson from the analysis of all high energy data.

In fact for energies above $b\bar{b}$ threshold and below ~ 100 GeV, the Higgs decays into $b\bar{b}$ in approximately 85% of the cases and into $\tau^+\tau^-$ in approximately 8% of the cases.

Figure 3 shows, as an example, the preliminary limit derived by DELPHI from all high energy data: $m_H > 84.4$ GeV at 95% CL and the expected sensitivity. Similar limits are obtained from the other LEP experiments.

In the years 1998 to 2000 LEP2 is expected to deliver a luminosity larger than 200 pb^{-1} per experiment at a centre-of-mass energy eventually as high as ~ 200 GeV. As shown in Figure 4 for $\sqrt{s} = 205$ GeV, such a luminosity should allow to discover or exclude a standard model Higgs lighter than approximately 105 GeV.

In the minimal super-symmetric extension of the standard model (MSSM), two Higgs doublets are introduced, in order to give masses to the up-type quarks on the one hand and to the down-type quarks and charged leptons on the other. The

Higgs particle spectrum therefore consists of five physical states: two CP-even neutral scalars (h,A), one CP-odd neutral pseudo-scalar (A) and a charged Higgs boson pair (H$^\pm$). Of these, h and A could be detectable at LEP2. In fact, at tree-level h is predicted to be lighter than the Z. However, radiative corrections to m_h are proportional to the fourth power of the top mass, shifting the upper limit of m_h to approximately 135 GeV, depending on the MSSM parameters.

The main production mechanisms are through the Higgs-strahlung process $e^+e^- \to HZ$, as for the standard model Higgs, and the associated pair production $e^+e^- \to hA$. The corresponding cross sections may be written in terms of the Higgs-strahlung cross section in the standard model, σ_{SM}, as:

$$\sigma(e^+e^- \to Zh) = \sin^2(\beta - \alpha)\, \sigma_{SM} \qquad (29)$$
$$\sigma(e^+e^- \to hA) = \cos^2(\beta - \alpha)\, \sigma_{SM}.$$

The parameter $\tan\beta$ gives the ratio of the vacuum expectation values of the two

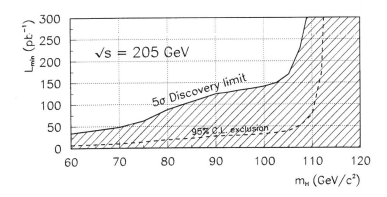

FIGURE 4. Minimum luminosity needed per experiment for a combined 5 σ discovery or a 95% CL exclusion as a function of m_H.

Higgs doublets and α is a mixing angle in the CP-even sector.

At small $\tan\beta$, the Higgs-strahlung h Z process dominates with a cross section of the order of 0.8 pb. Conversely, at large $\tan\beta$ the associated hA production becomes the dominant mechanism with rates similar to the previous case.

All the analyses developed for the standard model Higgs produced via the Higgs-strahlung mechanism can be used with no modification for the super-symmetric case, provided that the Higgs decays to standard model particles ($b\bar{b}$, $\tau^+\tau^-$). The results can then be reinterpreted in the MSSM context, by simply rescaling the number of expected events by the factor $\sin^2(\beta - \alpha)$.

For the pair production process, the signal consists of events with four b-quark jets or a $\tau^+\tau^-$ pair recoiling against a pair of b-quark jets. Figure 5 shows for DELPHI a preliminary limit on m_h as a function of $\tan\beta$. At all values of $\tan\beta$, m_h is larger than approximately 75 GeV. Similar limits are obtained on m_A.

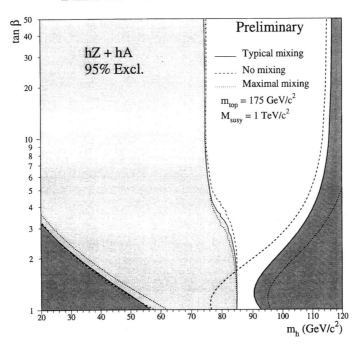

FIGURE 5. Preliminary DELPHI limit on m_h from the analysis of all high energy data.

OUTLOOK AND CONCLUSION

At present the data continue to support the standard model in a remarkable way, favouring a scenario with a light Higgs. Tight bounds on all conceivable forms of new physics are also obtained.

The LEP1 programme of precision tests of the standard model is close to its end. However some improvements can still be expected:

- The line-shape results as well as the measurements of the LEP beam energy have to be finalised.

- New revised R_b measurements employing improved techniques, as already done by ALEPH, DELPHI and SLD, should be available in the near future from the other experiments. In addition, the errors on R_b, as well as on R_c, $\mathcal{A}_b$ and $\mathcal{A}_c$, from SLD are expected to reduce by a factor of approximately 2.0 using the data expected by the end of 1998.

- The uncertainty on observables such as the τ polarization asymmetry, which is still statistics dominated, will be reduced, since not all the data have been analysed yet.

- A factor of ∼1.6 improvement can still be expected on the determination of $\sin^2\theta_{\text{eff}}^{\text{lept}}$ from A_{LR} at SLAC from the increased statistics.

- The W mass will be measured more precisely at LEP2 and at the TEVATRON.

Direct searches for new physics have been so far unsuccessful. The analysis of the LEP2 data has allowed to push the limit on the standard model Higgs mass from the 65 GeV lower bound obtained at LEP1 to approximately 88 GeV. By the end of the year 2000, it should be possible to find or exclude a Higgs of ∼ 105 GeV. This is a particularly interesting mass region to explore given the present indication for a light Higgs from the standard model fit of the electroweak precision data. Contrary to the Higgs search, which is still at an early phase, the search for supersymmetric particles has already allowed to exclude most of the allowed parameter space. For example, the chargino lower limit which was ∼ 45 Gev before LEP, is now ∼ 91 GeV, and it will reach ∼ 100 GeV by the year 2000.

ACKNOWLEDGEMENTS

I am especially grateful to my friends and colleagues of the LEP Electroweak Working Group and in particular to Christoph Paus for producing the combination of experimental results which are used in this paper and for several discussions and graphs. I would also like to thank Patrick Janot for providing many of the Higgs plots. I would also like to express my sincere gratitude to the mexican colleagues, and in particular to Marti Ruiz Altaba, Fernando Quevedo and Juan Carlos D'Olivo, for their invitation to Morelia and for the excellent organization and very lively and nice atmosphere of the Workshop.

REFERENCES

1. J. Timmermans, Proceedings of LP'97, Hamburg, 1997; S. Dong, ibidem; D. Ward, Proceedings of HEP97, Jerusalem,1997.
2. The LEP Collaborations, the LEP Electroweak Working Group and the SLD Heavy Flavour Group, CERN preprint CERN-PPE/97-154.
3. S. Jadach, E. Richter-Wąs, Z. Wąs and B.F.L. Ward, Phys. Lett. **B353** (1995) 362; Comput. Phys. Commun. **70** (1992) 305.
4. S. Jadach, W. Placzek, E. Richter-Wąs, B.F.L. Ward and Z. Wąs, *Upgrade of the Monte Carlo program BHLUMI for Bhabha scattering at low angles to version 4.04*, CERN-TH/96-158, June 1996, Submitted to Comp. Phys. Commun..
5. LEP Energy Working Group, R. Assmann et al., Z. Phys. **C66** (1995) 567.
6. G. Wilkinson, *The determination of the LEP energy in the 1995 Z^0 scan*, Proceedings of ICHEP96, Warsaw, 25-31 July 1996.
7. L. Arnaudon *et al*, Phys. Lett. **B284** (1992) 431.
8. LEP Energy Working Group, private communication. The contact person is Tiziano Camporesi, mail address T.Camporesi@cern.ch.
9. A. Leike, T. Riemann, J. Rose, Phys. Lett. **B 273** (1991) 513;
   T. Riemann, Phys. Lett. **B 293** (1992) 451;
   S. Kirsch, T. Riemann, Comp. Phys. Comm. **88** (1995) 89.
10. LEP Electroweak Working Group, *An Investigation of the Interference between Photon and Z-Boson Exchange*, Internal Note, LEPEWWG/LS/97-02, September 1997 http://www.cern.ch/LEPEWWG/smatrix/s9702.ps.gz.
11. TOPAZ Collaboration, K. Miyabayashi et al., Phys. Lett. **B 347** (1995) 171.
12. The LEP Heavy Flavour Group, *QCD Corrections to $A_{FB}^{b\bar{b}}$ and $A_{FB}^{c\bar{c}}$ Measurements at LEPI*, LEPHF/97-01, http://www.cern.ch/LEPEWWG/heavy/.
13. SLD Collaboration, P. Rowson, talk presented at Moriond 97.
14. S. Eidelmann and F. Jegerlehner, Z. Phys. **C67** (1995) 585.
15. SLD Collaboration, SLAC-PUB-7629, contributed paper to EPS-HEP-97, Jerusalem, **EPS-122**;
    SLD Collaboration, SLAC-PUB-7630, contributed paper to EPS-HEP-97, Jerusalem, **EPS-123**;
    SLD Collaboration, contributed paper to EPS-HEP-97, Jerusalem, **EPS-124**;

SLD Collaboration, SLAC-PUB-7595, contributed paper to EPS-HEP-97, Jerusalem, **EPS-126**;
E. Etzion, talk presented at EPS-HEP-97, Jerusalem, to appear in the proceedings.
16. SLD Collaboration, SLAC-PUB-7585, contributed paper to EPS-HEP-97, Jerusalem, **EPS-118**.
17. ALEPH Collaboration, R. Barate *et al*, Physics Letters **B 401** (1997) 150;
ALEPH Collaboration, R. Barate *et al*, Physics Letters **B 401** (1997) 163.
18. DELPHI Collaboration, *Measurement of the partial decay width* $R_b^0 = \Gamma_{b\bar{b}}/\Gamma_{had}$ *with the DELPHI detector at LEP*, DELPHI 97-106 CONF 88, contributed paper to the EPS-HEP-97, Jerusalem, **EPS-419**.
19. OPAL Collaboration, K. Ackerstaff *et al*, Z. Phys. **C74** (1997) 1.
20. Y.K. Kim, talk presented at the Lepton-Photon Symposium 1997, Hamburg, 28 July - 1 Aug, 1997, to appear in the proceedings.
21. M. Rijssenbeek, talk presented at ICHEP96, Warsaw, 25-31 July 1996, to appear in the proceedings.
22. CDF Collaboration, J. Lys, *Top Mass Measurements at CDF*, Proc. ICHEP96, Warsaw, 25-31 July 1996, 1196.
23. DØ Collaboration, S. Abachi *et al*, Phys. Rev. Lett. **79** (1997) 1197.
24. R. Raja, talk presented at XXXIInd Rencontres de Moriond, Les Arcs, 16-22 March 1997, to appear in the proceedings.
25. Electroweak libraries:
ZFITTER: D. Bardin *et al*, Z. Phys. **C44** (1989) 493; Comp. Phys. Comm. **59** (1990) 303; Nucl. Phys. **B351**(1991) 1; Phys. Lett. **B255** (1991) 290 and CERN-TH 6443/92 (May 1992);
BHM (G. Burgers, W. Hollik and M. Martinez): W. Hollik, Fortschr. Phys. **38** (1990) 3, 165; M. Consoli, W. Hollik and F. Jegerlehner: Proceedings of the Workshop on Z physics at LEP I, CERN Report 89-08 Vol.I,7 and G. Burgers, F. Jegerlehner, B. Kniehl and J. Kühn: the same proceedings, CERN Report 89-08 Vol.I, 55;
TOPAZ0: G. Montagna, O. Nicrosini, G. Passarino, F. Piccinnii and R. Pittau, Nucl. Phys. **B401** (1993) 3; Comp. Phys. Comm. **76** (1993) 328.
These computer codes are upgraded by including the results of [26] and references therein
26. *Reports of the working group on precision calculations for the Z resonance*, eds. D. Bardin, W. Hollik and G. Passarino, CERN Yellow Report 95-03, Geneva, 31 March 1995.
27. P. Janot, *Searches for New Particles at Present Colliders*, talk presented at EPS-HEP-97, Jerusalem, to appear in the proceedings.
28. Presentations by LEP Collaborations at LEPC Open Session, 11 November 1997.

Detection Techniques of Ultra High Energy Cosmic Rays

H. Salazar

Facultad de Ciencias Físico-Matemáticas
Universidad Autónoma de Puebla
Puebla, Pue., México

Abstract. Several detection techniques in current high-energy cosmic rays are examined. The results of some experiments are discussed and a review of the future projects is made, emphasizing their discovery potential in accordance with their aperture.

INTRODUCTION

Nowadays there exist a growing and strong interest on the subject of ultra high-energy cosmic-rays, which is reflected by the large number of existing and planned experiments. We can mention, among others, Hi-Res Fly's Eye, Telescope Array, Auger Observatory, OWL and Air-Watch, Pion Eyes, etc. All these projects share a common feature: they are very ambitious and they will collect a statistically significant number of events within a relatively short period of time. This can help to answer the typical questions with regards to the energy spectrum, composition, and direction of arrival of cosmic rays with energy above the GZK cut. The existence of cosmic rays of this sort is not questioned any longer; on the contrary this is precisely the motivation for the above mentioned projects.

In order to have elements to reflect on the scope and present situation of each of the projects, we review the state of the detection techniques in high-energy cosmic rays.

In the first section, the main working experiments and others that have finished their operating period are described, provided that they are able to detect cosmic rays with energies above 10^{17} eV. In the second section, the results for these experiments are presented. Finally, in the third section, the projects that are currently in the R&D stage are reviewed, emphasizing their capabilities in accordance with their aperture.

EXPERIMENTS AND TECHNIQUES IN ULTRA HIGH ENERGY COSMIC RAY DETECTION

Above about 100 Tev the Energy spectrum of cosmic rays can not be measured any longer by means of direct techniques by using balloons or satellites to fly or transport instrumentation to high altitudes. Among all this radiation there are the most energetic particles whose energy appears to reach beyond 10^{20} eV. The information of these particles can only be obtained indirectly from studies on the extensive air showers or atmospheric cascades, which are produced by primary particles. Because one has to infer the nature of the primary from the secondary cascade, the measurements become relatively difficult and the calculation of the characteristics of the primary flux is model dependent. Each model is an extrapolation of particle interaction measurements in man-made particle accelerators. For the most energetic cosmic rays, their extreme energy is seven orders of magnitude greater than those nucleons accelerated by humans on earth. When the cascade reaches the ground, it can cover an area of tens of square kilometers. Giant Ground Arrays of scintillation or water Cerenkov units registrate the density of secondary particles and arrival times to each of them. Energy of the primary cosmic ray is then calculated from either shower size or density of particles far from the core, s(600m). The arrival direction is calculted through relative time of detection at each unit, and composition comes from relative densities of secondary particles. On the other hand, the collision of the secondary particles with nitrogen molecula in the air produces a process of excitation and de-exitation in these molecula and consequently a flash of fluorescent light in the atmosphere. The air fluorescence detectors view this light as it propagates in the atmosphere; they determine the track geometry by either the photomultiplier tube trigger times or so-called "stereo" reconstruction, then they calculate the primary energy by an integral along the track lenght, and deduce the chemical composition by the shape of the longitudinal shower development. It is important to mention that besides the fluorescence light, there is a copious Cerenkov light produced along the shower axis by the charged particles, and a large area Cerenkov array can be used to detect that light. Alternative to the air fluorescence light detection, the total flux of the Cerenkov light is a good measure of the total particle track integral in space and thus it constitutes a good primary energy measurement. The angular and lateral distribution of Cerenkov light can be used to deduce the primary composition.

MIT Volcano Ranch Station.

The first giant array, which used scintillation detectors to find the direction(from pulse times) and size(from pulse amplitudes) of showers events, recorded events having energies greater than 5×10^{18} eV[1, 2]. This event collection allowed serious research on extremely high-energy cosmic rays.

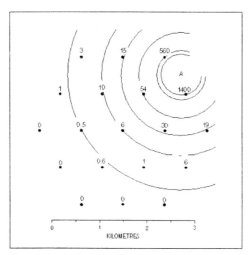

FIGURE 1. An extensive air shower recorded by Volcano Ranch array with energy above 10^{20} eV. The numbers indicate the observed particle densities (m^{-2}).

Following pioneering work by Bassi, Clark and Rosi at MIT for the reconstrution of the incidence direction of an EAS, and with the experience of the Agasa experiment which ran from 1954-1957, Linsley and Scarsi constructed the first giant array which started operation in 1959 at Volcano Ranch in New Mexico.

The Volcano Ranch array was a system of 19 scintillation detectors arranged in a hexagonal grid with spacing of approximately 884m between units. Each $3.3m^2 \times 9cm$ thick scintillator was linked to a central location by high bandwidth cable. The energy loss, correlated with the particle density at each detector as well as the arrival time of the shower front relative to the other detectors was recorded by oscilloscope photography.

In February 1962, this first giant array recorded a cosmic ray air shower that indicated a total energy of the primary particle of $1.0 \times 10^{20} ev$ and the zenith angle nearly vertical $(10 \pm 5°)$[2]. The shower was about twice the size of the largest they had report previously, recorded in March 1961. Fig.1 and 2 show the shower densities and lateral distribution of is event which was the most energetic ever recorded at Volcano Ranch.

The Haverah Park Array

The Haverah experiment was operated by the University of Leeds from 1968 to 1987. Although the Haverah Park array intended primarily to study the lower energy Extensive Air Showers, this giant array recorded a small number of ultra high energy showers. The Haverah Park experiment was an array of water Cerenkov

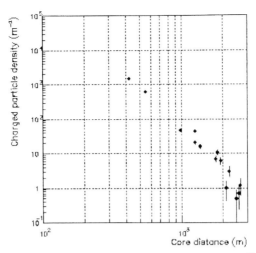

FIGURE 2. The first event recorded, for which an energy of $\geq 10^{20}$ eV was claimed, was registered in 1962 by Volcano Ranch system.

detectors, 1.2 m deep spread over $12 km^2$, located at an altitude of 200 m above sea level and at 53° 58.2'N, 1°38.22'W. The 225 stations were of a unit area of $2.25 m^2$ and deployed in clusters of up to $34 m^2$. In the late 1970's an upgrade to $30 \times 1 m^2$ units was built in a central area. The trigger requirement was fixed to produce an array threshold of approximately 6×10^{16} eV.

Haverah Park water Cerenkov array used extensively Hilla's idea of the use of S(600) as an energy estimator. Model calculations have shown that for a given primary energy, the particle density at large distances of the core has smaller fluctuations. In fact showers models and mass composition affect the primary energy estimates only weakly. Particle density is then a more robust energy parameter against total ground shower size.

We show in fig 3 a very well measured ultra-high energy event recorded by the Haverah Park array whose assigned energy was 1×10^{20} eV[3, 4].

Yakustk EAS Array

YAKUSTK array is located in Siberia at a distance of 50km from Yakustk and 100m above sea level. Its geographical coordinates are 61.7°N and 129.4°E. At present, Yakustk EAS surface array consists of 49 scintillation detectors of $2 \times 2 m^2$ area and 9 surface detectors of $2 m^2$ forming an array of $10 km^2$ area and spaced 500m. Yakustk array includes Cerenkov light detectors (27 with one 15 cm diameter photocathode PMT and 18 with 3 PMT's) and underground muon detectors too. The Yakustk array more effectively operates EAS after modernization in the energy of $2 \times 10^{17} - 10^{19}$ eV(ankle region).

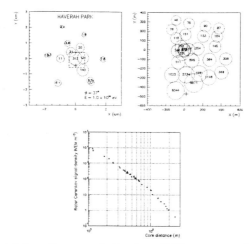

FIGURE 3. An exquisitely well measured ultra-large EAS event recorded by the Haverah Park system. This event fell close to an "infilled" array of smaller high dynamic range detectors, and hence it was measured in very great detail. Assigned event energy was 1×10^{20} eV. The commonly assumed EAS property of axial symetry is dramatically demonstrated, since there is a large number of redundant observations, each having very high statistical weight.

The first stage of the array was 13 surface units located in a circle of 1km radius recording data during 1969-1971. Similar techniques to the Volcano Ranch array were used to determine total energy, arrival direction and composition for primary particles (lateral distribution of particles, size of the shower and relative arrival time as function of the position).

The next stage of the array was completed in 1973 increasing the total area of the array up to $18km^2$. And begin starting a more sistematic study of cosmic rays with energy of $10^{17} - 10^{20}$ eV.

In May, 1989 a giant shower was detected at the Yakustk array. The energy calculated was $(1.5 \pm 0.7) \times 10^{20}$eV and zenith angle of $58.7°$. Fig 4 shows the lateral distribution for this event[5].

AGASA Array

The Akeno air shower experiment started in 1979 with an array covering an area of $1km^2$. In 1985 the array was upgraded to $20km^2$. From 1990, the array has been expanded gradually to a Giant Air Shower Array (AGASA) of $100km^2$ area. The AGASA consists of 111 scintillation detectors of units of $2.2m^2$ area and spacing of about 1km. Data transmition is carried out through two optical fiber cables which connect the central station of each branch and the unit detectors. The triggering requirement is a 5-fold coincidence of neighbouring detectors.

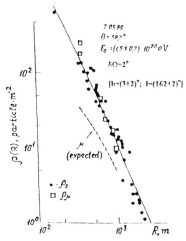

FIGURE 4. The giant shower detected at the Yakuts array (1989)

The whole area is divided into four branches, Akeno, Sudama, Takane and Nagasaka. Data acquisition started independently in each branch from 1990 and it was unified into one at the end of 1995. The technique used to evaluate the energy is the lateral distribution, particularly the densitiy at a distance of 600m (600).

AGASA, geographically located at 138° 30'E and 38° 47'N, has detected in 1993 an event whose energy exceeded considerably the GZK cutt-off energy. This event had 2.1×10^{20} eV and zenith angle 23°. Fig 5 shows the number of particles detected at each unit and the lateral distribution of charged particles for this event[6].

FLUORESCENCE DETECTOR TECHNIQUES

By the beginning of the 1960's, Chudakov and Suga were the first to realize that atmospheric nitrogen fluorescence might be used to detect extensive air showers. They pointed out that emission from N_2 had a convenient wave length for atmospheric transmission and for its detection through photomultiplier tubes. Greisen and his Cornell group [1] where the first to build an air shower detector, they used Fresnel lenses as light collectors, but the first succesful demonstration of the technique was performed in 1976 by the University of Utah group led by G. Cassiday. They built a mirror-based detection system at the same place as the Linsley's Volcano Ranch ground array in New Mexico. They could find events detected by in coincidence Linsley's ground array and their fluorescence detector. Then, they designed the Fly's Eye experiment in Utah in 1978. The Utah group published in 1983 the results from the highest energy cosmic ray ever detected. The calculated energy of this event is 3.2×10^{20} eV and 44°. Fig 6 shows the longitudinal profile. The currently running fluorescence experiment is called the High Resolution Eye

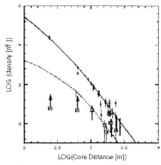 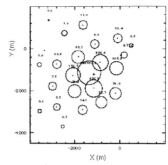

FIGURE 5. Left: The lateral distribution of charged particles (closed circles) and muons (squares) of the super energetic event recorded by AGASA. The large open circle is the density measured by a detector designed to study the arrival time distribution of particles in air shower. Right: Map of the density distribution of event. A cross shows the estimated location of the shower core.

Detector (HiRes), an international collaboration which is a big upgrade and improvement of the Fly's Eye (now defunct) experiment.

The Fluorescence Light and the Atmospheric Monitoring

The fluorescence detectors operate under dark, moonless skies, limiting their duty cycle to 10-12%. There are possibilities, not confirmed yet for operation under crescent moon, probably increasing the duty cycle to above 15%.

Some of the ionization energy loss experienced by charged particles in the atmosphere is applied to the exitation of N_2 molecules and N_2^+ ions. The excitation givesrise to fluorescence light if the excitation is not quenched by some other proceses like collision with other molecula. Several bands of emission are concentrated between 300 and 400nm[7]. About 88% of the total emission is found in this range with the three strongest bands at 337nm. and 391nm. Fluorescence from other molecula is not significant in this range. Certainly, this fluorescence light has a very low production efficiency. However if the showers are large enough, it is possible to detect them in a very spread area. For instance, the first version of the Fly's Eye detector had an aperture of $1000\,Km^2sr$. Its upgrated version, the HiRes detector; has now an aperture 10 times bigger for the same energy.

The technique of the measurement of cosmic rays showers by nitrogen fluorescence relies critically upon knowledge of the atmospheric absortion and scattering profile. With scale lenghts on the order of 10 Km, and new detectors observing events to distances of several scale lengths, regular measurements of the atmo-

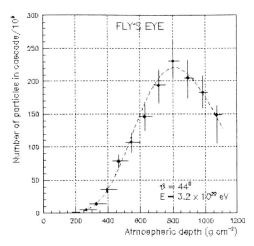

FIGURE 6. The highest energy cosmic ray ever detected. The calculated energy of this event is 3.2×10^{20} eV and zenith angle of $44°$. This figure shows the longituditudinal profile.

spheric aerosol concentration at the observatory site are very performed routinaly.

There are two effects which dominate the propagation of fluorescence light through the atmosphere. The first is the Rayleigh scattering by atmospheric molecules. This effect is largely constant with time and generally dependent on atmospheric effects such as temperature and relative humidity. The second, which is less understood is the Mie scattering. This is a scaterring of light by aerosol particles suspended in the atmosphere. It dependents on both the concentration of aerosol in the atmosphere and their size distribution. These are functions of temperature and relative humidity as well as other meterological conditions such as wind speed. Mie scaterring can vary greatly on relatively short time scales. MODTRAN and LOWTRAN programs have been found to reproduce atmospheric effects with reasonable accuracy. The desert aerosol model of these programs has been found to resemble closely the atmosphere at Dogway Proving Grounds, Utah, the site of Fly's Eye and HiRes.

Several methods have been used to measure the athmosferic conditions in a fluorescence detector site. The Fly's Eye experiment used 23 vertical argon flashers deployed around its detector. These flashers project colimated beams of ultraviolet light into the atmosphere at thirty minute intervals. These beams of light are scattered and viewed by the Fly's Eye detector. However, the position and trajectory of these flashers is fixed, allowing only a partial zone monitoring of the atmosphere [8].

More recently, a LIDAR system has been developed for use with the HiRes detector. This device has a variable trajectory, but it is restricted to a single location

(Fly's Eye 2) due to the physical size of the components and the precise optical alignments required. Furthermore, the trajectory information is only accurate to approximately 1 degree.

Finally, a LaserScope prototype has been designed and built at the University of Adelaine for use in calibrating the Hi Res detector and monitoring the atmosphere in the aperture of the detector. The heart of the LaserScope is a Continum MiniLite-10 YAG laser, with a frequency tripper. This is a very compact, water-cooled, solid state laser which has no difficulty in being held at any orientation and is capable of generating a 4 mJ pulse at 355nm. with a pulse width of 5 ns. The head of the laser and the associated optics are contained in a box weighing less than 4 kg [9].

The Fly's Eye Experiments

The Fly's Eye detector was the first successful air fluorescence shower detector and showed its power in both energy and in composition resolution. Because the detector viewed the shower development curves, its energy estimation is almost totally model independent.

The Fly's Eye detectors were located at Dogway, Utah ($40°N$, $113°W$, atmospheric depth 860 gcm^{-2}). The original detector, Fly's Eye I (completed in 1981), consisted of 67 spherical mirrors of 1.5 m diameter, each with 12 or 14 PMT's at the focus. The mirrors where arranged so that the entire night sky was imaged, which each phototube viewing a hexagonal region of the sky 5.5 degrees in diameter. In 1986 a second detector (Fly's Eye II) was completed 3.4 km. away, consisted of 36 mirrors of the same design. This detector viewed half of the night sky in the direction of the Fly's Eye I for a stereo view of a subset of the air showers. There were 880 PMT's in Fly's Eye I and 464 tubes in Fly's Eye II.

An important feature of the Fly's Eye detector was that those showers viewed by both Fly's Eye I and II (i.e. "stereo events") were measured with significant redundancy. This provided a model independent way of checking the energy and depth of shower maximum ($Xmax$) resolution. The energy and $Xmax$ values were independently reconstructed from each eye using the stereo geometry. By comparing the results, the stereo geometry resolution was determined to be 24% for events below $2 \times 10^{18} \, eV$. The ($Xmax$) resolution for the stereo events is 47 g/cm^2. The monocular energy resolution was calculated by comparing the monocular energy with the stereo energy event by event, and the FWHM is estimated to be 36% for events below $2 \times 10^{18} \, eV$. It should be pointed out that the monocular energy resolution is underestimated using this method since the stereo energy reconstruction also uses the Fly's Eye I phototubes intensities. In the stereo case, the energy resolution $((E_1 - E_2/Eavg)$ follows a Gaussian distribution, but in the monocular case, only

$log(E_{mono}/E_{stereo})$ is Gaussian. Here, E_1 and E_2 are the energies determined from Fly's Eye I and II using the stereo geometry independently. $Eavg$ is the average of E_1 and E_2. $Emono$ is the energy determined by the monocular reconstruction using Fly's Eye I information only, and $Estereo$ is the energy determined by stereo reconstruction using information from both eyes.

The HiRes Experiment

The High Resolution Fly's Eye Detector (High Res)[10] will be at least an order of magnitude more sensitive to cosmic rays above than the original Fly's Eye pair. HiRes is designed to study the cosmic ray flux above $3 \times 10^{18} eV$ and measure its composition and anisotropy with significantly improved resolution.

A set of fixed mirrors with 256 photomultipliers at the focus of each mirror is used to grid the sky to 1 degree resolution. HiRes currently consists of two sites on top of two mountains separated by 13km in Western Utah. At present, it consists of a 14-mirror prototype at the first site and a 4-mirror site at the second site. It is based on the original Fly's Eye experiment that was located at the same site. HiRes is building upon the experience gained in this experiment with a new detector that has greater sensitivity, larger aperture, and improved resolution in both energy and direction. The prototype also overlooks the Chicago AirShower Array (CASA) and the Michigan muon array (MIA) which both detect the passage of particles from airshowers that reachs ground level.

The HiRes Program is fully funded and scheduled for complete operation in 1999. Several important questions may be answered within 3 year of HigRes data. For example, HiRes will have approximately 40 events above 5×10^{19}eV with stereo mode reconstruction. This experiment should be able to answer the basic question- does the GKZ mechanism affect the energy spectrum of the cosmic rays at the highest energies? HiRes will certanly be the fastest route to the first glimpse at this type of physics.

The HiRes project is a collaboration between the University of Adelaide, Columbia University, University of Illinois at Urbana-Champaign and the University of Utah to study the highest energy cosmic rays.

THE ENERGY SPECTRUM AT THE HIGHEST ENERGY

Analysis of accumulated databanks point to a general agreement that the cosmic ray energy extends well beyond the GZK cut-off although there is no final agreement on the shape of the spectrum above 4×10^{19} eV.

The energy spectrum of high energy cosmic rays above the 10 GeV, energy where the magnetic field of the sun does not affect any more, is well represented by a power law form. In terms of the value of that power, the spectrum can be divided into three regions: two "knees" and one "ankle".

The first "Knee" appears around 3×10^{15} eV where the spectral power index changes from -2.7 to -3.0. Unfortunately, despite intensive effort, there is poor agreement about the position of the first knee in the energy spectrum. The differences betwen the estimates of the knee position are larger than what can be understood from likely systematic errors. Table 1 is an actualized version of the results of several experiments on the position of the bend in the primary spectrum[6].

TABLE 1. Where is the Knee in the Cosmic Ray spectrum?

1. Tibet(Amenomori et al (1995):the spectrum steepens continuosly.
2. EAS-TOP (V4:125):4×10^{15} eV.
3. KASCADE (V6:157):$(3-6) \times 10^{15}$ eV.
4. AKENO (Nagano et al 1984):3×10^{15} eV.
5. MSU (Fomin et al 1991):3.5×10^{15} eV.
6. Tunka Valley (V4:129):4×10^{15} eV.
7. DICE (oral presentation: Ap.J. Letters submitted):4×10^{15} eV.
8. HEGRA (Wuppertal)(V4:121):$(1-2) \times 10^{15}$ eV.
9. HEGRA (Munich-Madrid-Hamburg)(V4:69):2×10^{15} eV.

The second "knee" is somewhere between $(10^{17}-10^{18})$ eV where the spectral slope bends from -3.0 to around -3.3. Fig. 6 shows the measurements of the diferential flux reported at the 25 ICCR conference in the range from 10^5 to 10^8 GeV.

The "ankle" is seen in the region of 3×10^{18} eV. Above that energy, the spectral slope flattens out to about -2.7, however, there is large uncertainty due to poor statistics and resolution. Like the "knee" the exact shape of the "ankle" is very uncertain.

The best results[11] on the "ankle" structure come from the Fly's Eye stereo data of Fig 7, because of the well-controlled errors systematic of those data. The spectrum becomes steeper immediately beyond $10^{17.6}$ and flattens beyond $10^{18.5}$ eV. The change in the espectral slope forms a dip centered at $10^{18.5}$ eV.

The Fly's Eye experiment is not the only data to see the dip. AGASA, Haverah Park and Yakustsk have reported similar observations. See fig. 8 and 9.

Among all of these experiments, Fly's Eye gave the lowest spectral normalization, and Yakustsk gave the highest. The difference could be due to three potential problems: the absolute energy calibration, the exposure calculation, or the energy resolution[6].

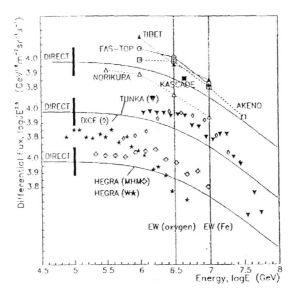

FIGURE 7. Measurements of the differential flux reported in the range 10^5 to 10^8 GeV. The vertical lines marked EW (oxigen and Fe) refer to the position of bumps in the spectrum proposed by Erlykin and Wolfendale (V4:85 and 161). They would wish to see some displacement in the energy scales of certain experiments. The spectra are displaced vertically and are grouped according to the methods used. The line drawn in each case passes through the direct estimate of the all particle spectrum at 10^5 GeV and is intended only to guide the eye.

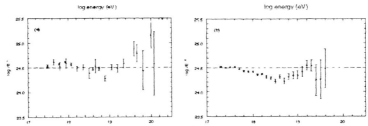

FIGURE 8. Differential energy spectrum ($\times E^3$) as observed by, a) The Haverah Park array, b) the Fly's Eye experiment. Note that the exhibited spectrum for the Fly's Eye is for events recorded by more than one eye ("stereo" events), and so does not include their largest event at $E = 3 \times 10^{20}$ eV.

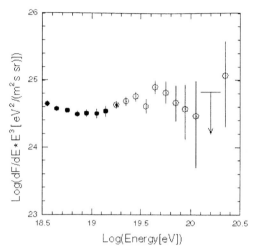

FIGURE 9. Differential energy spectrum ($\times E^3$) as observed by the AGASA experiment, for an integrated exposure of $630 km^2$ yr sr. The upper limit shown is at 90% confidence.

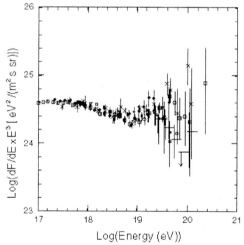

FIGURE 10. Combined differential energy spectra ($\times E^3$) from The Haverah Park ($\times$), Fly's Eye (stereo, •), Yakutsk ($\square$), and AkenoAGASA($\odot$) experiments. The energy scale of each experiment has been slightly shifted to match the AGASA result around 10^{18} eV.

COMING PROJECTS AND CONCLUSIONS

The Telescope Array Prototype

The Telescope Array[12] project is another undertaking of the University of Tokyo, Japan, now in collaboration with the HiRes group. The development work is going on at Akeno and the Fly's Eye site in Utah, where the prototype detectors have been installed. The seven-telescope array is under construction at Utah, and three are in operation. With these prototype detectors, they have carried out the observation of the Crab, AGN's and Marcarian nebulas[5] in the TeV gamma ray mode. Its geographical position is $40.33°N$, $113.02°W$, atmospheric depth 860 gcm^{-2}, at altitude of 1600m above the sea level. The prototype detector works in dual modes: The Cerenkov and the air fluorescence mode. Construction started in 1996, and currently three telescopes with respective separation of 120m are in operation. The seven telescopes will be arranged in a hexagonal grid with a separation of 70m to maximize the detection efficiency of Tev gamma rays. Each telescope has an alt-azimuth mount with a $6m^2$ main dish. The main dish consists of 19 hexagonal shape segmented mirrors coated with anodized aluminum. The reflectivity of the mirrors are about 90% at wave legth of 400nm. At the focal plane, a high resolution imaging camera of 256 PMT's is installed to measure detailed images of Cerenkov light from both gamma and cosmic rays. The typical cosmic ray rate is about 1000/min and the gamma ray rate from the Crab nebula is about 0.5/min. [astro-phys] with three telescopes. Therefore rejection of cosmic ray background events using the shape parameter of the Cerenkov light is essential to obtain a reasonable S/N ratio in this experiment. The absolute pointing accuracy of the telescopes is typically 1 arcmin which is frequently callibrated by imaging bright stars.

The signals from the PMTs are amplified in the PMT camera and then are fed to the data adquisition system. The pulse height and the pulse timing of the signals from PMTs are measured in each event. The rise time of signals is about 10ns and the TDC values are digitized with one ns accuracy. With the TDC information, the background photon noise can be efficiently eliminated in the image processing. The threshold level of discriminators corresponds to 4-photo-electrons and the single PMT tube rate is 5 khz. The triggering system requires 4 hit channels, so the resultant cosmic rate event rate in each telescope becomes about 4-5 Hz.

The Auger Project

The Auger Collaboration proposes[4] to build two arrays, each $3000\,km^2$ in area, one in the southern hemisphere (San Rafael Argentina) and one in the northern hemisphere (Millard Count, Utah). Detector arrays are located in each hemisphere to give full sky coverage. This feature allows a more sensitive search for galactic anisotropy, dipole and quadrupole moments of source distribution, and eliminates holes in the sky when cataloging points source distribution.

The ground array consists of water tanks and PMTs to detect large air showers and measure the energy and will provide some information on the particle type. In order to calibrate these ground array, Auger proposes to build three fluorescence detectors to detect events (about 10%) in hybrid mode. The fluorescent and extendend air shower array detector types are complementary and can be used to reduce systematic errors, particulary on energy determination.

The proposal results from a 5 year study and involves an international collaboration with physicists from about 18 countries. There are large numbers of ways by which the highest energy cosmic rays might be detected. The Auger group has chosen to conbine the two must established methods in a conservative experiment that is almost certainly guaranteed to achieve its predicted sensitivity. As noted above there is a very definitive merit in having two distinct but overlapping detectors, this is one of the real strengths of the Auger experiment. These are fairly simple techniques and there is little mystery associated to their use.

The Auger project moves beyond HiRes and the proposed Telescope Array project to provide both significantly increased statistics and full-sky coverage, Auger's hybrid design will permit showers to be caracterized with far greater confidence than previously possible. Observations by Auger will form the backdrop againts which the scientific goals and performance requirements for a potential future project such as OWL are stablished.

ACKNOWLEDGEMENTS

I am grateful to the organizers of this Workshop. I am indebted with the Mexican Auger Collaboration, specially to Arturo Fernández, Sergio Román Marciano Vargas and Galo De La Cueva for the ideas discussed in this work and for helping to undertake the experimental aspects of the Auger Project.

REFERENCES

1. J.Linsley and L. Scarsi, *Physical Review* **128**, 2384-2392 (1962).
2. J.Linsley, *Phys. Review Letters* **10**, 146-148 (1963).
3. C. L. Pryke, *Ph.D Thesis (University of Leeds)* (1996).
4. *Design Report. The Pierre Auger Project. Revised version, March 1997* (1997).
5. B.N. Afanasiev et al, *Proceedings of the International Symposium on Extremely High Energy Cosmic Rays: Astrophysics and Future Observation (Tokio)* 32-49 (1996).
6. N. Hayashida, et. al, *PREPRINT astro-ph/9804043* (1998).
7. S. Yoshida and H. Dai, *Journal Phys (To be published)* **G**, (1998).

8. A. A. Watson, *Rapporteaur talk: 25th International Cosmic Ray Conference(Durban)* (1997).
9. B. Dawson, *Pierre Auger Project tecnical notes GAP-96-017* (1996).
10. Bird D. J., *The Astrophysical Journal* **441**, 144-151 (1995).
11. Bird D. J., *(submitted to the Proceedings of the Auger Meeting on Nitrogen Fluorescence Cosmic Ray Detectors (Salt Lake City)*, 1996.
12. Au W. et. al, *Proceedings of the XXII Cosmic Ray Conference (Dublin)* **2**, 692 (1992).
13. Bird D. J., et. al, *Proceedings of the XXV Cosmic Ray Conference (Durban)* **5**, 548 (1997).
14. Aiso S. et.al, *Proceedings of the XXV Cosmic Ray Conference (Durban)* **4**, 548 (1997).

Precision Measurement of the Σ^0 Hyperon Mass

M.H.L.S. Wang, E.P. Hartouni, M.N. Kreisler, J. Uribe,
M.D. Church, E.E. Gottschalk, B.C. Knapp, B.J. Stern,
L.R. Wiencke, D.C. Christian, G. Gutierrez, A. Wehmann,
C. Avilez, J. Felix, G. Moreno,
M. Forbush, F.R. Huson, and J.T. White

The BNL E766 Collaboration

presented by
Michael N. Kreisler

I would like to begin by thanking the organizers of this conference for inviting me to present this paper. The research [1] that is described in this paper is part of a program to study strong interaction mechanisms in proton proton collisions. The program consists of two experiments: Brookhaven E766 in which we studied the reactions $pp \to p+$ all charged particles with 27.5 GeV/c incident protons and Fermilab E690 in which we studied the reactions $pp \to p+$ all charged particles with 800 GeV/c incident protons. In these experiments, we employed state-of-the-art data acquisition systems and acquired large samples of data: at Brookhaven we amassed 300 million high multiplicity events and at Fermilab, 5.5 billion events. This program is rich in physics topics and has resulted in 9 Ph.D. theses granted to date, three of which were students from Mexico.

The physics topics studied with these data samples include the polarization of Λ^0's, hyperon production mechanisms, light meson spectroscopy, correlations between like-sign pions, final state Coulomb interactions, precision mass measurements, and charm production. We expect many other topics to be studied and invite you, if you are interested, to join our collaboration.

The subject of this paper is the precision measurement of the invariant mass of an elementary particle. One might ask why anyone should care if the Σ^0 mass is measured more accurately. There are several answers. The first and most important is that one of the major goals of experimental physics is the improvement of our knowledge of the fundamental properties of nature. If a precision measurement does that, "no further justification is needed" [2]. In addition, this particular measurement also provides essential input to theories of constituent interactions [3] and improves our understanding of the baryon octet and decuplet mass relationships [4]. But an even stronger argument is that the previous measurements of this quantity are not very good.

In Table 1, we present the previous results of experiments measuring the Σ^0 mass. As one can see from an examination of this table, even the most accurate of the previous experiments had only a very small number of events (208). The newest of these previous measurements was performed more than 20 years ago. In addition, none of these measurements were direct measurements of the Σ^0 mass – that is, the experiments were unable to measure all of the decay products of the Σ^0 so that the mass could be determined by a direct kinematic calculation.

The measurement of the Σ^0 mass in this paper uses the main decay of the Σ^0:

$$\Sigma^0 \to \Lambda^0 + \gamma \qquad (1)$$
$$\Lambda^0 \to p + \pi^- \qquad (2)$$
$$\gamma \to e^+ + e^-. \qquad (3)$$

Naively, one might believe that the measurement of the invariant mass is easy. The steps are (1) find events in which there are Λ^0's and γ's; (2) measure the momenta of all particles; (3) calculate the invariant mass of the Λ^0 and γ; (4) fit the mass distribution to a signal plus background and (5) publish the results. None of the steps is either easy or quick.

In order to collect the large samples of data described above, we designed, constructed, and used a large spectrometer system. The proton beam interacted in a twelve inch long liquid hydrogen target. Charged particles produced in a pp interaction were detected and measured by a set of six multi-plane mini-drift chambers. The chambers were located inside the aperture of a large analyzing magnet. Particle identification was provided by a multi-element Cherenkov counter and two scintillation counter hodoscopes for time-of-flight. Further information about the spectrometer, the triggers used, and the details of the sophisticated computer algorithms developed to reconstruct the massive number of events can be found elsewhere [5]. In Figure 2, we present one view of the reconstruction of an event which is a candidate for a Σ^0 decay. The event is characterized by the existence of a two particle vertex (possibly the Λ^0 decay) clearly separated from the multi-particle vertex (the pp interaction point) and a two particle pair with a very small opening angle between the two particles. We shall return to a discussion of the selection process later in this paper.

A precision measurement requires that all of the small details are done correctly. Before describing the specific steps leading to the Σ^0 mass determination, I would like to give a flavor of the kinds of details we worried about.

First, we had to align the coordinate system in which we measured the magnetic field of the spectrometer analyzing magnet with the coordinate system of the drift chambers in which the trajectories of the particles were measured. Careful surveying does most of the job, but a precise tuning of the relationship between the two coordinate systems is necessary. We accomplished this by using a sample of 60,000 K_s^0 decays from the subset of our data which were exclusive events. An exclusive event is one in which all of the final state particles have been measured and identified. There is no ambiguity about the identification of the K_s^0's in this sample. The K_s^0 invariant mass distribution was examined as a function of small changes in the relative positions of the two coordinate systems to establish the best relative alignment.

Second, we had to correct for time dependent variations of the magnetic field. The invariant masses in this experiment are calculated from the momenta of the decay particles determined from the curvature of those particles in the magnetic field. This determination depends directly on the value of the magnetic field in the spectrometer which in turn is proportional to the current in the coils of the magnet. Since the regulation of that current is only 0.1%, it was necessary to use the following technique. Due to the speed of our data acquisition system, we recorded sufficient events every six minutes to yield approximately 5000 K_s^0 decays. With these events, we were able to measure the "K_s^0 mass" as a function of time and correct for the time dependent variations of the magnetic field.

We had to ensure that the fitting routines did not depend on the step size used to calculate the momenta. Also for K_s^0 and Λ^0 decays, we selected only those events in which the two decay particles bend towards each other as they pass through the magnetic field. The invariant mass in such an event is less sensitive to measurement errors than events in which the two decay particles bend away from each other. (We leave the proof of this as an exercise for the reader.)

Finally, we had to determine the absolute value of the magnetic field. We chose to do this by adjusting the field so that the invariant mass of the exclusive K_s^0 sample coincided with the current world average value of the K_s^0 mass ($497.672 MeV/c^2$). Figure 3 shows the invariant mass distribution for that sample. We note that the standard deviation of the K_s^0 mass distribution is only $1.28 MeV/c^2$.

Using this calibration of the magnetic field, we were able to measure and publish the most precise measurement of the lambda and anti-lambda hyperon masses [6]. The excellent mass resolution of the spectrometer (the standard deviation of the Λ^0 mass distribution is $0.5 MeV/c^2$) insures that the selection of Λ^0's for the measurement of the Σ^0 mass measurement has a very small background.

The selection of γ candidates deserves some discussion. γ candidates were identified by a pair of oppositely charged particles with a small opening angle between the two particles. The selection of real e^+e^- pairs from this subset proceeded as follows.

We define the parameter q_T, the perpendicular momentum between the two particles as

$$q_T = 2\frac{|\vec{p}^+ \times \vec{p}^-|}{|\vec{p}^+ + \vec{p}^-|} \qquad (4)$$

To determine the actual conversion point of the γ, we moved the location of the vertex of the pair until q_T^2 was minimized. In Figure 4, we show the distribution of q_T^2 for all candidates and the cut used to select the final γ candidates.

The details of the event selection are shown in Table 2, starting from total data sample of 300,000,000 events which yield 3,327 Σ^0 candidates within $\pm 6.9 MeV/c^2$

of the current world average value of $1192.55 MeV/c^2$. It is important to note that we excluded any event that could be a candidate for the decay $\Sigma^0 \to \Lambda^0 e^+ e^-$.

In order to determine the value if the mass, it is necessary to know the shape of the invariant mass distribution which depends on the details of both the uncertainties introduced in the measurement process and in the energy loss mechanisms affecting the decay particles. We chose to determine this shape by using a hybrid Monte Carlo technique [7]. In this technique, all of the parameters of the events are taken directly from the real data except those that describe the parameters of direct concern to the measurement. Thus, we began by taking events with real Σ^0's. We kept all of the information in the detector (wire hits, etc.) except those from the decay particles of the Σ^0 (the p, π^-, e^+ and e^-). Knowing the momentum vector of the Σ^0, we used standard Monte Carlo techniques to generate the decay of the Σ^0. Each "real" event was used as the seed to generate many such Monte Carlo events. We included all physical processes such as the Moliere description of multiple Coulomb scattering, energy loss by ionization and energy loss by bremsstrahlung. The particles were then propagated thorough the spectrometer. The resulting event was analyzed by the same program which had analyzed the original data sample.

The invariant mass distribution resulting from this hybrid Monte Carlo technique was then fit to the invariant mass distribution from the data. The fit determined the values of the Σ^0 invariant mass, the number of events, and included a linear background. The results of the fit are shown in Figure 5.

Before quoting our result it is necessary to note that we had to consider possible systematic errors in addition to the statistical uncertainty. This measurement depends on the value of both the Λ^0 hyperon mass and the K_s^0 meson mass. The contribution to the Σ^0 mass systematic uncertainty from the uncertainties in these two masses are 0.006 and $0.0077 MeV/c^2$ respectively. The systematic uncertainty from the hybrid Monte Carlo technique is less than $0.0005 MeV/c^2$ and our uncertainty (at the 5% level) in the amount of material in the spectrometer contributes at most $0.01 MeV/c^2$ to the Σ^0 mass systematic uncertainty.

Adding the systematic errors in quadrature we find for our results on the Σ^0 mass and the $\Sigma^0 - \Lambda^0$ mass difference to be:

$$M_{\Sigma^0} = 1192.65 \pm 0.020 \pm 0.014 MeV/c^2 \tag{5}$$

and:

$$M_{\Sigma^0} - M_{\Lambda^0} = 76.966 \pm 0.020 \pm 0.013 MeV/c^2. \tag{6}$$

Let me close by noting that this is a significant result. Our uncertainty in the Σ^0 mass is more than 7 times smaller than the best previous result and was based on 16 times the statistics. Likewise, the $\Sigma^0 - \Lambda^0$ mass difference is more than 14 times more accurate than the previous best result. Finally, we note that this measurement is the first direct measurement of the Σ^0 mass. Based on these observations, we felt it imperative to redo the world averages which are presented in Table 2.

TABLE 1. Previous Results on Σ^0 Mass

value (MeV/c^2)	# events	technique
1191.83 ± 0.55	109	heavy liquid bubble chamber [8]
1192.41 ± 0.14	208	H2 bubble chamber [9]
1192.25 ± 0.23	18	H2 bubble chamber [10]
1190.3 + 1.2 -2.0	8	LH2 bubble chamber [11]
1184.7 ± 3.6	3	propane bubble chamber [12]

TABLE 2. New World Averages (MeV/c^2)

Σ^-	1197.451 ± 0.031	$\Sigma^- - \Lambda^0$	*81.694 ± 0.066*
Σ^0	*1192.65 ± 0.025*	$\Sigma^0 - \Lambda^0$	*76.96 ± 0.03*
Σ^+	1189.37 ± 0.06	$\Sigma^- - \Sigma^0$	4.86 ± 0.07
Λ^0	1115.683 ± 0.006	$\Sigma^- - \Sigma^+$	8.10 ± 0.11

We look forward to other interesting results coming from our data sample and encourage any of you who wish to work in this experiment to contact the members of our collaboration.

REFERENCES

1. M.H.L.S. Wang et al., Phys. Rev. D **56**, 2544 (1997).
2. V. Fitch, private communication.
3. A. DeRujula, H. Georgi, and S.L. Glashow, Phys. Rev. D **12**, 147 (1975); N. Isgur and G. Karl, Phys. Rev. D **20**, 1191 (1979); G. S. Adkins, C. R. Nappi, and E. Witten, Nucl. Phys. B **228**, 552 (1983).
4. Y. Dong, J. Su, and S. Wu, J. Phys. G: Nucl. Part. Phys. **20**, 73 (1994); W-Y. P. Hwang and K. C. Yang, Phys. Rev. D **49**, 460 (1994); I. Duck, Z. Phys. A **350**, 71 (1994); E. Jenkins and R. Lebed, Phys. Rev. D **52**, 282 (1995); M. K. Banerjee and J. Milana, Phys. Rev. D **52**, 6451 (1995).
5. J. Uribe et al., Phys. Rev. D **49**, 4373 (1994); E. P. Hartouni et al., IEEE Trans. Nucl. Sci. **36**, 1480 (1989); D. C. Christian et al., Nucl. Instr. and Meth. **A345**, 62 (1994); L. R. Wiencke et al., Phys. Rev. D **46**, 3708 (1992); E. Gottschalk et al., Phys. Rev. D **53**, 4756 (1996)
6. E. P. Hartouni et al., Phys. Rev. Lett. **72**, 1322 (1994).
7. G. Bunce, Nucl. Instr. and Meth. **172**, 553 (1980).
8. J. Colas et al., Nucl. Phys. **B91**, 253 (1975).
9. P. Schmidt, Phys. Rev. **140**, B1328 (1965).
10. R. A. Burnstein et al., Phys. Rev. Lett. **13**, 66, (1964).
11. F. Eisler et al., Phys. Rev. **110**, 1, (1958).
12. R. Plano et al., Nuovo Cimento **5**, 216, (1957).

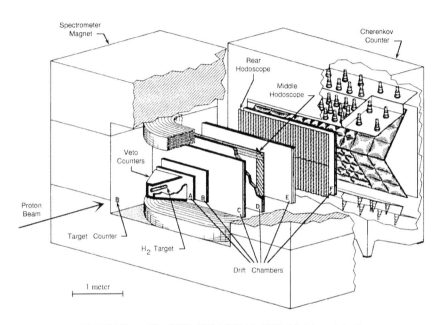

FIGURE 1. The BNL E766/FNAL E690 Multiparticle Spectrometer.

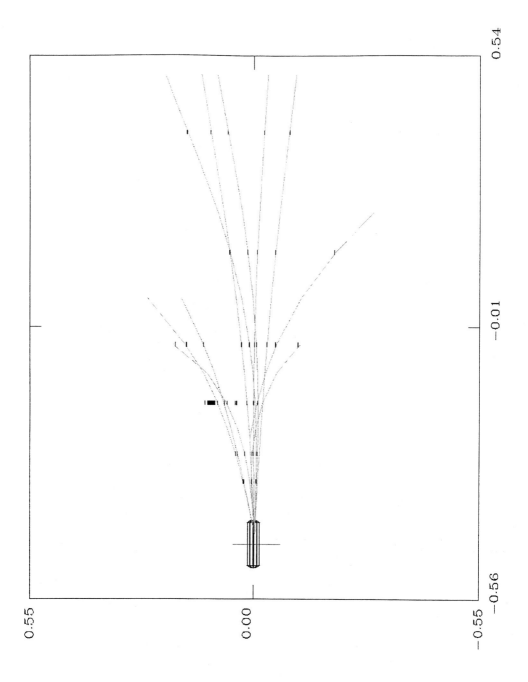

FIGURE 2. One view of a $\Sigma^0 \to \Lambda^0 + \gamma$ candidate event.

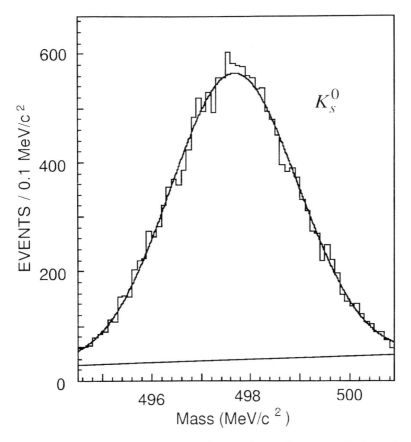

FIGURE 3. Invariant mass for the K_S^0 events fit to a Gaussian and a linear background shown explicitly. The standard deviation is $1.28 MeV/c^2$.

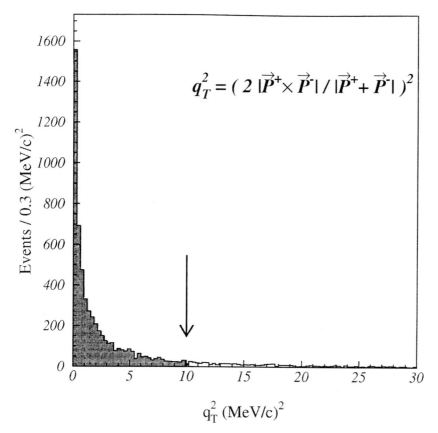

FIGURE 4. q_T^2 distribution of candidate e^+e^- pairs from the γ conversion of $\Sigma^0 \to \Lambda^0 + \gamma$ events. The arrow shows the location of the cut used.

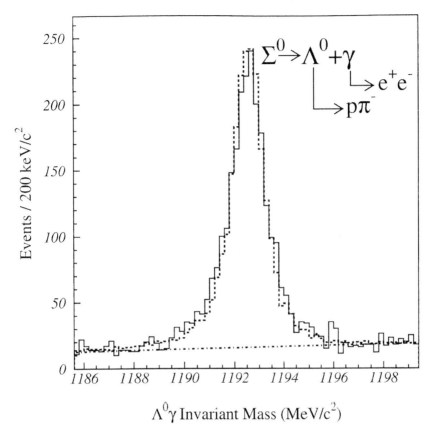

FIGURE 5. Invariant mass for $\Sigma^0 \rightarrow \Lambda^0 + \gamma$ events (solid line) fit to the Monte Carlo distribution (dashed line) added to a linear background (dot-dashed line).

II. Phenomenology

The Standard Model and Beyond

G. Altarelli

Theoretical Physics Division, CERN
CH - 1211 Geneva 23
and
Università di Roma Tre, Rome, Italy

Contents

1. Introduction
2. Status of the Data
3. Precision Electroweak Data and the Standard Model
4. A more General Analysis of Electroweak Data
    4.1 Basic Definitions and Results
    4.2 Experimental Determination of the Epsilon Variables
    4.3 Comparing the Data with the Minimal Supersymmetric Standard Model
5. The Case for Physics beyond the Standard Model
6. The Search for the Higgs
7. New Physics at HERA?
    7.1 Introduction
    7.2 Structure Functions
    7.3 Contact Terms
    7.4 Leptoquarks
    7.5 S-quarks with R-Parity Violation
    7.6 Charged Current Events
8. Conclusion

1 Introduction

In recent years new powerful tests of the Standard Model (SM) have been performed mainly at LEP but also at SLC and at the Tevatron. The running of LEP1 was terminated in 1995 and close-to-final results of the data analysis are now available [1],[2]. The experiments at the Z_0 resonance have enormously improved the accuracy in the electroweak neutral current sector. The top quark has been at last found and the errors on m_Z and $\sin^2\theta_{eff}$ went down by two and one orders of magnitude, respectively, since the start of LEP in 1989. The LEP2 programme is in progress. The validity of the SM has been confirmed to a level that we can say was unexpected at the beginning. In the present data there is no significant evidence for departures from the SM, no convincing hint of new physics (also including the first results from LEP2) [3]. The impressive success of the SM poses strong limitations on the possible forms of new physics. Favoured are models of the Higgs sector and of new physics that preserve the SM structure and only very delicately improve it, as is the case for fundamental Higgs(es) and Supersymmetry. Disfavoured are models with a nearby strong non perturbative regime that almost inevitably would affect the radiative corrections, as for composite Higgs(es) or technicolor and its variants[4]-[6].

2 Status of the Data

The relevant electro-weak data together with their SM values are presented in table 1 [1]-[2]. The SM predictions correspond to a fit of all the available data (including the directly measured values of m_t and m_W) in terms of m_t, m_H and $\alpha_s(m_Z)$, described later in sect.3, table 4 (last column).

Other important derived quantities are, for example, N_ν the number of light neutrinos, obtained from the invisible width: $N_\nu = 2.993(11)$, which shows that only three fermion generations exist with $m_\nu < 45$ GeV, or the leptonic width Γ_l, averaged over e, μ and τ: $\Gamma_l = 83.91(10) MeV$, or the hadronic width: $\Gamma_h = 1743.2(2.3)$ MeV.

For indicative purposes, in table 1 the "pulls" are also shown, defined as: pull = (data point- fit value)/(error on data point). At a glance we see that the agreement with the SM is quite good. The distribution of the pulls is statistically normal. The presence of a few $\sim 2\sigma$ deviations is what is to be expected. However it is maybe worthwhile to give a closer look at these small discrepancies.

Perhaps the most annoying feature of the data is the persistent difference between the values of $\sin^2\theta_{eff}$ measured at LEP and at SLC. The value of $\sin^2\theta_{eff}$ is obtained from a set of combined asymmetries. From asymmetries one derives the ratio $x = g_V^l/g_A^l$ of the vector and axial vector couplings of the Z_0, averaged over the charged leptons. In turn $\sin^2\theta_{eff}$ is defined by $x = 1 - 4\sin^2\theta_{eff}$. SLD obtains x from the single measurement of A_{LR}, the left-right asymmetry, which requires longitudinally polarized beams. The distribution of the present measurements of $\sin^2\theta_{eff}$ is shown in fig. 1. The LEP average, $\sin^2\theta_{eff} = 0.23199(28)$, differs by 2.9σ from the SLD value $\sin^2\theta_{eff} = 0.23055(41)$. The most precise individual measurement at LEP is from A_b^{FB}: the combined LEP error on this quantity is about the same as the SLD error, but the two values are 3.1σ's away. One might attribute this to the fact that the b measurement is more delicate and affected by a complicated systematics. In fact one notices from fig. 1 that the value obtained at LEP from A_l^{FB}, the average for l=e, μ and τ, is somewhat low (indeed quite in agreement with the SLD value). However the statement that LEP and SLD agree on leptons while they only disagree when the b quark is considered is not quite right. First, the value of A_e, a quantity essentially identical to A_{LR}, measured at LEP from the angular distribution of the τ polarization, differs by 1.8σ from the SLD value. Second, the low value of $\sin^2\theta_{eff}$ found at LEP from A_l^{FB} turns out to be entirely due to the τ lepton channel which leads to a central value different than that of e and μ [2]. The e and μ asymmetries, which are experimentally simpler, are perfectly on top of the SM fit. Suppose we take only e and μ asymmetries at LEP and disregard the b and τ measurements: the LEP average becomes $\sin^2\theta_{eff} = 0.23184(55)$, which is still 1.9σ away from the SLD value.

In conclusion, it is difficult to find a simple explanation for the SLD-LEP discrepancy on $\sin^2\theta_{eff}$. In view of this, the error on the nominal SLD-LEP average, $\sin^2\theta_{eff} = 0.23152(23)$, should perhaps be enlarged, for example, by a factor $S = \sqrt{\chi^2/N_{df}} \sim 1.4$, according to the recipe adopted by the Particle Data Group in such cases. Accordingly, in the following we will often use the average

$$\sin^2\theta_{eff} = 0.23152 \pm 0.00032 \qquad (1)$$

Thus the LEP-SLC discrepancy results in an effective limitation of the experimental precision on $\sin^2\theta_{eff}$. The data-taking by the SLD experiment is still in progress and also at LEP seizable improvements on A_τ and A_b^{FB} are foreseen as soon as the corresponding analyses will be completed. We hope to see the difference to decrease or to be understood.

From the above discussion one may wonder if there is evidence for something special in the τ channel, or equivalently if lepton universality is really supported by the data. Indeed this is the case: the hint of a difference in A_τ^{FB} with respect to the corresponding e and μ asymmetries is not confirmed by the measurements of A_τ and Γ_τ which appear normal [1],[2],[7]. In principle the fact that an anomaly shows up in A_τ^{FB} and not in A_τ and Γ_τ is not inconceivable because the FB lepton asymmetries are very small and very precisely measured. For example, the extraction of A_τ^{FB} from the data on the angular distribution of τ's could be biased if the imaginary part of the continuum was altered by some non universal new physics effect [8]. But a more trivial experimental problem is at the moment quite plausible.

A similar question can be asked for the b couplings. We have seen that the measured value of A_b^{FB} is about 2σ's below the SM fit. At the same time R_b which used to show a major discrepancy is now only about 1.4σ's away from the SM fit (as a result of the more sophisticated second generation experimental techniques). It is often stated that there is a -2.5σ deviation on the measured value of A_b vs the SM expectation [1],[2]. But in fact that depends on how the data are combined. In our opinion one should rather talk of a -1.8σ effect. Let us discuss this point in detail. A_b can be measured directly at SLC by taking advantage of the beam longitudinal polarization. At LEP one measures $A_b^{FB} = 3/4\, A_e A_b$. One can then derive A_b by inserting a value for A_e. The question is what to use for A_e: the LEP value obtained,

Table 1: The electroweak data and the SM values obtained from a global fit.

Quantity	Data (August '97)	Standard Model Fit	Pull
m_Z (GeV)	91.1867(20)	91.1866	0.0
Γ_Z (GeV)	2.4948(25)	2.4966	-0.7
σ_h (nb)	41.486(53)	41.467	0.4
R_h	20.775(27)	20.756	0.7
R_b	0.2170(9)	0.2158	1.4
R_c	0.1734(48)	0.1723	-0.1
A_{FB}^l	0.0171(10)	0.0162	0.9
A_τ	0.1411(64)	0.1470	-0.9
A_e	0.1399(73)	0.1470	-1.0
A_{FB}^b	0.0983(24)	0.1031	-2.0
A_{FB}^c	0.0739(48)	0.0736	0.0
A_b (SLD direct)	0.900(50)	0.935	-0.7
A_c (SLD direct)	0.650(58)	0.668	-0.3
$\sin^2\theta_{eff}$(LEP-combined)	0.23199(28)	0.23152	1.7
$A_{LR} \to \sin^2\theta_{eff}$	0.23055(41)	0.23152	-2.4
m_W (GeV) (LEP2+$p\bar{p}$)	80.43(8)	80.375	0.7
$1 - \frac{m_W^2}{m_Z^2}$ (νN)	0.2254(37)	0.2231	0.6
Q_W (Atomic PV in Cs)	-72.11(93)	-73.20	1.2
m_t (GeV)	175.6(5.5)	173.1	0.4

using lepton universality, from the measurements of A_l^{FB}, A_τ, A_e: $A_e = 0.1461(33)$, or the combination of LEP and SLD etc. The LEP electroweak working group adopts for A_e the SLD+LEP average value which also includes A_{LR} from SLD: $A_e = 0.1505(23)$. This procedure leads to a -2.5σ deviation. However, in this case, the well known $\sim 2\sigma$ discrepancy of A_{LR} with respect to A_e measured at LEP and also to the SM fit, which is not related to the b couplings, further contributes to inflate the number of σ's. Since we are here concerned with the b couplings it is perhaps wiser to obtain A_b from LEP by using the SM value for A_e (that is the pull-zero value of table 1): $A_e^{SM} = 0.1467(16)$. With the value of A_b derived in this way from LEP we finally obtain

$$A_b = 0.895 \pm 0.022 \quad (\text{LEP + SLD}, A_e = A_e^{SM}: -1.8) \quad (2)$$

In the SM A_b is so close to 1 because the b quark is almost purely left-handed. A_b only depends on the ratio $r = (g_R/g_L)^2$ which in the SM is small: $r \sim 0.033$. To adequately decrease A_b from its SM value one must increase r by a factor of about 1.6, which appears large for a new physics effect. Also such a large change in r must be compensated by decreasing g_L^2 by a small but fine-tuned amount in order to counterbalance the corresponding large positive shift in R_b. In view of this the most likely way out is that A_b^{FB} and A_b have been a bit underestimated at LEP and actually there is no anomaly in the b couplings. Then the LEP value of $\sin^2 \theta_{eff}$ would slightly move towards the SLD value, but, as explained above, by far not enough to remove the SLD-LEP discrepancy (for example, if the LEP average for $\sin^2 \theta_{eff}$ is computed by moving the central value of A_b^{FB} to the pull-zero value in table 1 with the same figure for the error, one finds $\sin^2 \theta_{eff} = 0.23162(28)$, a value still 2.2$\sigma$'s away from the SLD result).

3 Precision Electroweak Data and the Standard Model

For the analysis of electroweak data in the SM one starts from the input parameters: some of them, α, G_F and m_Z, are very well measured, some other ones, $m_{f_{light}}$, m_t and $\alpha_s(m_Z)$ are only approximately determined while m_H is largely unknown. With respect to m_t the situation has much improved since the CDF/D0 direct measurement of the top quark mass [9]. From the input parameters one computes the radiative corrections [10] to a sufficient precision to match the experimental capabilities. Then one compares the theoretical predictions and the data for the numerous observables which have been measured, checks the consistency of the theory and derives constraints on m_t, $\alpha_s(m_Z)$ and hopefully also on m_H.

Some comments on the least known of the input parameters are now in order. The only practically relevant terms where precise values of the light quark masses, $m_{f_{light}}$, are needed are those related to the hadronic contribution to the photon vacuum polarization diagrams, that determine $\alpha(m_Z)$. This correction is of order 6%, much larger than the accuracy of a few permil of the precision tests. Fortunately, one can use the actual data to in principle solve the related ambiguity. But the leftover uncertainty is still one of the main sources of theoretical error. In recent years there has been a lot of activity on this subject and a number of independent new estimates of $\alpha(m_Z)$ have appeared in the literature [11],(see also [12]). A consensus has been

Table 2: Measurements of $\alpha_s(m_Z)$. In parenthesis we indicate if the dominant source of errors is theoretical or experimental. For theoretical ambiguities our personal figure of merit is given.

Measurements	$\alpha_s(m_Z)$	
R_τ	0.122 ± 0.006	(Th)
Deep Inelastic Scattering	0.116 ± 0.005	(Th)
Y_{decay}	0.112 ± 0.010	(Th)
Lattice QCD	0.117 ± 0.007	(Th)
$Re^+e^-(\sqrt{s} < 62 \text{ GeV})$	0.124 ± 0.021	(Exp)
Fragmentation functions in e^+e^-	0.124 ± 0.012	(Th)
Jets in e^+e^- at and below the Z	0.121 ± 0.008	(Th)
Z line shape (Assuming SM)	0.120 ± 0.004	(Exp)

established and the value used at present is

$$\alpha(m_Z)^{-1} = 128.90 \pm 0.09 \tag{3}$$

As for the strong coupling $\alpha_s(m_Z)$ the world average central value is by now quite stable. The error is going down because the dispersion among the different measurements is much smaller in the most recent set of data. The most important determinations of $\alpha_s(m_Z)$ are summarized in table 2 [13]. For all entries, the main sources of error are the theoretical ambiguities which are larger than the experimental errors. The only exception is the measurement from the electroweak precision tests, but only if one assumes that the SM electroweak sector is correct. Our personal views on the theoretical errors are reflected in the table 2. The error on the final average is taken by all authors between ± 0.003 and ± 0.005 depending on how conservative one is. Thus in the following our reference value will be

$$\alpha_s(m_Z) = 0.119 \pm 0.004 \tag{4}$$

Finally a few words on the current status of the direct measurement of m_t. The present combined CDF/D0 result is [9]

$$m_t = 175.6 \pm 5.5 \text{ GeV} \tag{5}$$

The error is so small by now that one is approaching a level where a more careful investigation of the effects of colour rearrangement on the determination of m_t is needed. One wants to determine the top quark mass, defined as the invariant mass of its decay products (i.e. b+W+ gluons + γ's). However, due to the need of colour rearrangement, the top quark and its decay products cannot be really isolated from the rest of the event. Some smearing of the mass distribution is induced by this colour crosstalk which involves the decay products of the top, those of the antitop and also the fragments of the incoming (anti)protons. A reliable quantitative computation of the smearing effect on the m_t determination is difficult because of the importance of non perturbative effects. An induced error of the order of one GeV on m_t is reasonably expected. Thus further progress on the m_t determination demands tackling this problem in more depth.

Table 3: Errors from different sources: Δ^{exp}_{now} is the present experimental error; $\Delta\alpha^{-1}$ is the impact of $\Delta\alpha^{-1} = \pm 0.09$; Δ_{th} is the estimated theoretical error from higher orders; Δm_t is from $\Delta m_t = \pm 6$ GeV; Δm_H is from $\Delta m_H = 60$–1000 GeV; $\Delta\alpha_s$ corresponds to $\Delta\alpha_s = \pm 0.003$. The epsilon parameters are defined in sect. 4.1 [18].

Parameter	Δ^{exp}_{now}	$\Delta\alpha^{-1}$	Δ_{th}	Δm_t	Δm_H	$\Delta\alpha_s$
Γ_Z (MeV)	± 2.5	± 0.7	± 0.8	± 1.4	± 4.6	± 1.7
σ_h (pb)	53	1	4.3	3.3	4	17
$R_h \cdot 10^3$	27	4.3	5	2	13.5	20
Γ_l (keV)	100	11	15	55	120	3.5
$A^l_{FB} \cdot 10^4$	10	4.2	1.3	3.3	13	0.18
$\sin^2\theta \cdot 10^4$	~ 3.2	2.3	0.8	1.9	7.5	0.1
m_W (MeV)	80	12	9	37	100	2.2
$R_b \cdot 10^4$	9	0.1	1	2.1	0.25	0
$\epsilon_1 \cdot 10^3$	1.2		~ 0.1			0.2
$\epsilon_3 \cdot 10^3$	1.4	0.5	~ 0.1			0.12
$\epsilon_b \cdot 10^3$	2.1		~ 0.1			1

In order to appreciate the relative importance of the different sources of theoretical errors for precision tests of the SM, we report in table 3 a comparison for the most relevant observables, evaluated using refs. [10],[14]. What is important to stress is that the ambiguity from m_t, once by far the largest one, is by now smaller than the error from m_H. We also see from table 3 that the error from $\Delta\alpha(m_Z)$ is especially important for $\sin^2\theta_{eff}$ and, to a lesser extent, is also sizeable for Γ_Z and ϵ_3, to be defined later on.

The most important recent advance in the theory of radiative corrections is the calculation of the $o(g^4 m_t^2/m_W^2)$ terms in $\sin^2\theta_{eff}$ and m_W (not yet in $\delta\rho$) [15]. The result implies a small but visible correction to the predicted values but expecially a seizable decrease of the ambiguity from scheme dependence (a typical effect of truncation).

We now discuss fitting the data in the SM. Similar studies based on older sets of data are found in refs.[16]. As the mass of the top quark is finally rather precisely known from CDF and D0 one must distinguish two different types of fits. In one type one wants to answer the question: is m_t from radiative corrections in agreement with the direct measurement at the Tevatron? For answering this interesting but somewhat limited question, one must clearly exclude the CDF/D0 measurement of m_t from the input set of data. Fitting all other data in terms of m_t, m_H and $\alpha_s(m_Z)$ one finds the results shown in the third column of table 4 [2]. Other similar fits where also m_W direct data are left out are shown. The extracted value of m_t is typically a bit too low. There is a strong correlation between m_t and m_H. The results on $\sin^2\theta_{eff}$ and m_W [17] drive the fit to small values of m_H. This can be seen from figs.2 and 3 (note that in fig. 2 the value of $\sin^2\theta_{eff}$ found by SLD would be too low to be shown on the scale of the plot). Then, at small m_H, the widths drive the fit to small m_t (see fig. 4). In this context it is important to remark that fixing m_H at 300 GeV, as was often done in the past, is by now completely obsolete, because it introduces too strong a bias on the fitted value of m_t.

Table 4: SM fits from different sets of data (with and without the direct measurements of m_W and m_t).

Parameter	LEP(incl.m_W)	All but m_W, m_t	All but m_t	All Data
m_t (GeV)	158+14 − 11	157+10 − 9	161+10 − 8	173.1 ± 5.4
m_H (GeV)	83+168 − 49	41+64 − 21	42+75 − 23	115+116 − 66
$log[m_H(\text{GeV})]$	1.92+0.48 − 0.39	1.62+0.41 − 0.31	1.63+0.44 − 0.33	2.06+0.30 − 0.37
$\alpha_s(m_Z)$	0.121 ± 0.003	0.120 ± 0.003	0.120 ± 0.003	0.120 ± 0.003
χ^2/dof	8/9	14/12	16/14	17/15

The change induced on the fitted value of m_t when moving m_H from 300 to 65 or 1000 GeV is in fact larger than the error on the direct measurement of m_t.

In a more general type of fit, e.g. for determining the overall consistency of the SM or the best present estimate for some quantity, say m_W, one should of course not ignore the existing direct determinations of m_t and m_W. Then, from all the available data, by fitting m_t, m_H and $\alpha_s(m_Z)$ one finds the values shown in the last column of table 4. This is the fit also referred to in table 1. The corresponding fitted values of $\sin^2\theta_{eff}$ and m_W are:

$$\sin^2\theta_{eff} = 0.23152 \pm 0.00022,$$
$$m_W = 80.375 \pm 0.030 \text{GeV} \qquad (6)$$

The fitted value of $\sin^2\theta_{eff}$ is identical to the LEP+SLD average and the caution on the error expressed in the previous section applies. The error of 30 MeV on m_W clearly sets up a goal for the direct measurement of m_W at LEP2 and the Tevatron.

As a final comment we want to recall that the radiative corrections are functions of $log(m_H)$. It is truly remarkable that the fitted value of $log(m_H)$ (the decimal logarithm) is found to fall right into the very narrow allowed window around the value 2 specified by the lower limit from direct searches, $m_H > 77$ GeV, and the theoretical upper limit in the SM $m_H < 600 - 800$ GeV (see sect.6). The fulfilment of this very stringent consistency check is a beautiful argument in favour of a fundamental Higgs (or one with a compositeness scale much above the weak scale).

4 A More General Analysis of Electroweak Data

We now discuss an update of the epsilon analysis [18] which is a method to look at the data in a more general context than the SM. The starting point is to isolate from the data that part which is due to the purely weak radiative corrections. In fact the epsilon variables are defined in such a way that they are zero in the approximation when only effects from the SM at tree level plus pure QED and pure QCD corrections are taken into account. This very simple version of improved Born approximation is a good first approximation according to the data and is independent of m_t and m_H. In fact the whole m_t and m_H dependence arises from weak loop corrections and therefore is only contained in the epsilon variables. Thus the epsilons are extracted from the data without need of specifying m_t and m_H. But their predicted value in the SM or in any extension of it depend on m_t and m_H. This is to be compared with the competitor method based on the S, T, U variables [19],[20]. The latter cannot be obtained from the data without specifying m_t and m_H because they are defined as deviations from the complete SM prediction for specified m_t and m_H. Of course there are very many variables that vanish if pure weak loop corrections are neglected, at least one for each relevant observable. Thus for a useful definition we choose a set of representative observables that are used to parametrize those hot spots of the radiative corrections where new physics effects are most likely to show up. These sensitive weak correction terms include vacuum polarization diagrams which being potentially quadratically divergent are likely to contain all possible non decoupling effects (like the quadratic top quark mass dependence in the SM). There are three independent vacuum polarization contributions. In the same spirit, one must add the $Z \to b\bar{b}$ vertex which also includes a large top mass dependence. Thus altogether we consider four defining observables: one asymmetry, for example A^l_{FB}, (as representative of the set of measurements that lead to the determination of $\sin^2\theta_{eff}$), one width (the leptonic width Γ_l is particularly suitable because it is practically independent of α_s), m_W and R_b. Here lepton universality has been taken for granted, because the data show that it is verified within the present accuracy. The four variables, ϵ_1, ϵ_2, ϵ_3 and ϵ_b are defined in ref.[18] in one to one correspondence with the set of observables A^{FB}_l, Γ_l, m_W, and R_b. The definition is so chosen that the quadratic top mass dependence is only present in ϵ_1 and ϵ_b, while the m_t dependence of ϵ_2 and ϵ_3 is logarithmic. The definition of ϵ_1 and ϵ_3 is specified in terms of A^{FB}_l and Γ_l only. Then adding m_W or R_b one obtains ϵ_2 or ϵ_b. We now specify the relevant definitions in detail.

4.1 Basic Definitions and Results

We start from the basic observables m_W/m_Z, Γ_l and A^{FB}_l and Γ_b. From these four quantities one can isolate the corresponding dynamically significant corrections Δr_W, $\Delta\rho$, Δk and ϵ_b, which contain the small effects one is trying to disentangle and are defined in the following. First we introduce Δr_W as obtained from m_W/m_Z by the relation:

$$(1 - \frac{m_W^2}{m_Z^2})\frac{m_W^2}{m_Z^2} = \frac{\pi\alpha(m_Z)}{\sqrt{2}G_F m_Z^2(1 - \Delta r_W)} \qquad (7)$$

Here $\alpha(m_Z) = \alpha/(1 - \Delta\alpha)$ is fixed to the central value $1/128.90$ so that the effect of the running of α due to known physics is extracted from $1 - \Delta r = (1 - \Delta\alpha)(1 - \Delta r_W)$. In fact, the error on $1/\alpha(m_Z)$, as given in eq.(32) would then affect Δr_W. In order to define $\Delta\rho$ and Δk we consider the effective vector and axial-vector couplings g_V and g_A of the on-shell Z to charged leptons, given by the formulae:

$$\Gamma_l = \frac{G_F m_Z^3}{6\pi\sqrt{2}}(g_V^2 + g_A^2)(1 + \frac{3\alpha}{4\pi}),$$

$$A^{FB}_l(\sqrt{s} = m_Z) = \frac{3g_V^2 g_A^2}{(g_V^2 + g_A^2)^2} = \frac{3x^2}{(1+x^2)^2}. \qquad (8)$$

Note that Γ_l stands for the inclusive partial width $\Gamma(Z \to l\bar{l}+\text{photons})$. We stress the following points. First, we have extracted from $(g_V^2 + g_A^2)$ the factor $(1 + 3\alpha/4\pi)$ which is induced in Γ_l from final state radiation. Second, by the asymmetry at the peak in eq.(8) we mean the quantity which is commonly referred to by the LEP experiments (denoted as A_{FB}^0 in ref.[2]), which is corrected for all QED effects, including initial and final state radiation and also for the effect of the imaginary part of the γ vacuum polarization diagram. In terms of g_A and $x = g_V/g_A$, the quantities $\Delta\rho$ and Δk are given by:

$$g_A = -\frac{\sqrt{\rho}}{2} \sim -\frac{1}{2}(1 + \frac{\Delta\rho}{2}),$$

$$x = \frac{g_V}{g_A} = 1 - 4\sin^2\theta_{eff} = 1 - 4(1 + \Delta k)s_0^2. \quad (9)$$

Here s_0^2 is $\sin^2\theta_{eff}$ before non pure-QED corrections, given by:

$$s_0^2 c_0^2 = \frac{\pi\alpha(m_Z)}{\sqrt{2}G_F m_Z^2} \quad (10)$$

with $c_0^2 = 1 - s_0^2$ ($s_0^2 = 0.231095$ for $m_Z = 91.188$ GeV).

We now define ϵ_b from Γ_b, the inclusive partial width for $Z \to b\bar{b}$ according to the relation

$$\Gamma_b = \frac{G_F m_Z^3}{6\pi\sqrt{2}}\beta(\frac{3-\beta^2}{2}g_{bV}^2 + \beta^2 g_{bA}^2)N_C R_{QCD}(1 + \frac{\alpha}{12\pi}) \quad (11)$$

where $N_C = 3$ is the number of colours, $\beta = \sqrt{1 - 4m_b^2/m_Z^2}$, with $m_b = 4.7$ GeV, R_{QCD} is the QCD correction factor given by

$$R_{QCD} = 1 + 1.2a - 1.1a^2 - 13a^3; \quad a = \frac{\alpha_s(m_Z)}{\pi} \quad (12)$$

and g_{bV} and g_{bA} are specified as follows

$$g_{bA} = -\frac{1}{2}(1 + \frac{\Delta\rho}{2})(1 + \epsilon_b),$$

$$\frac{g_{bV}}{g_{bA}} = \frac{1 - 4/3\sin^2\theta_{eff} + \epsilon_b}{1 + \epsilon_b}. \quad (13)$$

This is clearly not the most general deviation from the SM in the $Z \to b\bar{b}$ vertex but ϵ_b is closely related to the quantity $-Re(\delta_{b-vertex})$ defined in the last of refs.[21] where the large m_t corrections are located.

As is well known, in the SM the quantities Δr_W, $\Delta\rho$, Δk and ϵ_b, for sufficiently large m_t, are all dominated by quadratic terms in m_t of order $G_F m_t^2$. As new physics can more easily be disentangled if not masked by large conventional m_t effects, it is convenient to keep $\Delta\rho$ and ϵ_b while trading Δr_W and Δk for two quantities with no contributions of order $G_F m_t^2$. We thus introduce the following linear combinations:

$$\epsilon_1 = \Delta\rho,$$

$$\epsilon_2 = c_0^2\Delta\rho + \frac{s_0^2 \Delta r_W}{c_0^2 - s_0^2} - 2s_0^2\Delta k,$$

$$\epsilon_3 = c_0^2\Delta\rho + (c_0^2 - s_0^2)\Delta k. \quad (14)$$

The quantities ϵ_2 and ϵ_3 no longer contain terms of order $G_F m_t^2$ but only logarithmic terms in m_t. The leading terms for large Higgs mass, which are logarithmic, are contained in ϵ_1 and ϵ_3. In the Standard Model one has the following "large" asymptotic contributions:

$$\begin{align}
\epsilon_1 &= \frac{3 G_F m_t^2}{8\pi^2 \sqrt{2}} - \frac{3 G_F m_W^2}{4\pi^2 \sqrt{2}} \tan^2\theta_W \ln\frac{m_H}{m_Z} + ..., \\
\epsilon_2 &= -\frac{G_F m_W^2}{2\pi^2 \sqrt{2}} \ln\frac{m_t}{m_Z} + ..., \\
\epsilon_3 &= \frac{G_F m_W^2}{12\pi^2 \sqrt{2}} \ln\frac{m_H}{m_Z} - \frac{G_F m_W^2}{6\pi^2 \sqrt{2}} \ln\frac{m_t}{m_Z} ...., \\
\epsilon_b &= -\frac{G_F m_t^2}{4\pi^2 \sqrt{2}} + ...
\end{align} \tag{15}$$

The relations between the basic observables and the epsilons can be linearised, leading to the approximate formulae

$$\begin{align}
\frac{m_W^2}{m_Z^2} &= \frac{m_W^2}{m_Z^2}\Big|_B (1 + 1.43\epsilon_1 - 1.00\epsilon_2 - 0.86\epsilon_3), \\
\Gamma_l &= \Gamma_l|_B (1 + 1.20\epsilon_1 - 0.26\epsilon_3), \\
A_l^{FB} &= A_l^{FB}|_B (1 + 34.72\epsilon_1 - 45.15\epsilon_3), \\
\Gamma_b &= \Gamma_b|_B (1 + 1.42\epsilon_1 - 0.54\epsilon_3 + 2.29\epsilon_b).
\end{align} \tag{16}$$

The Born approximations, as defined above, depend on $\alpha_s(m_Z)$ and also on $\alpha(m_Z)$. Defining

$$\delta\alpha_s = \frac{\alpha_s(m_Z) - 0.119}{\pi}; \quad \delta\alpha = \frac{\alpha(m_Z) - \frac{1}{128.90}}{\alpha}, \tag{17}$$

we have

$$\begin{align}
\frac{m_W^2}{m_Z^2}\Big|_B &= 0.768905(1 - 0.40\delta\alpha), \\
\Gamma_l|_B &= 83.563(1 - 0.19\delta\alpha)\text{MeV}, \\
A_l^{FB}|_B &= 0.01696(1 - 34\delta\alpha), \\
\Gamma_b|_B &= 379.8(1 + 1.0\delta\alpha_s - 0.42\delta\alpha).
\end{align} \tag{18}$$

Note that the dependence on $\delta\alpha_s$ for $\Gamma_b|_B$, shown in eq.(18), is not simply the one loop result for $m_b = 0$ but a combined effective shift which takes into account both finite mass effects and the contribution of the known higher order terms.

The important property of the epsilons is that, in the Standard Model, for all observables at the Z pole, the whole dependence on m_t (and m_H) arising from one-loop diagrams only enters through the epsilons. The same is actually true, at the relevant level of precision, for all higher order m_t-dependent corrections. Actually, the only residual m_t dependence of the various observables not included in the epsilons is in the terms of order $\alpha_s^2(m_Z)$ in the pure QCD correction factors to the hadronic widths [22]. But this one is quantitatively irrelevant,

Table 5: Values of the epsilons in the SM as functions of m_t and m_H as obtained from recent versions[14] of ZFITTER and TOPAZ0 (also including the new results of ref.[15]). These values (in 10^{-3} units) are obtained for $\alpha_s(m_Z) = 0.119$, $\alpha(m_Z) = 1/128.90$, but the theoretical predictions are essentially independent of $\alpha_s(m_Z)$ and $\alpha(m_Z)$ [18].

m_t (GeV)	ϵ_1 m_H (GeV) =			ϵ_2 m_H (GeV) =			ϵ_3 m_H (GeV) =			ϵ_b All m_H
	70	300	1000	70	300	1000	70	300	1000	
150	3.55	2.86	1.72	−6.85	−6.46	−5.95	4.98	6.22	6.81	−4.50
160	4.37	3.66	2.50	−7.12	−6.72	−6.20	4.96	6.18	6.75	−5.31
170	5.26	4.52	3.32	−7.43	−7.01	−6.49	4.94	6.14	6.69	−6.17
180	6.19	5.42	4.18	−7.77	−7.35	−6.82	4.91	6.09	6.61	−7.08
190	7.18	6.35	5.09	−8.15	−7.75	−7.20	4.89	6.03	6.52	−8.03
200	8.22	7.34	6.04	−8.59	−8.18	−7.63	4.87	5.97	6.43	−9.01

especially in view of the errors connected to the uncertainty on the value of $\alpha_s(m_Z)$. The theoretical values of the epsilons in the SM from state of the art radiative corrections [10, 14], also including the recent development of ref.[15], are given in table 5. It is important to remark that the theoretical values of the epsilons in the SM, as given in table 2, are not affected, at the percent level or so, by reasonable variations of $\alpha_s(m_Z)$ and/or $\alpha(m_Z)$ around their central values. By our definitions, in fact, no terms of order $\alpha_s^n(m_Z)$ or $\alpha \ln m_Z/m$ contribute to the epsilons. In terms of the epsilons, the following expressions hold, within the SM, for the various precision observables

$$\Gamma_T = \Gamma_{T0}(1 + 1.35\epsilon_1 - 0.46\epsilon_3 + 0.35\epsilon_b),$$
$$R = R_0(1 + 0.28\epsilon_1 - 0.36\epsilon_3 + 0.50\epsilon_b),$$
$$\sigma_h = \sigma_{h0}(1 - 0.03\epsilon_1 + 0.04\epsilon_3 - 0.20\epsilon_b),$$
$$x = x_0(1 + 17.6\epsilon_1 - 22.9\epsilon_3),$$
$$R_b = R_{b0}(1 - 0.06\epsilon_1 + 0.07\epsilon_3 + 1.79\epsilon_b). \qquad (19)$$

where $x = g_V/g_A$ as obtained from A_l^{FB}. The quantities in eqs.(16,19) are clearly not independent and the redundant information is reported for convenience. By comparison with the codes of ref.[14] (we also added the complete results of ref.[15]) we obtain

$$\Gamma_{T0} = 2489.46(1 + 0.73\delta\alpha_s - 0.35\delta\alpha) \ MeV,$$
$$R_0 = 20.8228(1 + 1.05\delta\alpha_s - 0.28\delta\alpha),$$
$$\sigma_{h0} = 41.420(1 - 0.41\delta\alpha_s + 0.03\delta\alpha) \ nb,$$
$$x_0 = 0.075619 - 1.32\delta\alpha,$$
$$R_{b0} = 0.2182355. \qquad (20)$$

Note that the quantities in eqs.(20) should not be confused, at least in principle, with the corresponding Born approximations, due to small "non universal" electroweak corrections. In practice, at the relevant level of approximation, the difference between the two corresponding quantities is in any case significantly smaller than the present experimental error.

Table 6: Experimental values of the epsilons in the SM from different sets of data. These values (in 10^{-3} units) are obtained for $\alpha_s(m_Z) = 0.119 \pm 0.003$, $\alpha(m_Z) = 1/128.90 \pm 0.09$, the corresponding uncertainties being included in the quoted errors.

$\epsilon\ 10^3$	Only def. quantities	All asymmetries	All High Energy	All Data
$\epsilon_1\ 10^3$	4.0 ± 1.2	4.3 ± 1.2	4.1 ± 1.2	3.9 ± 1.2
$\epsilon_2\ 10^3$	-8.3 ± 2.3	-9.1 ± 2.2	-9.3 ± 2.2	-9.4 ± 2.2
$\epsilon_3\ 10^3$	2.9 ± 1.9	4.3 ± 1.4	4.1 ± 1.4	3.9 ± 1.4
$\epsilon_b\ 10^3$	-3.2 ± 2.3	-3.3 ± 2.3	-3.9 ± 2.1	-3.9 ± 2.1

In principle, any four observables could have been picked up as defining variables. In practice we choose those that have a more clear physical significance and are more effective in the determination of the epsilons. In fact, since Γ_b is actually measured by R_b (which is nearly insensitive to α_s), it is preferable to use directly R_b itself as defining variable, as we shall do hereafter. In practice, since the value in eq.(20) is practically indistinguishable from the Born approximation of R_b, this determines no change in any of the equations given above but simply requires the corresponding replacement among the defining relations of the epsilons.

4.2 Experimental Determination of the Epsilon Variables

The values of the epsilons as obtained, following the specifications in the previous sect.4.1, from the defining variables m_W, Γ_l, A_l^{FB} and R_b are shown in the first column of table 6. To proceed further and include other measured observables in the analysis we need to make some dynamical assumptions. The minimum amount of model dependence is introduced by including other purely leptonic quantities at the Z pole such as A_τ, A_e (measured from the angular dependence of the τ polarization) and A_{LR} (measured by SLD). For this step, one is simply assuming that the different leptonic asymmetries are equivalent measurements of $\sin^2 \theta_{eff}$ (for an example of a peculiar model where this is not true, see ref.[23]). We add, as usual, the measure of A_b^{FB} because this observable is dominantly sensitive to the leptonic vertex. We then use the combined value of $\sin^2 \theta_{eff}$ obtained from the whole set of asymmetries measured at LEP and SLC with the error increased according to eq.(1) and the related discussion. At this stage the best values of the epsilons are shown in the second column of table 6. In figs. 5-8 we report the 1σ ellipses in the indicated ϵ_i-ϵ_j planes that correspond to this set of input data. In fig. 9, for example, we also give a graphical representation in the ϵ_3-ϵ_b plane, of the uncertainties due to $\alpha(m_Z)$ and $\alpha_s(m_Z)$.

All observables measured on the Z peak at LEP can be included in the analysis provided that we assume that all deviations from the SM are only contained in vacuum polarization diagrams (without demanding a truncation of the q^2 dependence of the corresponding functions) and/or the $Z \to b\bar{b}$ vertex. From a global fit of the data on m_W, Γ_T, R_h, σ_h, R_b and $\sin^2 \theta_{eff}$ (for LEP data, we have taken the correlation matrix for Γ_T, R_h and σ_h given by the LEP experiments [2], while we have considered the additional information on R_b and $\sin^2 \theta_{eff}$ as independent) we obtain the values shown in the third column of table 6. The comparison of theory and experiment at this stage is also shown in figs. 5-8. More detailed information is shown in figs.

10-11, which both refer to the level when also hadronic data are taken into account. But in fig. 10 we compare the results obtained if $\sin^2\theta_{eff}$ is extracted in turn from different asymmetries among those listed in fig. 1. The ellipse marked "average" is the same as the one labeled "All high en." in fig. 6 and corresponds to the value of $\sin^2\theta_{eff}$ which is shown on the figure (and in eq.(1)). We confirm that the value from A_{LR} is far away from the SM given the experimental value of m_t and the bounds on m_H and would correspond to very small values of ϵ_3 and of ϵ_1. We see also that while the τ FB asymmetry is also on the low side, the combined e and μ FB asymmetry are right on top of the average. Finally the b FB asymmetry is on the high side. An analogous plot is presented in fig. 11. In this case the defining width Γ_l is replaced in turn with either the total or the hadronic or the invisible width. The important conclusion that one obtains is that the widths are indeed well consistent among them even with respect to this new criterium of leading to the same epsilons.

To include in our analysis lower energy observables as well, a stronger hypothesis needs to be made: vacuum polarization diagrams are allowed to vary from the SM only in their constant and first derivative terms in a q^2 expansion [19]-[20]. In such a case, one can, for example, add to the analysis the ratio R_ν of neutral to charged current processes in deep inelastic neutrino scattering on nuclei [24], the "weak charge" Q_W measured in atomic parity violation experiments on Cs [25] and the measurement of g_V/g_A from $\nu_\mu e$ scattering [26]. In this way one obtains the global fit given in the fourth column of table 6 and shown in figs. 5-8. For completeness, we also report the corresponding values of Δr_W and Δk (defined in eqs. (7,9)): $10^3 \times \Delta r_W = -27.0 \pm 4.3$, $10^3 \times \Delta k = 16.7 \pm 18.4$. With the progress of LEP the low energy data, while important as a check that no deviations from the expected q^2 dependence arise, play a lesser role in the global fit. Note that the present ambiguity on the value of $\delta\alpha^{-1}(m_Z) = \pm 0.09$ [11] corresponds to an uncertainty on ϵ_3 (the other epsilons are not much affected) given by $\Delta\epsilon_3 \, 10^3 = \pm 0.6$ [18]. Thus the theoretical error is still comfortably less than the experimental error. In fig. 12 we present a summary of the experimental values of the epsilons as compared to the SM predictions as functions of m_t and m_H, which shows agreement within 1σ. However the central values of ϵ_1, ϵ_2 and ϵ_3 are all somewhat low, while the central value of ϵ_b is shifted upward with respect to the SM as a consequence of the still imperfect matching of R_b.

A number of interesting features are clearly visible from figs.5-12. First, the good agreement with the SM and the evidence for weak corrections, measured by the distance of the data from the improved Born approximation point (based on tree level SM plus pure QED or QCD corrections). There is by now a solid evidence for departures from the improved Born approximation where all the epsilons vanish. In other words a clear evidence for the pure weak radiative corrections has been obtained and LEP/SLC are now measuring the various components of these radiative corrections. For example, some authors [27] have studied the sensitivity of the data to a particularly interesting subset of the weak radiative corrections, i.e. the purely bosonic part. These terms arise from virtual exchange of gauge bosons and Higgses. The result is that indeed the measurements are sufficiently precise to require the presence of these contributions in order to fit the data. Second, the general results of the SM fits are reobtained from a different perspective. We see the preference for light Higgs manifested by the tendency for ϵ_3 to be rather on the low side. Since ϵ_3 is practically independent of m_t, its low value demands m_H small. If the Higgs is light then the preferred value of m_t is somewhat lower than the Tevatron result (which in the epsilon analysis is not included among the input data). This is because also the value of $\epsilon_1 \equiv \delta\rho$, which is determined by the widths, in particular by the leptonic width, is somewhat low. In fact ϵ_1 increases with m_t and, at fixed m_t, decreases with m_H, so that for small m_H the low central value of ϵ_1 pushes m_t down. Note that also the central value of ϵ_2 is on the low side, because the experimental value of m_W is a little bit too large. Finally, we see that adding the hadronic quantities or the low energy observables hardly makes a difference in the ϵ_i-ϵ_j plots with respect to the case with only the leptonic variables being included (the ellipse denoted by "All Asymm."). But, take for example the ϵ_1-ϵ_3 plot: while the leptonic ellipse contains the same information as one could obtain from a $\sin^2\theta_{eff}$ vs Γ_l plot, the content of the other two ellipses is much larger because it shows that the hadronic as well as the low energy quantities match the leptonic variables without need of any new physics. Note that the experimental values of ϵ_1 and ϵ_3 when the hadronic quantities are included also depend on the input value of α_s given in eq.(4).

4.3 Comparing the Data with the Minimal Supersymmetric Standard Model

The MSSM [40] is a completely specified, consistent and computable theory. There are too many parameters to attempt a direct fit of the data to the most general framework. So we consider two significant limiting cases: the "heavy" and the "light" MSSM.

The "heavy" limit corresponds to all sparticles being sufficiently massive, still within the limits of a natural explanation of the weak scale of mass. In this limit a very important result holds [29]: for what concerns the precision electroweak tests, the MSSM predictions tend to reproduce the results of the SM with a light Higgs, say $m_H \sim 100$ GeV. So if the masses of SUSY partners are pushed at sufficiently large values the same quality of fit as for the SM is guaranteed. Note that for $m_t = 175.6$ GeV and $m_H \sim 70$ GeV the values of the four epsilons computed in the SM lead to a fit of the corresponding experimental values with $\chi^2 \sim 4$, which is reasonable for $d.o.f = 4$. This value corresponds to the fact that the central values of $\epsilon_1, \epsilon_2, \epsilon_3$ and $-\epsilon_b$ are all below the SM value by about 1σ, as can be seen from fig. 12.

In the "light" MSSM option some of the superpartners have a relatively small mass, close to their experimental lower bounds. In this case the pattern of radiative corrections may sizeably deviate from that of the SM [30]. The potentially largest effects occur in vacuum polarisation amplitudes and/or the $Z \to b\bar{b}$ vertex. In particular we recall the following contributions :

i) a threshold effect in the Z wave function renormalisation [29] mostly due to the vector coupling of charginos and (off-diagonal) neutralinos to the Z itself. Defining the vacuum polarisation functions by $\Pi_{\mu\nu}(q^2) = -ig_{\mu\nu}[A(0) + q^2 F(q^2)] + q_\mu q_\nu$ terms, this is a positive contribution to $\epsilon_5 = m_Z^2 F'_{ZZ}(m_Z^2)$, the prime denoting a derivative with respect to q^2 (i.e. a contribution to a higher derivative term not included in the naive epsilon formalism, but compatible with the scheme described in Sect. 4.1). The ϵ_5 correction shifts ϵ_1, ϵ_2 and ϵ_3 by $-\epsilon_5$, $-c^2\epsilon_5$ and $-c^2\epsilon_5$ respectively, where $c^2 = \cos^2\theta_W$, so that all of them are reduced by a comparable amount. Correspondingly all the Z widths are reduced without affecting the asymmetries. This effect falls down particularly fast when the lightest chargino mass increases from a value close to $m_Z/2$. Now that we know, from the LEP2 runs, that the chargino mass is probably not smaller than m_Z its possible impact is drastically reduced.

ii) a positive contribution to ϵ_1 from the virtual exchange of split multiplets of SUSY partners, for example of the scalar top and bottom superpartners [31], analogous to the contribution of the top-bottom left-handed quark doublet. From the experimental value of m_t not much space is left for this possibility, and the experimental value of ϵ_1 is an important constraint on the spectrum. This is especially true now that the rather large lower limits on the chargino mass reduce the size of a possible compensation from ϵ_5. For example, if the stop is light then it must be mainly a right-handed stop. Also large values of $\tan\beta$ are disfavoured because they tend to enhance the splittings among SUSY partner multiplets. In general it is simpler to decrease the predicted values of ϵ_2 and ϵ_3 by taking advantage of ϵ_5 than to decrease ϵ_1, because the negative shift from ϵ_5 is most often counterbalanced by the increase from the effect of split SUSY multiplets.

iii) a negative contribution to ϵ_b due to the virtual exchange of a charged Higgs [32]. If one defines, as customary, $\tan\beta = v_2/v_1$ (v_1 and v_2 being the vacuum expectation values of the Higgs doublets giving masses to the down and up quarks, respectively), then, for negligible bottom Yukawa coupling or $\tan\beta << m_t/m_b$, this contribution is proportional to $m_t^2/\tan^2\beta$.

iv) a positive contribution to ϵ_b due to virtual chargino-stop exchange [33] which in this case is proportional to $m_t^2/\sin^2\beta$ and prefers small $\tan\beta$. This effect again requires the chargino and the stop to be light in order to be sizeable.

With the recent limits set by LEP2 on the masses of SUSY partners the above effects are small enough that other contributions from vertex diagrams could be comparable. Thus in the following we will only consider the experimental values of the epsilons obtained at the level denoted by "All Asymmetries" which only assumes lepton universality.

We have analysed the problem of what configurations of masses in the "light" MSSM are favoured or disfavoured by the present data (updating ref.[34]). We find that no lower limits on the masses of SUSY partners are obtained which are better than the direct limits. One exception is the case of stop and sbottom masses, which are severely constrained by the ϵ_1 value and also, at small $\tan\beta$, by the increase at LEP2 of the direct limit on the Higgs mass. Charged Higgs masses are also constrained. Since the central values of ϵ_1,ϵ_2 and ϵ_3 are all below the SM it is convenient to make ϵ_5 as large as possible. For this purpose light gaugino and slepton masses are favoured. We find that for $m_{\chi_1^+} \sim 90-120$ GeV the effect is still sizeable. Also favoured are small values of $\tan\beta$ that allow to put slepton masses relatively low, say, in the range 100-500 GeV, without making the split in the isospin doublets too large for ϵ_1. Charged Higgses must be heavy because they contribute to ϵ_b with the wrong sign. A light right-handed stop could help on R_b for a Higgsino-like chargino. But one needs small mixing (the right-handed stop must be close to the mass eigenstate) and beware of the Higgs mass constraint at small $\tan\beta$ (a Higgs mass above ~ 80 GeV, the range of LEP2 for susy Higgses at $\sqrt{s} = 183$ GeV, starts being a strong constraint at small $\tan\beta$). So we prefer in the following to keep the stop mass large. The limits on $b \to s\gamma$ also prefer heavy charged Higgs and stop

[35].

The scatter plots obtained in the planes ϵ_1-ϵ_3 and ϵ_2-ϵ_3 for $-200 < \mu < 200$ GeV, $0 < M < 250$ GeV, $\tan \beta = 1.5 - 2.5$, $m_{\tilde{l}} = 100 - 500$ GeV and $m_{\tilde{q}} = 1$ TeV are shown in figs. 13 and 14, together with the SM prediction for $m_t = 175.6$ GeV and $m_h \sim 70$ GeV. We see that in most cases the χ^2 is not improved. If we restrict to the small area, marked with a small star in fig. 13, where both ϵ_1 and ϵ_3 are improved we can check that also ϵ_2 is improved, as is seen from fig. 14 where the same values of the parameters have been employed in the region marked with the star. This region where the χ^2 is decreased by slightly more than one unity is included in the hypervolume $\mu = 133 - 147$ GeV, $M = 212 - 250$ GeV, ($m_{\chi_1^+} = 90 - 105$ GeV, $m_{\chi_1^0} = 58 - 72$ GeV, $m_{\chi_2^0} = 129 - 147$ GeV) with $\tan \beta \sim 1.5$, $m_{\tilde{q}} \sim 1$ TeV and $m_{\tilde{l}} = 100$ GeV. In this configuration ϵ_b is unchanged. We see that the advantage with respect to the SM is at most of the order of 1 in χ^2.

5 The Case for Physics beyond the Standard Model

Given the striking success of the SM why are we not satisfied with that theory? Why not just find the Higgs particle, for completeness, and declare that particle physics is closed? The main reason is that there are strong conceptual indications for physics beyond the SM.

It is considered highly unplausible that the origin of the electro-weak symmetry breaking can be explained by the standard Higgs mechanism, without accompanying new phenomena. New physics should be manifest at energies in the TeV domain. This conclusion follows fron an extrapolation of the SM at very high energies. The computed behaviour of the $SU(3) \otimes SU(2) \otimes U(1)$ couplings with energy clearly points towards the unification of the electro-weak and strong forces (Grand Unified Theories: GUTS) at scales of energy $M_{GUT} \sim 10^{14} - 10^{16}$ GeV which are close to the scale of quantum gravity, $M_{Pl} \sim 10^{19}$ GeV [36],[37]. One can also imagine a unified theory of all interactions also including gravity (at present superstrings [38] provide the best attempt at such a theory). Thus GUTS and the realm of quantum gravity set a very distant energy horizon that modern particle theory cannot anymore ignore. Can the SM without new physics be valid up to such large energies? This appears unlikely because the structure of the SM could not naturally explain the relative smallness of the weak scale of mass, set by the Higgs mechanism at $m \sim 1/\sqrt{G_F} \sim 250$ GeV with G_F being the Fermi coupling constant. This so-called hierarchy problem [39] is related to the presence of fundamental scalar fields in the theory with quadratic mass divergences and no protective extra symmetry at m=0. For fermions, first, the divergences are logarithmic and, second, at m=0 an additional symmetry, i.e. chiral symmetry, is restored. Here, when talking of divergences we are not worried of actual infinities. The theory is renormalisable and finite once the dependence on the cut off is absorbed in a redefinition of masses and couplings. Rather the hierarchy problem is one of naturalness. If we consider the cut off as a manifestation of new physics that will modify the theory at large energy scales, then it is relevant to look at the dependence of physical quantities on the cut off and to demand that no unexplained enormously accurate cancellation arise.

According to the above argument the observed value of $m \sim 250 \ GeV$ is indicative of the existence of new physics nearby. There are two main possibilities. Either there exist fundamental scalar Higgses but the theory is stabilised by supersymmetry, the boson-fermion symmetry that would downgrade the degree of divergence from quadratic to logarithmic. For approximate supersymmetry the cut off is replaced by the splitting between the normal particles and their supersymmetric partners. Then naturalness demands that this splitting (times the size of the weak gauge coupling) is of the order of the weak scale of mass, i.e. the separation within supermultiplets should be of the order of no more than a few TeV. In this case the masses of most supersymmetric partners of the known particles, a very large managerie of states, would fall, at least in part, in the discovery reach of the LHC. There are consistent, fully formulated field theories constructed on the basis of this idea, the simplest one being the MSSM [40]. Note that all normal observed states are those whose masses are forbidden in the limit of exact $SU(2) \otimes U(1)$. Instead for all SUSY partners the masses are allowed in that limit. Thus when supersymmetry is broken in the TeV range but $SU(2) \otimes U(1)$ is intact only s-partners take mass while all normal particles remain massless. Only at the lower weak scale the masses of ordinary particles are generated. Thus a simple criterium exists to understand the difference between particles and s-particles.

The other main avenue is compositeness of some sort. The Higgs boson is not elementary but either a bound state of fermions or a condensate, due to a new strong force, much stronger than the usual strong interactions, responsible for the attraction. A plethora of new "hadrons", bound by the new strong force would exist in the LHC range. A serious problem for this idea is that nobody sofar has been able to build up a realistic model along these lines, but that could eventually be explained by a lack of ingenuity on the theorists side. The most appealing examples are technicolor theories [4]-[5]. These models where inspired by the breaking of chiral symmetry in massless QCD induced by quark condensates. In the case of the electroweak breaking new heavy techniquarks must be introduced and the scale analogous to Λ_{QCD} must be about three orders of magnitude larger. The presence of such a large force relatively nearby has a strong tendency to clash with the results of the electroweak precision tests [6].

The hierarchy problem is certainly not the only conceptual problem of the SM. There are many more: the proliferation of parameters, the mysterious pattern of fermion masses and so on. But while most of these problems can be postponed to the final theory that will take over at very large energies, of order M_{GUT} or M_{Pl}, the hierarchy problem arises from the unstability of the low energy theory and requires a solution at relatively low energies.

A supersymmetric extension of the SM provides a way out which is well defined, computable and that preserves all virtues of the SM. The necessary SUSY breaking can be introduced through soft terms that do not spoil the good convergence properties of the theory. Precisely those terms arise from supergravity when it is spontaneoulsly broken in a hidden sector [41]. But alternative mechanisms of SUSY breaking are also being considered [42]. In the most familiar approach SUSY is broken in a hidden sector and the scale of SUSY breaking is very large of order $\Lambda \sim \sqrt{G_F^{-1/2} M_P}$ where M_P is the Planck mass. But since the hidden sector only communicates with the visible sector through gravitational interactions the splitting of the SUSY multiplets is much smaller, in the TeV energy domain, and the Goldstino is practically decoupled. In an alternative scenario the (not so much) hidden sector is connected to the visible one by ordinary gauge interactions. As these are much stronger than the gravitational interactions, Λ can be much smaller, as low as 10-100 TeV. It follows that the Goldstino is very light in these models (with mass of order or below 1 eV typically) and is the lightest, stable SUSY particle, but its couplings are observably large. The radiative decay of the lightest neutralino into the Goldstino leads to detectable photons. The signature of photons comes out naturally in this SUSY breaking pattern: with respect to the MSSM, in the gauge mediated model there are typically more photons and less missing energy. Gravitational and gauge mediation are extreme alternatives: a spectrum of intermediate cases is conceivable. The main appeal of gauge mediated models is a better protection against flavour changing neutral currents. In the gravitational version even if we accept that gravity leads to degenerate scalar masses at a scale near M_{Pl} the running of the masses down to the weak scale can generate mixing induced by the large masses of the third generation fermions [37].

At present the most direct phenomenological evidence in favour of supersymmetry is obtained from the unification of couplings in GUTS. Precise LEP data on $\alpha_s(m_Z)$ and $\sin^2 \theta_W$ confirm what was already known with less accuracy: standard one-scale GUTS fail in predicting $\sin^2 \theta_W$ given $\alpha_s(m_Z)$ (and $\alpha(m_Z)$) while SUSY GUTS [43] are in agreement with the present, very precise, experimental results. According to the recent analysis of ref.[44], if one starts from the known values of $\sin^2 \theta_W$ and $\alpha(m_Z)$, one finds for $\alpha_s(m_Z)$ the results:

$$\begin{aligned} \alpha_s(m_Z) &= 0.073 \pm 0.002 \quad \text{(Standard GUTS)} \\ \alpha_s(m_Z) &= 0.129 \pm 0.010 \quad \text{(SUSY GUTS)} \end{aligned} \quad (21)$$

to be compared with the world average experimental value $\alpha_s(m_Z)$ =0.119(4).

Some experimental hints for new physics beyond the SM come from the sky and from cosmology. I refer to solar and atmospheric neutrinos, dark matter and baryogenesis. It seems to me that by now it is difficult to imagine that the neutrino anomalies can all disappear or be explained away [45],[46]. Probably they are a manifestation of neutrino oscillations, hence neutrino masses. Massive neutrinos are natural in most GUT's. The see-saw mechanism [47] provides an elegant explanation of the smallness of neutrino masses, inversely proportional to the large scale of L violation. Minimal SU(5) is disfavoured by neutrino masses, because there is no ν_R and B-L is conserved by gauge interactions, while SO(10) provides a very natural context for them [36]. A number of different observations show that most of the matter in the universe is non luminous and that both cold and hot dark matter forms are needed [48]. Neutrinos with a mass of a few eV's could provide the hot dark matter (particles that are relativistic when they drop out of equilibrium). But neutrinos cannot be the totality of dark matter because they are too faintly interacting and have no enough clumping at galactic distances. Cold dark matter particles do not exist in the SM. Plausible candidates are axions or neutralinos. In this respect SUSY in the MSSM version scores a good point while other SUSY options, like the gauge mediated models and the R-parity violating versions, do not offer such a simple possibility. Baryogenesis is interesting because it could occur at the weak scale [49] but not in the SM. For baryogenesis one needs the three famous Sakharov conditions [50]: B violation, CP violation and no termal equilibrium. In principle these conditions could be verified in the SM.

B is violated by instantons when kT is of the order of the weak scale (but B-L is conserved). CP is violated by the CKM phase and out of equilibrium conditions could be verified during the electroweak phase transition. So the conditions for baryogenesis appear superficially to be present for it to occur at the weak scale in the SM. However, a more quantitative analysis [51],[52] shows that baryogenesis is not possible in the SM because there is not enough CP violation and the phase transition is not sufficiently strong first order, unless $m_H < 40\ GeV$, which is by now excluded by LEP. Certainly baryogenesis could also occur below the GUT scale, after inflation. But only that part with $|B - L| > 0$ would survive and not be erased at the weak scale by instanton effects. Thus baryogenesis at $kT \sim 10^{12} - 10^{15}\ GeV$ needs B-L violation at some stage like for m_ν. The two effects could be related if baryogenesis arises from leptogenesis [54] then converted into baryogenesis by instantons. While baryogenesis at a large energy scale is thus not excluded it is interesting that recent studies have shown that baryogenesis at the weak scale could be possible in the MSSM [53]. In fact, in this model there are additional sources of CP violations and the bound on m_H is modified by a sufficient amount by the presence of scalars with large couplings to the Higgs sector, typically the s-top. What is required is that $m_H \lesssim 80 - 100\ GeV$ (in the LEP2 range!), a s-top not heavier than the top quark and, preferentially, a small $\tan\beta$.

6 The Search for the Higgs

The SM works with remarkable accuracy. But the experimental foundation of the SM is not completed if the electroweak symmetry breaking mechanism is not experimentally established. Experiments must decide what is true: the SM Higgs or Higgs plus SUSY or new strong forces and Higgs compositeness.

The theoretical limits on the Higgs mass play an important role in the planning of the experimental strategy. The large experimental value of m_t has important implications on m_H both in the minimal SM [55]-[57] and in its minimal supersymmetric extension[58],[59].

It is well known[55]-[57] that in the SM with only one Higgs doublet a lower limit on m_H can be derived from the requirement of vacuum stability. The limit is a function of m_t and of the energy scale Λ where the model breaks down and new physics appears. Similarly an upper bound on m_H (with mild dependence on m_t) is obtained [60] from the requirement that up to the scale Λ no Landau pole appears. If one demands vacuum stability up to a very large scale,of the order of M_{GUT} or M_{Pl} then the resulting bound on m_H in the SM with only one Higgs doublet is given by [56]:

$$m_H(GeV) > 138 + 2.1\,[m_t - 175.6] - 3.0\,\frac{\alpha_s(m_Z) - 0.119}{0.004}\ . \qquad (22)$$

In fact one can show that the discovery of a Higgs particle at LEP2, or $m_H \lesssim 100\ GeV$, would imply that the SM breaks down at a scale Λ of the order of a few TeV. Of course, the limit is only valid in the SM with one doublet of Higgses. It is enough to add a second doublet to avoid the lower limit. The upper limit on the Higgs mass in the SM is important for assessing the chances of success of the LHC as an accelerator designed to solve the Higgs problem. The upper limit [60] has been recently reevaluated [61]. For $m_t \sim 175\ GeV$ one finds $m_H \lesssim 180\ GeV$ for $\Lambda \sim M_{GUT} - M_{Pl}$ and $m_H \lesssim 0.5 - 0.8\ TeV$ for $\Lambda \sim 1\ TeV$.

A particularly important example of theory where the bound is violated is the MSSM, which we now discuss. As is well known [40], in the MSSM there are two Higgs doublets, which implies three neutral physical Higgs particles and a pair of charged Higgses. The lightest neutral Higgs, called h, should be lighter than m_Z at tree-level approximation. However, radiative corrections [62] increase the h mass by a term proportional to m_t^4 and logarithmically dependent on the stop mass . Once the radiative corrections are taken into account the h mass still remains rather small: for $m_t = 174$ GeV one finds the limit (for all values of tg β) $m_h < 130$ GeV [59]. Actually there are reasons to expect that m_h is well below the bound. In fact, if h_t is large at the GUT scale, which is suggested by the large observed value ot m_t and by a natural onsetting of the electroweak symmetry breaking induced by m_t, then at low energy a fixed point is reached in the evolution of m_t. The fixed point corresponds to $m_t \sim 195 \sin \beta$ GeV (a good approximate relation for tg $\beta = v_{up}/v_{down} < 10$). If the fixed point situation is realized, then m_h is considerably below the bound, $m_h \lesssim 100\ GeV$ [59].

In conclusion, for $m_t \sim 175$ GeV, we have seen that, on the one hand, if a Higgs is found at LEP the SM cannot be valid up to M_{Pl}. On the other hand, if a Higgs is found at LEP, then the MSSM has good chances, because this model would be excluded for $m_h > 130$ GeV.

7 New Physics at HERA?

7.1 Introduction

The HERA experiments H1 [63] and ZEUS [64], recently updated in ref. [65], have reported an excess of deep-inelastic e^+p scattering events at large values of $Q^2 \gtrsim 1.5 \times 10^4$ GeV^2, in a domain not previously explored by other experiments. The total e^+p integrated luminosity was of 14.2 +9.5 = 23.7 pb^{-1}, at H1 and of 20.1+13.4 = 33.5 pb^{-1} at ZEUS. The first figure refers to the data before the '97 run [63],[64], while the second one refers to part of the continuing '97 run, whose results were presented at the LP'97 Symposium [65]. Both experiments collected in the past about 1 pb^{-1} each with an e^- beam. A very schematic description of the situation is as follows. At $Q^2 \gtrsim 15\ 10^4$ in the neutral current channel (NC), H1 observes 12+6 = 18 events while about 5+3 = 8 were expected and ZEUS observes 12 + 6 = 18 events with about 9 + 6 =15 expected. In the charged current channel (CC), in the same range of Q^2, H1 observes 4+2 = 6 events while about 1.8+1.2 = 3 were expected and ZEUS observes 3 + 2 = 5 events with about 1.2 + 0.8 = 2 expected. The distribution of the first H1 data suggested a resonance in the NC channel. In the interval $187.5 < M < 212.5\ GeV$, which corresponds to $x \simeq 0.4$, and $y > 0.4$, H1 in total finds 7 + 1 = 8 events with about 1 + 0.5 = 1.5 expected. But in correspondence of the H1 peak ZEUS observes a total of 3 events, just about the number of expected events. In the domain $x > 0.55$ and $y > 0.25$ ZEUS observes 3 + 2 events with about 1.2 + 0.8 = 2 expected. But in the same domain H1 observes only 1 event in total more or less as expected.

We see that with new statistics the evidence for the signal remain meager. The bad features of the original data did not improve. First, there is a problem of rates. With more integrated luminosity than for H1, ZEUS sees about the same number of events in both the NC and CC channels. Second, H1 is suggestive of a resonance (although the evidence is now less than it was) while ZEUS indicates a large x continuum (here also the new data are not more encouraging). The difference could in part, but apparently not completely [66], be due to the different methods of mass reconstruction used by the two experiments, or to fluctuations in the event characteristics. Of course, at this stage, due to the limited statistics, one cannot exclude the possibility that the whole effect is a statistical fluctuation. All these issues will hopefully be clarified by the continuation of data taking. Meanwhile, it is important to explore possible interpretations of the signal, in particular with the aim of identifying additional signatures that might eventually be able to discriminate between different explanations of the reported excess.

7.2 Structure Functions

Since the observed excess is with respect to the SM expectation based on the QCD-improved parton model, the first question is whether the effect could be explained by some inadequacy of the conventional analysis without invoking new physics beyond the SM. In the somewhat analogous case of the apparent excess of jet production at large transverse energy E_T recently observed by the CDF collaboration at the Tevatron [67], it has been argued [68] that a substantial decrease in the discrepancy can be obtained by modifying the gluon parton density at large values of x where it has not been measured directly. New results [69] on large p_T photons appear to cast doubts on this explanation because these data support the old gluon density and not the newly proposed one. In the HERA case, a similar explanation appears impossible, at least for the H1 data. Here quark densities are involved and they are well known at the same x but smaller Q^2 [70],[71], and indeed the theory fits the data well there. Since the QCD evolution is believed to be safe in the relevant region of x, the proposed strategy is to have a new component in the quark densities at very large x, beyond the measured region, and small Q^2 which is driven at smaller x by the evolution and contributes to HERA when Q^2 is sufficiently large [70]. However it turns out that a large enough effect is only conceivable at very large x, $x \gtrsim 0.75$, which is too large even for ZEUS. The compatibility with the Tevatron is also an important constraint. This is because ep scattering is linear in the quark densities, while $p\bar{p}$ is quadratic, so that a factor of 1.5-2 at HERA implies a large effect also at the Tevatron. In addition, many possibilities including intrinsic charm [72] (unless $\bar{c} \neq c$ at the relevant x values [73]) are excluded from the HERA data in the CC channel [74]. More in general, if only one type of density is modified, then in the CC channel one obtains too large an effect in the $\bar{u}$ and d cases and no effect at all in the $\bar{d}$ and u cases [74]. In conclusion, it is a fact that nobody sofar was able to even roughly fit the data. This possibility is to be kept in mind if eventually the data will drift towards the SM and only a small excess at particularly large x and Q^2 is left in NC channel.

7.3 Contact Terms

Still considering the possibility that the observed excess is a non-resonant continuum, a rather general approach in terms of new physics is to interpret the HERA excess as due to an effective four-fermion $\bar{e}e\bar{q}q$ contact interaction [75] with a scale Λ of order a few TeV. It is interesting that a similar contact term of the $\bar{q}q\bar{q}q$ type, with a scale of exactly the same order of magnitude, could also reproduce the CDF excess in jet production at large E_T [67]. (Note, however, that this interpretation is not strengthened by more recent data on the dijet angular distribution [76]). One has studied in detail [77],[78] vector contact terms of the general form

$$\Delta L = \frac{4\pi \eta_{ij}}{(\Lambda_{ij}^\eta)^2} \, \bar{e}_i \gamma^\mu e_i \, \bar{q}_j \gamma_\mu q_j. \tag{23}$$

with $i,j = L, R$ and η a $\pm$ sign. Strong limits on these contact terms are provided by LEP2 [79] (LEP1 limits also have been considered but are less constraining [80]), Tevatron [81] and atomic parity violation (APV) experiments [25]. The constraints are even more stringent for scalar or tensor contact terms. APV limits essentially exclude all relevant $A_e V_q$ component. The CDF limits on Drell-Yan production are particularly constraining. Data exist both for electron and muon pairs up to pair masses of about 500 GeV and show a remarkable $e - \mu$ universality and agreement with the SM. New LEP limits (especially from LEP2) have been presented [79]. In general it would be possible to obtain a reasonably good fit of the HERA data, consistent with the APV and the LEP limits, if one could skip the CDF limits [82]. But, for example, a parity conserving combination $(\bar{e}_L \gamma^\mu e_L)(\bar{u}_R \gamma_\mu u_R) + (\bar{e}_R \gamma^\mu e_R)(\bar{u}_L \gamma_\mu u_L)$ with $\Lambda_{LR}^+ = \Lambda_{RL}^+ \sim 4$ TeV still leads to a marginal fit to the HERA data and is compatible with all existing limits [82]·[83] (see fig.15 [84]). Because we expect contact terms to satisfy $SU(2) \otimes U(1)$, because they reflect physics at large energy scales, the above phenomenological form is to be modified into $\bar{L}_L \gamma_\mu L_L (\bar{u}_R \gamma^\mu u_R + \bar{d}_R \gamma^\mu d_R) + \bar{e}_R \gamma_\mu e_R \bar{Q}_L \gamma^\mu Q_L)$, where L and Q are doublets [85]. This form is both gauge invariant and parity conserving. Here one has taken into account the requirement that contact terms corresponding to CC are too constrained to appear. More sophisticated fits have also been performed [82].

In conclusion, contact terms are severely constrained but not excluded. The problem of generating the phenomenologically required contact terms from some form of new physics at larger energies is far from trivial [85, 86]. Note also that contact terms require values of $g^2/\Lambda^2 \sim 4\pi/(3-4 \text{ TeV})^2$, which would imply a very strong nearby interaction. Indeed for g^2 of the order of the $SU(3) \otimes SU(2) \otimes U(1)$ couplings, Λ would fall below 1 TeV, where the contact term description is inadequate. We recall that the effects of contact terms should be present in both the e^+ and the e^- cases with comparable intensity. Definitely contact terms cannot produce a CC signal [87], as we shall see, and no events with isolated muons and missing energy.

7.4 Leptoquarks

I now focus on the possibility of a resonance with e^+q quantum numbers, namely a leptoquark [77, 88, 89, 90, 91, 92, 93], of mass $M \sim 190 - 210\ GeV$, according to H1. The most obvious possibility is that the production at HERA occurs from valence u or d quarks, since otherwise the coupling would need to be quite larger, and more difficult to reconcile with existing limits. However production from the sea is also considered. Assuming an S-wave state, one may have either a scalar or a vector leptoquark. I only consider here the first option, because vector leptoquarks are more difficult to reconcile with their apparent absence at the Tevatron. The coupling λ for a scalar ϕ is defined by $\lambda \phi \bar{e}_L q_R$ or $\lambda \phi \bar{e}_R q_L$, The corresponding width is given by $\Gamma = \lambda^2 M_\phi / 16\pi$, and the production cross section on a free quark is given in lowest order by $\sigma = \frac{\pi}{4s} \lambda^2$.

Including also the new '97 run results, the combined H1 and ZEUS data, interpreted in terms of scalar leptoquarks lead to the following list of couplings [95, 77, 94]:

$$e^+u \to \lambda\sqrt{B} \sim 0.017 - 0.025; \quad e^+d \to \lambda\sqrt{B} \sim 0.025 - 0.033; \quad e^+s \to \lambda\sqrt{B} \sim 0.15 - 0.25 \quad (24)$$

where B is the branching ratio into the e-q mode. By s the strange sea is meant. For comparison note that the electric charge is $e = \sqrt{4\pi\alpha} \sim 0.3$. Production via $e^+\bar{u}$ or $e^+\bar{d}$ is excluded by the fact that in these cases the production in e^-u or e^-d would be so copious that it should have shown up in the small luminosity already collected in the e^-p mode. The estimate of λ in the strange sea case is merely indicative due to the large uncertainties on the value of the small sea densities at the relatively large values relevant to the HERA data. The width is in all cases narrow: for $B \sim 1/2$ we have $\Gamma \sim 4 - 16\ MeV$ for valence and $350 - 1000\ MeV$ for sea densities.

It is important to notice that improved data from the CDF and D0[69] on one side and from APV [25] and LEP [79] on the other considerably reduce the window for leptoquarks. Consistency with the Tevatron, where scalar leptoquarks are produced via model-independent (and λ-independent) QCD processes with potentially large rates, demands a value of B sizeably smaller than 1. In fact, the most recent NLO estimates of the squark and leptoquark production cross sections [96, 97] allow to estimate that at 200 GeV approximately 6-7 events with e^+e^-jj final states should be present in the combined CDF and D0 data sets. For $B = 1$ the CDF limit is 210 GeV, the latest D0 limit is 225 GeV at 95%CL. The combined CDF+D0 limit is 240 GeV at 95%CL [69]. We see that for consistency one should impose:

$$B \lesssim 0.5 - 0.7 \quad (25)$$

Finally, the case of a 200 GeV vector leptoquark is most likely totally ruled out by the Tevatron data, since the production rate can be as much as a factor of 10 larger than that of scalar leptoquarks.

There are also lower limits on B, different for production off valence or sea quarks, so that only a definite window for B is left in all cases. For production off valence the limit arises from APV [25], while for the sea case it is obtained from recent LEP2 data [79].

One obtains a limit from APV because the s-channel echange amplitude for a leptoquark is equivalent at low energies to an $(\bar{e}q)(\bar{q}e)$ contact term with amplitude proportional to λ^2/M^2. After Fierz rearrangement a component on the relevant APV amplitude $A_e V_q$ is generated, hence the limit on λ. The results are [87]

$$e^+ u \to \lambda \lesssim 0.058; \quad e^+ d \to \lambda \lesssim 0.055 \qquad (26)$$

The above limits are for $M = 200\ GeV$ (they scale in proportion to M) and are obtained from the quoted error on the new APV measurement on Cs. This error being mainly theoretical, one could perhaps take a more conservative attitude and somewhat relax the limit. Comparing with the values for $\lambda\sqrt{B}$ indicated by HERA, given in eq.(24), one obtains lower limits on B:

$$e^+ u \to B \gtrsim 0.1 - 0.2; \quad e^+ d \to B \gtrsim 0.2 - 0.4 \qquad (27)$$

For production off the strange sea quark the upper limit on λ is obtained from LEP2 [79], in that the t-channel exchange of the leptoquark contributes to the process $e^+ e^- \to s\bar{s}$ (similar limits for valence quarks are not sufficiently constraining, because the values of λ required by HERA are considerably smaller). Recently new results have been presented by ALEPH, DELPHI and OPAL [79]. The best limits are around $\lambda \lesssim 0.6 - 0.7$ This, given eq.(24), corresponds to

$$e^+ s \to B \gtrsim 0.05 - 0.2 \qquad (28)$$

Recalling the Tevatron upper limits on B, given in eq.(25), we see that only a definite window for B is left in all cases.

Note that one given leptoquark cannot be present both in $e^+ p$ and in $e^- p$ (unless it is produced from strange quarks).

7.5 S-quarks with R-parity Violation

I now consider specifically leptoquarks and SUSY [99, 100, 101, 102, 103, 77]. In general, in SUSY one could consider leptoquark models without R-parity violation. It is sufficient to introduce together with scalar leptoquarks also the associated spin-1/2 leptoquarkinos [98]. In this way one has not to give up the possibility that neutralinos provide the necessary cold dark matter in the universe. We find it more attractive to embed a hypothetical leptoquark in the minimal supersymmetric extension of the Standard Model [40] with violation of R parity [104]. The connection with the HERA events has been more recently invoked in ref. [102]. The corresponding superpotential can be written in the form

$$W_R \equiv \mu_i H L_i + \lambda_{ijk} L_i L_j E_k^c + \lambda'_{ijk} L_i Q_j D_k^c + \lambda''_{ijk} U_i^c D_j^c D_k^c, \qquad (29)$$

where $H, L_i, E_j^c, Q_k, (U,D)_l^c$ denote superfields for the $Y = 1/2$ Higgs doublet, left-handed lepton doublets, lepton singlets, left-handed quark doublets and quark singlets, respectively. The indices i, j, k label the three generations of quarks and leptons. Furthermore, we assume the absence of the λ'' couplings, so as to avoid rapid baryon decay, and the λ couplings are not directly relevant in the following.

The squark production mechanisms permitted by the λ' couplings in (29) include e^+d collisions to form $\tilde{u}_L, \tilde{c}_L$ or $\tilde{t}_L$, which involve valence d quarks, and various collisions of the types e^+d_i ($i = 2, 3$) or $e^+\bar{u}_i$ ($i = 1, 2, 3$) which involve sea quarks. A careful analysis leads to the result that the only processes that survive after taking into account existing low energy limits are

$$e_R^+ d_R \to \tilde{c}_L; \quad e_R^+ d_R \to \tilde{t}_L; \quad e_R^+ s_R \to \tilde{t}_L \qquad (30)$$

For example $e_R^+ d_R \to \tilde{u}_L$ is forbidden by data on neutrinoless double beta decay which imply [105]

$$|\lambda'_{111}| < 7 \times 10^{-3} \left(\frac{m_{\tilde{q}}}{200 \text{ GeV}}\right)^2 \left(\frac{m_{\tilde{g}}}{1 \text{ TeV}}\right)^{\frac{1}{2}}. \qquad (31)$$

where $m_{\tilde{q}}$ is the mass of the lighter of $\tilde{u}_L$ and $\tilde{d}_R$, and $m_{\tilde{g}}$ is the gluino mass.

It is interesting to note [100] that the left s-top could be a superposition of two mass eigenstates $\tilde{t}_1, \tilde{t}_2$, with a difference of mass that can be large as it is proportional to m_t:

$$\tilde{t}_L = \cos\theta_t \, \tilde{t}_1 + \sin\theta_t \, \tilde{t}_2 \qquad (32)$$

where θ_t is the mixing angle. With $m_1 \sim 200$ GeV, $m_2 \sim 230$ GeV and $\sin^2\theta_t \sim 2/3$ one can obtain a broad mass distribution, more similar to the combined H1 and ZEUS data. (But with the present data one has to swallow that H1 only observes $\tilde{t}_1$ while ZEUS only sees $\tilde{t}_2$!). However, the presence of two light leptoquarks makes the APV limit more stringent. In fact it becomes

$$B < B_\infty [1 + \tan^2\theta_t \frac{m_1^2}{m_2^2}] \qquad (33)$$

Thus, for the above mass and mixing choices, the above quoted APV limit B_∞ must be relaxed invoking a larger theoretical uncertainty on the Cs measurement.

Let us now discuss [77] if it is reasonable to expect that $\tilde{c}$ and $\tilde{t}$ decay satisfy the bounds on the branching ratio B. A virtue of s-quarks as leptoquark is that competition of R-violating and normal decays ensures that in general $B < 1$.

In the case of $\tilde{c}_L$, the most important possible decay modes are the R-conserving $\tilde{c}_L \to c\chi_i^0$ ($i = 1, .., 4$) and $\tilde{c}_L \to s\chi_j^+$ ($j = 1, 2$), and the R-violating $\tilde{c}_L \to de^+$, where χ_i^0, χ_j^+ denote neutralinos and charginos, respectively. In this case it has been shown that, if one assumes that $m_{\chi_j^+} > 200$ GeV, then, in a sizeable domain of the parameter space, the neutralino mode can be sufficiently suppressed so that $B \sim 1/2$ as required (for example, the couplings of a higgsino-like neutralino are suppressed by the small charm mass).

In the case of $\tilde{t}_L$, it is interesting to notice that the neutralino decay mode $\tilde{t}_L \to t\chi_i^0$ is kinematically closed in a natural way. In order to obtain a large value of B in the case of s-top production off d-quarks, in spite of the small value of λ, it is sufficient to require that all charginos are heavy enough to forbid the decay $\tilde{t}_L \to b\chi_j^+$. However, we do not really want to obtain B too close to 1, so that in this case some amount of fine tuning is required. Or, with charginos heavy, one could invoke other decay channels as, for example, $\tilde{t} \to \tilde{b}W^+$ [108]. But the large splitting needed between $\tilde{t}$ and $\tilde{b}$ implies problems with the ρ-parameter of electroweak precision tests, unless large mixings in both the s-top and s-bottom sectors are involved and their values suitably chosen. To obtain $B \sim 1/2$ is more natural in the case of s-top production off s-quarks, because of the larger value of λ, which is of the order of the gauge couplings.

The interpretation of HERA events in terms of s-quarks with R-parity violation requires a very peculiar family and flavour structure [106]. The flavour problem is that there are very strong limits on products of couplings from absence of FCNC. The unification problem is that nucleon stability poses even stronger limits on products of λ couplings that differ by the exchange of quarks and leptons which are treated on the same footing in GUTS. However it was found that the unification problem can be solved and the required pattern can be embedded in a grand unification framework [106]. The already intricated problem of the mysterious texture of masses and couplings is however terribly enhanced in these scenarios.

7.6 Charged Current Events

We have mentioned that in the CC channel at $Q^2 \gtrsim 15\ 10^4$ H1 and ZEUS see a total of 11 events with 5 expected. The statistics is even more limited than in the NC case, so one cannot at the moment derive any firm conclusion on the existence and on the nature of an excess in that channel. However, the presence or absence of a simultaneous CC signal is extremely significant for the identification of the underlying new physics (as it would also be the case for the result of a comparable run with an e^- beam, which however is further away in time). It is found that in most of the cases the CC signal is not expected to arise [87, 107, 109, 108]. But if it is present at a comparable rate as for the NC signal, the corresponding indications are very selective. In fact the following results are found. Due to the existing limits on charged current processes, it is not possible to find a set of contact terms that satisfy $SU(2) \otimes U(1)$ invariance and lead to a significant production of CC events. For leptoquarks, we recall that a leptoquark with branching ratio equal to 1 in e^+q is excluded by the recent Tevatron limits. Therefore on one hand some branching fraction in the CC channel is needed. On the other hand, one finds that there is limited space for the possibility that a leptoquark can generate a CC signal at HERA with one single parton quark in the final state. This occurrence would indicate $SU(2) \otimes U(1)$ violating couplings or couplings to a current containing the charm quark. A few mechanisms for producing CC final states from $\tilde{c}$ or $\tilde{t}$ have been proposed [87, 109]. In all cases $\tilde{c}$ or $\tilde{t}$ lead to multiparton final states. Since apparently the CC candidates are all with one single jet, some strict requirements on the masses of the partecipating particles must be imposed so that some partons are too soft to be visible while others coelesce into a single visible jet. So, s-quarks with R-parity violating decays could indeed produce CC events or events with charged leptons and missing energy. But the observation of such events would make the model much more constrained.

8 Conclusion

The HERA anomaly is an interesting feature that deserves further attention and more experimental effort. But at the moment it does not represent a convincing evidence of new physics. The same is true for the other few possible discrepancies observed here and there in the data. The overall picture remains in impressive agreement with the SM. Yet, for conceptual reasons, we remain confident that new physics will eventually appear at the LHC if not before.

Acknowledgments

It is for me a pleasure to thank Juan Carlos D'Olivo, Fernando Quevedo and Marti Ruiz-Altaba for their kind invitation and warm hospitality in Mexico.

References

[1] J.Timmermans, Proceedings of LP'97, Hamburg, 1997;
S. Dong, ibidem;
D. Ward, Proceedings of HEP97, Jerusalem, 1997.

[2] The LEP Electroweak Working Group, LEPEWWG/97-02.

[3] C. Dionisi, Proceedings of LP'97, Hamburg, 1997;
P. Janot, Proceedings of HEP97, Jerusalem, 1997.

[4] S. Weinberg, *Phys. Rev.* **D13** (1976) 974 and *Phys. Rev.* **D19** (1979) 1277;
L. Susskind, *Phys. Rev.* **D20** (1979) 2619;
E. Farhi and L. Susskind, *Phys. Rep.* **74** (1981) 277.

[5] R. Casalbuoni et al., *Phys. Lett.* **B258** (1991) 161;
R.N. Cahn and M. Suzuki, LBL-30351 (1991);
C. Roiesnel and Tran N. Truong, *Phys. Lett.* **B253** (1991) 439;
T. Appelquist and G. Triantaphyllou, *Phys. Lett.* **B278** (1992) 345;
T. Appelquist, Proceedings of the Rencontres de la Vallé e d'Aoste, La Thuile, Italy, 1993;
R. Chivukula, hep-ph/9701322.

[6] J. Ellis, G.L. Fogli and E. Lisi, *Phys. Lett.* **B343** (1995) 282.

[7] W. Li, Proceedings of LP'97, Hamburg, 1997.

[8] F. Caravaglios, *Phys.Lett.* **B394** (1997) 359.

[9] P. Giromini, Proceedings of LP'97, Hamburg, 1997;
A. Yagil, Proceedings of HEP97, Jerusalem, 1997.

[10] G. Altarelli, R. Kleiss and C. Verzegnassi (eds.), Z Physics at LEP 1 (CERN 89-08, Geneva, 1989), Vols. 1–3;
Precision Calculations for the Z Resonance, ed. by D. Bardin, W. Hollik and G. Passarino, CERN Rep 95-03 (1995);
M.I. Vysotskii, V.A. Novikov, L.B. Okun and A.N. Rozanov, hep-ph/9606253.

[11] F. Jegerlehner, *Z.Phys.* **C32** (1986) 195;
B.W. Lynn, G.Penso and C.Verzegnassi, *Phys. Rev.* **D35** (1987) 42;
H. Burkhardt et al., *Z.Phys.* **C43** (1989) 497;
F. Jegerlehner, *Progr. Part. Nucl. Phys.* **27** (1991) 32;
M.L. Swartz, Preprint SLAC-PUB-6710, 1994;
M.L. Swartz, *Phys. Rev.* **D53** (1996) 5268;
A.D. Martin and D. Zeppenfeld, *Phys. Lett.* **B345** (1995) 558;
R.B. Nevzorov, A.V. Novikov and M.I. Vysotskii, hep-ph/9405390;
H. Burkhardt and B. Pietrzyk, *Phys. Lett.* **B356** (1995) 398;
S. Eidelman and F. Jegerlehner, *Z. Phys.* **C67** (1995) 585.

[12] B. Pietrzyk, Proceedings of the Symposium on Radiative Corrections, Cracow, 1996.

[13] S. Catani, Proceedings of LP'97, Hamburg, 1997;
Yu.L. Dokshitser, Proceedings of HEP97, Jerusalem, 1997.

[14] ZFITTER: D. Bardin et al., CERN-TH. 6443/92 and refs. therein;
TOPAZ0: G. Montagna et al., *Nucl. Phys.* **B401** (1993) 3, *Comp. Phys. Comm.* **76** (1993) 328;
BHM: G.Burgers et al.,
LEPTOP: A.V. Novikov, L.B. Okun and M.I. Vysotsky, *Mod. Phys. Lett.* **A8** (1993) 2529;
WOH, W:Hollik : see ref. [10].

[15] G. Degrassi, P. Gambino and A. Vicini, *Phys. Lett.* **B383** (1996) 219;
G. Degrassi, P. Gambino and A. Sirlin, *Phys. Lett.* **B394** (1997) 188;
G. Degrassi, P. Gambino, M. Passera and A. Sirlin, hep-ph/9708311.

[16] J. Ellis, G.L. Fogli and E. Lisi, *Phys. Lett.* **B389** (1996) 321;
G. Altarelli, hep-ph/9611239;
AA. Gurtu, *Phys. Lett.* **B385** (1996) 415;
P. Langacker and J. Erler, hep-ph/9703428;
J.L. Rosner, hep-ph/9704331;
K. Hagiwara, D. Haidt and S. Matsumoto, hep-ph/9706331.

[17] Y.Y. Kim, Proceedings of LP'97, Hamburg, 1997.

[18] G. Altarelli, R. Barbieri and S. Jadach, *Nucl. Phys.* **B369** (1992) 3;
G. Altarelli, R. Barbieri and F. Caravaglios, *Nucl. Phys.* **B405** (1993) 3; *Phys. Lett.* **B349** (1995) 145.

[19] M.E. Peskin and T. Takeuchi, *Phys. Rev. Lett.* **65** (1990) 964 and *Phys. Rev.* **D46** (1991) 381.

[20] G. Altarelli and R. Barbieri, *Phys. Lett.* **B253** (1990) 161;
B.W. Lynn, M.E. Peskin and R.G. Stuart, SLAC-PUB-3725 (1985), in Physics at LEP, CERN 86-02, Vol. I, p. 90;
B. Holdom and J. Terning, *Phys. Lett.* **B247** (1990) 88;
D.C. Kennedy and P. Langacker, *Phys. Rev. Lett.* **65** (1990) 2967.

[21] A.A. Akundov et al., *Nucl. Phys.* **B276** (1988) 1;
F. Diakonov and W. Wetzel, HD-THEP-88-4 (1988);
W. Beenakker and H. Hollik, *Z. Phys.* **C40** (1988) 569;
B.W. Lynn and R.G. Stuart, *Phys. Lett.* **B252** (1990) 676;
J. Bernabeu, A. Pich and A. Santamaria, *Phys. Lett.* **B200** (1988) 569; *Nucl. Phys.* **B363** (1991) 326.

[22] B.A. Kniehl and J.H. Kuhn, *Phys. Lett.* **B224** (1989) 229; *Nucl.Phys.* **B329** (1990) 547.

[23] F. Caravaglios and G.G. Ross, *Phys. Lett.* **B346** (1995) 159.

[24] CHARM Collaboration, J.V. Allaby et al., *Phys. Lett.* **B177** (1986) 446; *Z. Phys.* **C36** (1987) 611;

CDHS Collaboration, H. Abramowicz et al., *Phys. Rev. Lett.* **57** (1986) 298;
A. Blondel et al., *Z. Phys.* **C45** (1990) 361;
CCFR Collaboration, K. McFarland, hep-ex/9701010.

[25] C.S. Wood et al., *Science* **275** (1997) 1759.

[26] CHARM II Collaboration, P. Vilain et al., *Phys. Lett.* **B335** (1997) 246.

[27] S. Dittmaier, D. Schildknecht, K. Kolodziej, M. Kuroda *Nucl. Phys.* **B426** (1994) 249; **B446** (1995) 334;
A. Sirlin, and P. Gambino, *Phys. Rev. Lett.* **73** (1994) 621;
S. Dittmaier, D. Schildknecht and G.Weiglein, *Nucl. Phys.* **B465** (1996) 3.

[28] H.P. Nilles, *Phys. Rep.* **C110** (1984) 1;
H.E. Haber and G.L. Kane, *Phys. Rep.* **C117** (1985) 75;
R. Barbieri, *Riv. Nuovo Cim.* **11** (1988) 1.

[29] R. Barbieri, F. Caravaglios and M. Frigeni,*Phys. Lett.* **B279** (1992) 169.

[30] S. Pokorski, Proceedings of ICHEP'96, Warsaw, 1996;
see also P. Chankowski, J. Ellis and S. Pokorski, CERN preprint TH/97-343, hep-ph/9712234.

[31] R. Barbieri and L. Maiani, *Nucl. Phys.* **B224** (1983) 32;
L. Alvarez-Gaumé, J. Polchinski and M. Wise, *Nucl. Phys.* **B221** (1983) 495.

[32] W. Hollik, *Mod. Phys. Lett.* **A5** (1990) 1909.

[33] A. Djouadi et al., *Nucl. Phys.* **B349** (1991) 48;
M. Boulware and D. Finnell, *Phys. Rev.* **D44** (1991) 2054. The sign discrepancy between these two papers appears now to be solved in favour of the second one.

[34] G. Altarelli, R. Barbieri and F. Caravaglios, *Phys. Lett.* **B314** (1993) 357.

[35] A. Brignole, F. Feruglio and F. Zwirner *Z. Phys.* **C71** (1996) 679.

[36] See, for example, G.G. Ross, *Grand Unified Theories*, Benjamin, New York, 1984;
R. Mohapatra, *Prog. Part. Nucl. Phys.* **26**, 1.

[37] R. Barbieri, Proceedings of LP'97, Hamburg, 1997.

[38] W. Lerche, Proceedings of LP'97, Hamburg, 1997.

[39] E. Gildener, *Phys.Rev.* **14**, 1667 (1976);
E. Gildener and S. Weinberg, *Phys.Rev.* **15**, 3333 (1976).

[40] H.P. Nilles, *Phys. Rep.* C **110**,1 (1984);
H.E. Haber and G.L. Kane, *Phys. Rep.* C **117**, 75 (1985);
R. Barbieri, *Riv. Nuovo Cim.* **11**, 1 (1988).

[41] A. Chamseddine, R. Arnowitt and P. Nath, *Phys. Rev. Lett.* **49**, 970 (1982);
R. Barbieri, S. Ferrara and C. Savoy, *Phys. Lett.* **110B**, 343 (1982);
E. Cremmer et al., *Phys. Lett.* **116B**, 215 (1982).

[42] For a review, see S. Dimopoulos, S. Thomas and J.D. Wells, hep-ph/9609434 and references therein.

[43] S. Dimopoulos, S. Raby and F. Wilczek, *Phys.Rev.* **24**, 1681 (1981);
S. Dimopoulos and H. Georgi, *Nucl. Phys.* **B193**, 150 (1981);
L.E. Ibáñez and G.G. Ross, *Phys. Lett.* **105B**, 439 (1981).

[44] P. Langacker and N.Polonsky, hep-ph/9503214.

[45] A. Rubbia, Proceedings of LP'97, Hamburg, 1997.

[46] Y. Totsuka, Proceedings of LP'97, Hamburg, 1997.

[47] M. Gell-Mann, P. Ramond and R. Slansky, in Supergravity, ed. D. Freedman et al., North Holland, 1979;
T. Yanagida, *Prog. Theo. Phys.* **B135**, 66 (1978).

[48] N.Turok, Proceedings of LP'97, Hamburg, 1997.

[49] V.A. Kuzmin, V.A. Rubakov and M.E. Shaposhnikov, *Phys. Lett.* **155B**, 36 (1985);
M.E. Shaposhnikov, *Nucl. Phys.* **B287**, 757 (1987); *Nucl. Phys.* **B299**, 797 (1988).

[50] A.D. Sakharov, *JETP Lett.* **91B**, 24 (1967).

[51] A.G. Cohen, D.B. Kaplan and A.E.Nelson, *Annu. Rev. Part. Sci.* **43**, 27 (1993);
M. Quiros, *Helv.Phys. Acta* **67**, 451 (1994)4;
V.A. Rubakov and M.E. Shaposhnikov, hep-ph/9603208.

[52] M. Carena and C.E.M. Wagner, hep-ph/9704347.

[53] M. Carena, M. Quiros, A. Riotto, I. Vilja and C.E.M. Wagner, hep-ph/9702409;
J. McDonald, hep-ph/9707290.

[54] See, for example, M. Fukugita and T.Yanagida, *Phys. Lett.* **B174**, 45 (1986);
G. Lazarides and Q. Shafi, *Phys. Lett.* **B258**, 305 (1991).

[55] M. Sher, *Phys. Rep.* **179**, 273 (1989); *Phys. Lett.* **B317**, 159 (1993).

[56] G. Altarelli and G. Isidori, *Phys. Lett.* **B337**, 141 (1994).

[57] J.A. Casas, J.R. Espinosa and M. Quiros, *Phys. Lett.* **B342**, 171 (1995).

[58] J.A. Casas et al., *Nucl. Phys.* **B436**, 3 (1995); E**B439**, 466 (1995).

[59] M. Carena and C.E.M. Wagner, *Nucl. Phys.* **B452**, 45 (1995).

[60] See, for example, M. Lindner, *Z. Phys.* **31** (1986) 295 and references therein.

[61] T. Hambye and K. Riesselmann, *Phys. Rev.* **D55**, 7255 (1997).

[62] H. Haber and R. Hempfling, *Phys. Rev. Lett.* **66**, 1815 (1991);
J. Ellis, G. Ridolfi and F. Zwirner, *Phys. Lett.* **B257**, 83 (1991);
Y. Okado, M. Yamaguchi and T. Yanagida, *Progr. Theor. Phys. Lett.* **85** (1991) 1;
R. Barbieri, F. Caravaglios and M. Frigeni, *Phys. Lett.* **B258**, 167 (1991);
For a 2-loop improvement, see also: R. Memplfling and A.H. Hoang, *Phys. Lett.* **B331**, 99 (1994).

[63] C. Adloff *et al.*, H1 collaboration, DESY 97-24, hep-ex/9702012.

[64] J. Breitweg *et al.*, ZEUS collaboration, DESY 97-25, hep-ex/9702015.

[65] B. Straub, Proceedings of LP'97, Hamburg, 1997.

[66] M. Drees, hep-ph/9703332, APCTP 97-03(1997)5;
U. Bassler, G. Bernardi, hep-ph/9707024, DESY 97-136.

[67] F. Abe et al., CDF collaboration, *Phys.Rev.Lett.* **77**, 438 (1996).

[68] J. Huston *et al.*, *Phys.Rev.Lett.* **77**, 444 (1996);
H.L. Lai *et al.*, *Phys. Rev.* **D55**, 1280 (1997).

[69] H. Schellman, Talk presented at LP'97, Hamburg, July 1997.

[70] S. Kuhlmann, H.L. Lai and W.K. Tung, hep-ph/9704338.

[71] S. Rock and P. Bosted, hep-ph/9706436.

[72] S.J. Brodsky *et al.*, *Phys. Lett.* **B93**, 451 (1980);
J.F. Gunion and R. Vogt, hep-ph/9706252.

[73] W. Melnitchouk and A.W. Thomas, hep-ph/9707387.

[74] K.S. Babu, C. Kolda and J. March-Russell, hep-ph/9705399.

[75] E. Eichten, K. Lane and M. Peskin, *Phys. Rev. Lett* **50**, 811 (1983).

[76] F. Abe et al., CDF collaboration, *Phys.Rev.Lett.* **77**, 5336 (1996).

[77] G. Altarelli, J. Ellis, G.F. Giudice, S. Lola and M.L. Mangano, CERN TH/97-40, hep-ph/9703276.

[78] K.S. Babu, C. Kolda, J. March-Russell and F. Wilczek, IASSNS-HEP-97-04, hep-ph/9703299;
V. Barger, K. Cheung, K. Hagiwara and D. Zeppenfeld, MADPH-97-991, hep-ph/9703311;
A.E. Nelson, hep-ph/9703379;

N.G. Deshpande, B. Dutta and Xiao-Gang He, hep-ph/9705236;
Gi-Chol Cho, K. Hagiwara and S. Matsumoto, hep-ph/9707334.

[79] ALEPH, DELPHI,OPAL Collaborations, submitted at LP'97, Hamburg, July 1997.

[80] M.C. Gonzalez-Garcia and S.F. Novaes, IFT-P-024-97, hep-ph/9703346;
J. Ellis, S. Lola and K. Sridhar, hep-ph/9705416.

[81] F. Abe et al., CDF collaboration, FERMILAB-PUB-97/171-E.

[82] V. Barger, K. Cheung, K. Hagiwara and D. Zeppenfeld, MADPH-97-999, hep-ph/9707412.

[83] N. Di Bartolomeo and M. Fabbrichesi, SISSA-34-97-EP, hep-ph/9703375.

[84] Updated from Ref.[83].

[85] F. Caravaglios, hep-ph/9706288.

[86] L. Giusti and A. Strumia, hep-ph/9706298.

[87] G. Altarelli, G.F. Giudice and M.L. Mangano, hep-ph/9705287.

[88] W. Buchmüller and D. Wyler, *Phys. Lett.* **BB177**, 377 (1986).

[89] M. Leurer, *Phys. Rev.* **D49**, 333 (1994);
S. Davidson, D. Bailey and B.A. Campbell, *Z. Phys.* **C61**, 613 (1993) and references therein.

[90] J.L. Hewett and T.G. Rizzo, SLAC-PUB-7430, hep-ph/9703337.

[91] J. Blümlein, E. Boos and A. Kryukov, hep-ph/9610408;
J. Blümlein, hep-ph/9703287.

[92] I. Montvay, hep-ph/9704280.

[93] M.A. Doncheski and R.W. Robinett, hep-ph/9707328.

[94] Updated with the new HERA data, M.L. Mangano, private communication.

[95] Z. Kunszt and W.J. Stirling, DTP-97-16, hep-ph/9703427;
T. Plehn, H. Spiesberger, M. Spira and P.M. Zerwas, DESY-97-043, hep-ph/9703433.

[96] W. Beenakker, R. Hopker, M. Spira and P.M. Zerwas, hep-ph/9610490.

[97] M. Kramer, T. Plehn, M. Spira and P.M. Zerwas,RAL-97-017, hep-ph/9704322.

[98] B. Dutta, R.N. Mohapatra and S. Nandi, hep-ph/9704428.

[99] J.L. Hewett, Proc. 1990 Summer Study on *High Energy Physics*, Snowmass, Colorado.

[100] T. Kon and T. Kobayashi, *Phys. Lett.* B**270**, 81 (1991);
T. Kon, T. Kobayashi, S. Kitamura, K. Nakamura and S. Adachi, *Z. Phys.* C**61**, 239 (1994);
T. Kobayashi, S. Kitamura, T. Kon, *Int. J. Mod. Phys.* **A11**, 1875 (1996). T. Kon and T. Kobayashi, ITP-SU-97-02, hep-ph/9704221.

[101] J. Butterworth and H. Dreiner, *Nucl. Phys.* B**397**, 3 (1993);
H. Dreiner and P. Morawitz, *Nucl. Phys.* B**428**, 31 (1994);
E. Perez, Y. Sirois and H. Dreiner, contribution to Beyond the Standard Model Group, 1995-1996 Workshop on Future Physics at HERA, see also the Summary by H. Dreiner, H.U. Martyn, S. Ritz and D. Wyler, hep-ph/9610232.

[102] D. Choudhury and S. Raychaudhuri, CERN TH/97-26, hep-ph/9702392.

[103] H. Dreiner and P. Morawitz, hep-ph/9703279;
J. Kalinowski, R. Ruckl, H. Spiesberger, and P.M. Zerwas, BI-TP-97-07, hep-ph/9703288.

[104] G. Farrar and P. Fayet, *Phys. Lett.* B**76**, 575 (1978);
S. Weinberg, *Phys.Rev.* **26**, 287 (1982);
N. Sakai and T. Yanagida, *Nucl. Phys.* B**197**, 133 (1982).

[105] M. Hirsch, H.V. Klapdor-Kleingrothaus and S.G. Kovalenko, *Phys. Rev. Lett.* **75**, 17 (1995), and *Phys. Rev.* D**53**, 1329 (1996).

[106] R. Barbieri, A. Strumia and Z. Berezhiani, IFUP-TH-13-97, hep-ph/9704275;
G.F. Giudice and R. Rattazzi, CERN-TH-97-076, hep-ph/9704339.

[107] K.S. Babu, C. Kolda and J. March-Russell, hep-ph/9705414.

[108] T. Kon, T.Matsushita and T. Kobayashi, hep-ph/9707355.

[109] M. Carena, D. Choudhury, S. Raychaudhuri and C.E.M. Wagner, hep-ph/9707458.

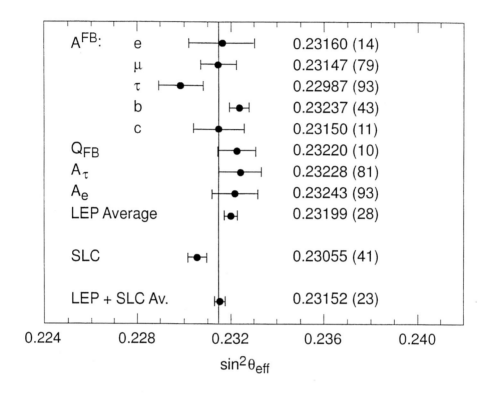

Fig. 1

The collected measurements of $\sin^2 \theta_{eff}$. The resulting value for the χ^2 is given by $\chi^2/d.o.f = 1.87$. As a consequence the error on the average is enlarged in the text by a factor $\sqrt{1.87}$ with respect to the formal average shown here.

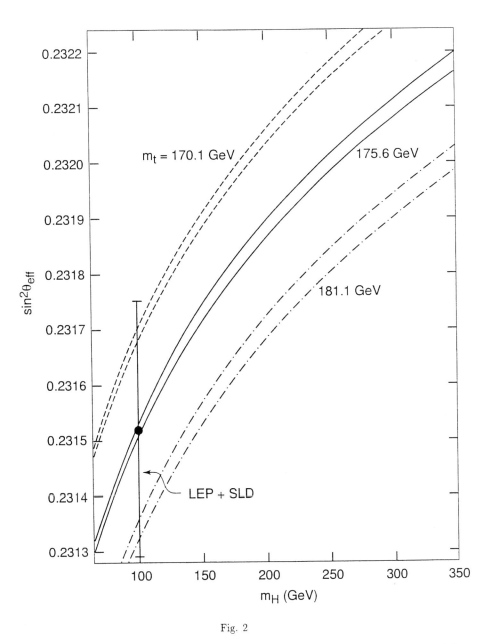

Fig. 2

SM prediction for $\sin^2\theta_{eff}$ as function of m_H for $m_t = 175.6 \pm 5.5$ as computed from updated radiative corrections [15]. The theoretical error bands from neglect of higher order terms, estimated from scheme and scale dependence, are shown. The combined LEP+SLD experimental value is indicatively plotted for $m_H \sim 100$ GeV (the SLD value would be very low, out of the plot scale). Small values of m_H are preferred.

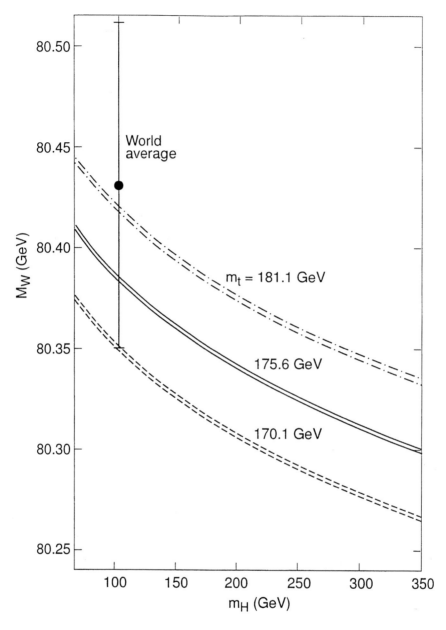

Fig. 3

SM prediction for m_W as function of m_H for $m_t = 175.6 \pm 5.5$ as computed from updated radiative corrections [15]. The theoretical error bands from neglect of higher order terms, estimated from scheme and scale dependence, are shown. The combined LEP+hadron colliders experimental value is indicatively plotted for $m_H \sim 100$ GeV. Small values of m_H are preferred.

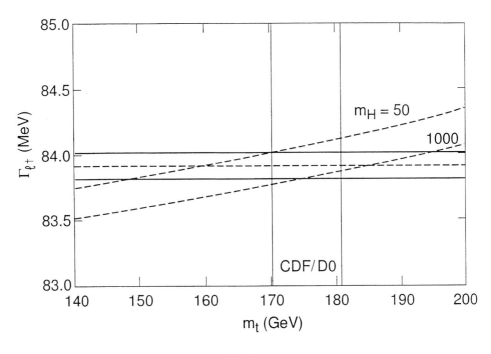

Fig. 4

SM prediction for the leptonic width as function of m_t. For small Higgs mass a value of m_t slightly smaller than the CDF/D0 experimental value is indicated.

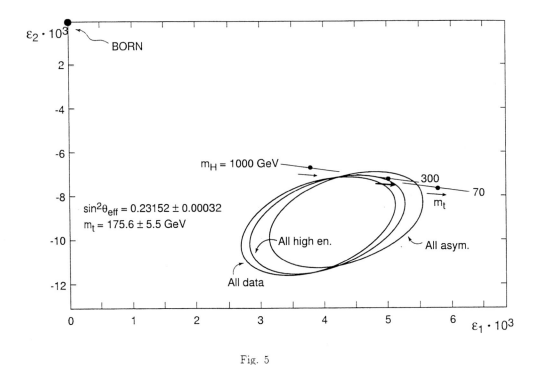

Fig. 5

Data vs theory in the ϵ_2-ϵ_1 plane. The origin point corresponds to the "Born" approximation obtained from the SM at tree level plus pure QED and pure QCD corrections. The predictions of the full SM (also including the improvements of ref.[15]) are shown for $m_H = 70$, 300 and 1000 GeV and $m_t = 175.6 \pm 5.5$ GeV (a segment for each m_H with the arrow showing the direction of m_t increasing from -1σ to $+1\sigma$). The three $1-\sigma$ ellipses (38% probability contours) are obtained from a) "All Asymm.": Γ_l, m_W and $\sin^2\theta_{eff}$ as obtained from the combined asymmetries (the value and error used are shown); b) "All High En.": the same as in a) plus all the hadronic variables at the Z; c) "All Data": the same as in b) plus the low energy data.

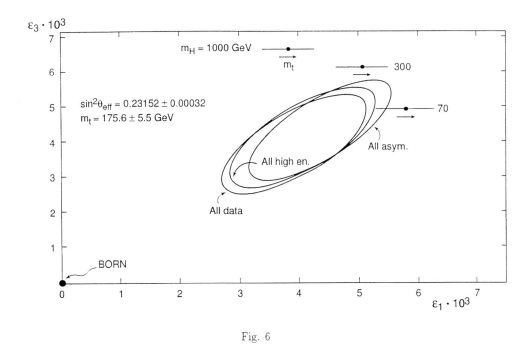

Fig. 6

Data vs theory in the ϵ_3-ϵ_1 plane (notations as in fig. 5).

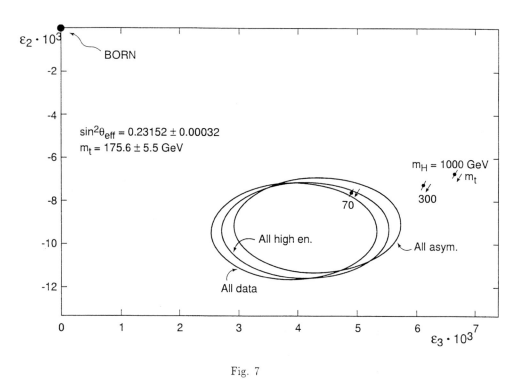

Fig. 7

Data vs theory in the ϵ_2-ϵ_3 plane (notations as in fig. 5).

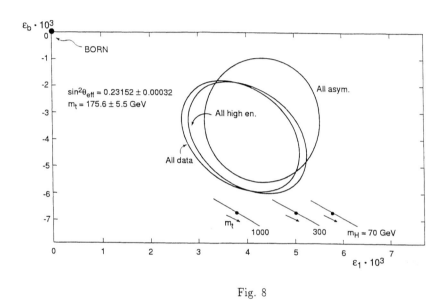

Fig. 8

Data vs theory in the ϵ_b-ϵ_1 plane (notations as in fig. 5).

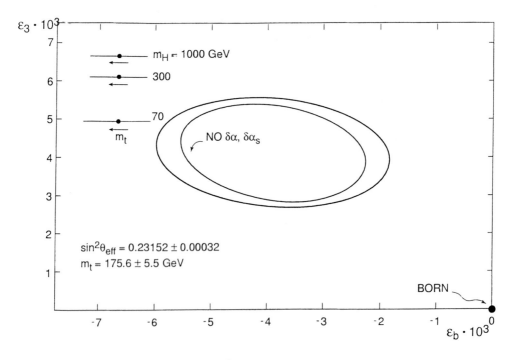

Fig. 9

Data vs theory in the ϵ_3-ϵ_b plane (notations as in fig. 5, except that both ellipses refer to the case b)) The inner $1-\sigma$ ellipse is without the errors induced by the uncertainties on $\alpha(m_Z)$ and $\alpha_s(m_Z)$.

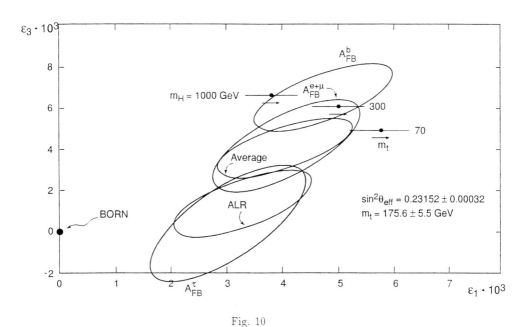

Fig. 10

Data vs theory in the ϵ_3-ϵ_1 plane (notations as in fig. 5). The ellipse indicated with "Average" corresponds to the case "All high en" of fig. 6 and is obtained from the average value of $sin^2\theta_{eff}$ displayed on the figure. The other ellipses are obtained by replacing the average $sin^2\theta_{eff}$ with the values obtained in turn from each individual asymmetry as shown by the labels.

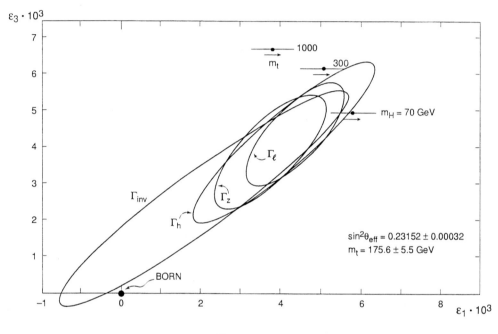

Fig. 11

Data vs theory in the ϵ_3-ϵ_1 plane (notations as in fig. 5). The different ellipses are obtained from m_W, the average value of $sin^2\theta_{eff}$ displayed on the figure, R_b plus one width among Γ_l, Γ_{inv}, the total width Γ_Z and Γ_h.

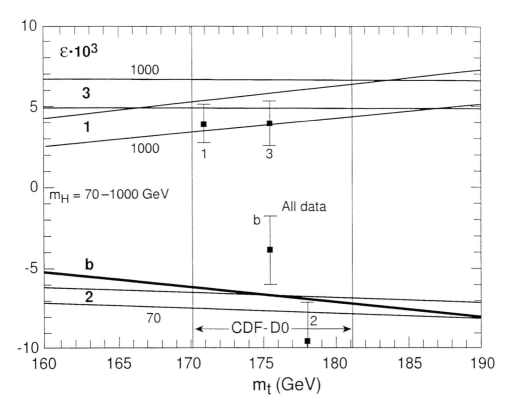

Fig. 12

The bands (labeled by the ϵ index) are the predicted values of the epsilons in the SM as functions of m_t for $m_H = 70-1000$ GeV (the m_H value corresponding to one edge of the band is indicated). The CDF/D0 experimental 1-σ range of m_t is shown. The experimental results for the epsilons from all data are displayed (from the last column of table 6). The position of the data on the m_t axis has been arbitrarily chosen and has no particular meaning.

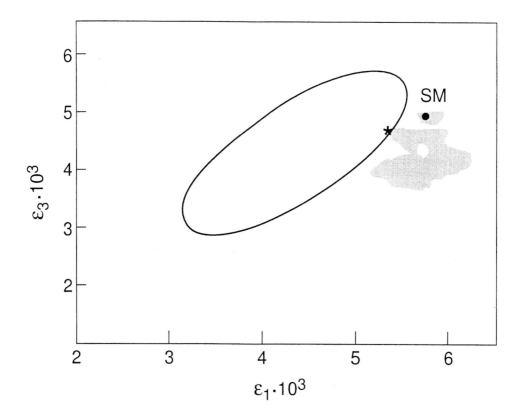

Fig. 13

Scatter plot obtained in the plane ϵ_1-ϵ_3 for $-200 < \mu < 200$ GeV, $0 < M < 250$ GeV, $\tan\beta = 1.5 - 2.5$, $m_{\tilde{l}} = 100 - 500$ GeV and $m_{\tilde{q}} = 1\ TeV$, together with the SM prediction for $m_t = 175.6$ GeV and $m_H \sim 70$ GeV. The separation $\mu > 0$ or $\mu < 0$ is clearly visible. The small star indicates the region with minimum χ^2.

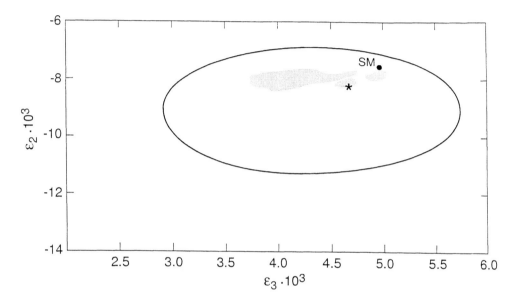

Fig. 14

Scatter plot obtained in the plane ϵ_2-ϵ_3 for $-200 < \mu < 200$ GeV, $0 < M < 250$ GeV, $\tan\beta = 1.5 - 2.5$, $m_{\tilde{l}} = 100 - 500$ GeV and $m_{\tilde{q}} = 1$ TeV, together with the SM prediction for $m_t = 175.6$ GeV and $m_H \sim 70$ GeV. The small star corresponds to the same values of the parameters of the region (marked with a star) in fig. 13. For these values of parameters the fit of ϵ_1, ϵ_2 and ϵ_3 improves while ϵ_b is unchanged.

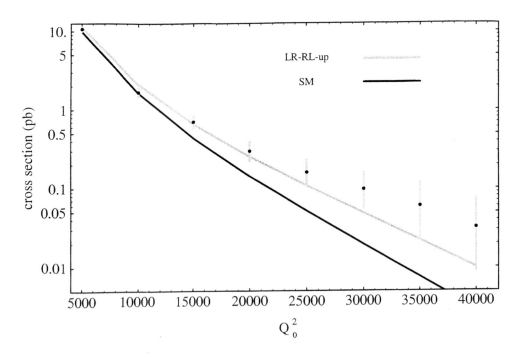

Fig. 15

Example of a fit to the HERA data presented at LP'97 from a LR+RL contact term with only u-quarks.

Cosmological phase transitions and baryogenesis

Mariano Quirós[†]

[†]*Instituto de Estructura de la Materia (CSIC), Serrano 123*
E–28006 Madrid, Spain

Abstract. We review various aspects of field theory at finite temperature related to the theory of phase transitions. In particular, the real and imaginary time formalisms are discussed, showing their equivalence in simple examples. Bubble nucleation by thermal tunneling, and the subsequent development of the phase transition is described in some detail. Some attention is also devoted to the breakdown of the perturbative expansion and the infrared problem in the finite temperature field theory. Finally the application to baryogenesis at the electroweak phase transition is done in the Standard Model and in the Minimal Supersymmetric Standard Model. In all cases we have translated the condition of not washing out any previously generated baryon asymmetry by upper bounds on the Higgs mass.

FIELD THEORY AT FINITE TEMPERATURE

The formalism used in conventional quantum field theory is suitable to describe observables (*c.g.* cross-sections) measured in empty space-time, as particle interactions in an accelerator. However, in the early stages of the universe, at high temperature, the environment had a non-negligible matter and radiation density, making the hypotheses of conventional field theories impracticable. For that reason, under those circumstances, the methods of conventional field theories are no longer in use, and should be replaced by others, closer to thermodynamics, where the background state is a thermal bath. This field has been called field theory at finite temperature and it is extremely useful to study all phenomena which happened in the early universe: phase transitions, inflationary cosmology, ... Excellent articles [1,2], review articles [3–5] and textbooks [6] exist which discuss different aspects of these issues. In this section we will review the main methods which will be useful for the theory of phase transitions at finite temperature.

Grand-canonical ensemble

In this section we shall give some definitions borrowed from thermodynamics and statistical mechanics. The **microcanonical ensemble** is used to describe an isolated system with fixed energy E, particle number N and volume V. The **canonical ensemble** describes a system in contact with a heat reservoir at temperature T: the energy can be exchanged between them and T, N and V are fixed. Finally, in the **grand canonical ensemble** the system can exchange energy and particles with the reservoir: T, V and the chemical potentials are fixed.

Consider now a dynamical system characterized by a hamiltonian [1] H and a set of conserved (mutually commuting) charges Q_A. The equilibrium state of the system at rest in the large volume V is described by the **grand-canonical density operator**

$$\rho = \exp(-\Phi) \exp\left\{-\sum_A \alpha_A Q_A - \beta H\right\} \tag{1}$$

where $\Phi \equiv \log Tr \exp\{-\sum_A \alpha_A Q_A - \beta H\}$ is called the Massieu function (Legendre transform of the entropy), α_A and β are Lagrange multipliers given by $\beta = T^{-1}$, $\alpha_A = -\beta\mu_A$, T is the temperature and μ_A are the chemical potentials.

Using (1) one defines the **grand canonical average** of an arbitrary operator $\mathcal{O}$, as

$$\langle \mathcal{O} \rangle \equiv Tr(\mathcal{O}\rho) \tag{2}$$

satisfying the property $\langle \mathbf{1} \rangle = 1$.

In the following of this section we will always consider the case of zero chemical potential. It will be re-introduced when necessary.

Generating functionals

We will start considering the case of a real scalar field $\phi(x)$, carrying no charges ($\mu_A = 0$), with hamiltonian H, i.e.

$$\phi(x) = e^{itH} \phi(0, \vec{x}) e^{-itH} \tag{3}$$

where the time $x^0 = t$ is analytically continued to the complex plane.

We define the thermal Green function as the grand canonical average of the ordered product of the n field operators

$$G^{(C)}(x_1, \ldots, x_n) \equiv \langle T_C \phi(x_1), \ldots, \phi(x_n) \rangle \tag{4}$$

where the T_C ordering means that fields should be ordered along the path C in the complex t-plane. For instance the product of two fields is defined as,

---

[1] All operators will be considered in the Heisenberg picture.

$$T_C\phi(x)\phi(y) = \theta_C(x^0 - y^0)\phi(x)\phi(y) + \theta_C(y^0 - x^0)\phi(y)\phi(x) \tag{5}$$

If we parameterize C as $t = z(\tau)$, where τ is a real parameter, T_C ordering means standard ordering along τ. Therefore the step and delta functions can be given as $\theta_C(t) = \theta(\tau)$, $\delta_C(t) = (\partial z/\partial \tau)^{-1} \delta(\tau)$.

The rules of the functional formalism can be applied as usual, with the prescription $\delta j(y)/\delta j(x) = \delta_C(x^0 - y^0)\delta^{(3)}(\vec{x} - \vec{y})$, and the generating functional $Z^\beta[j]$ for the full Green functions,

$$Z^\beta[j] = \sum_{n=0}^{\infty} \frac{i^n}{n!} \int_C d^4x_1 \ldots d^4x_n j(x_1) \ldots j(x_n) G^{(C)}(x_1, \ldots, x_n) \tag{6}$$

can also be written as,

$$Z^\beta[j] = \left\langle T_C \exp\left\{i \int_C d^4x j(x)\phi(x)\right\} \right\rangle \tag{7}$$

which is normalized to $Z^\beta[0] = \langle \mathbf{1} \rangle = 1$, as in (2), and where the integral along t is supposed to follow the path C in the complex plane.

Similarly, the generating functional for connected Green functions $W^\beta[j]$ is defined as $Z^\beta[j] \equiv \exp\{iW^\beta[j]\}$, and the generating functional for 1PI Green functions $\Gamma^\beta[\overline{\phi}]$, by the Legendre transformation,

$$\Gamma^\beta[\overline{\phi}] = W^\beta[j] - \int_C d^4x \frac{\delta W^\beta[j]}{\delta j(x)} j(x) \tag{8}$$

where the current $j(x)$ is eliminated in favor of the classical field $\overline{\phi}(x)$ as $\overline{\phi}(x) = \delta W^\beta[j]/\delta j(x)$. It follows that $\delta \Gamma^\beta[\overline{\phi}]/\delta \overline{\phi}(x) = -j(x)$, and $\overline{\phi}(x) = \langle \phi(x) \rangle$ is the grand canonical average of the field $\phi(x)$.

Symmetry violation is signaled by

$$\left. \frac{\delta \Gamma^\beta[\overline{\phi}]}{\delta \overline{\phi}} \right|_{j=0} = 0 \tag{9}$$

for a value of the field different from zero.

As in field theory at zero temperature, in a translationally invariant theory $\overline{\phi}(x) = \phi_c$ is a constant. In this case, by removing the overall factor of space-time volume arising in each term of $\Gamma^\beta[\phi_c]$, we can define the effective potential at finite temperature as,

$$\Gamma^\beta[\phi_c] = -\int d^4x V_{\text{eff}}^\beta(\phi_c) \tag{10}$$

and symmetry breaking occurs when

$$\frac{\partial V_{\text{eff}}^\beta(\phi_c)}{\partial \phi_c} = 0 \tag{11}$$

for $\phi_c \neq 0$.

Green functions

Scalar fields

Not all the contours are allowed if we require Green functions to be analytic with respect to t. Using (5) we can write the two-point Green function as,

$$G^{(C)}(x-y) = \theta_C(x^0 - y^0)G_+(x-y) + \theta_C(y^0 - x^0)G_-(x-y) \qquad (12)$$

where

$$G_+(x-y) = \langle \phi(x)\phi(y) \rangle, \quad G_-(x-y) = G_+(y-x) \qquad (13)$$

Now, take the complete set of states $|n\rangle$ with eigenvalues E_n: $H|n\rangle = E_n|n\rangle$. One can readily compute (13) at the point $\vec{x} = \vec{y} = 0$ as

$$G_+(x^0 - y^0) = e^{-\Phi} \sum_{m,n} |\langle m|\phi(0)|n\rangle|^2 \, e^{-iE_n(x^0 - y^0)} e^{iE_m(x^0 - y^0 + i\beta)} \qquad (14)$$

so that the convergence of the sum implies that $-\beta \leq Im(x^0 - y^0) \leq 0$ which requires $\theta_C(x^0 - y^0) = 0$ for $Im(x^0 - y^0) > 0$. From (13) it follows that the similar property for the convergence of $G_-(x^0 - y^0)$ is that $0 \leq Im(x^0 - y^0) \leq \beta$, which requires $\theta_C(y^0 - x^0) = 0$ for $Im(x^0 - y^0) < 0$, and the final condition for the convergence of the complete Green function on the strip

$$-\beta \leq Im(x^0 - y^0) \leq \beta \qquad (15)$$

is that we define the function $\theta_C(t)$ such that $\theta_C(t) = 0$ for $Im(t) > 0$. The latter condition implies that C must be such that *a point moving along it has a monotonically decreasing or constant imaginary part*.

A very important periodicity relation affecting Green functions can be easily deduced from the very definition of $G_+(x)$ and $G_-(x)$, Eq. (13). By using the definition of the grand canonical average and the cyclic permutation property of the trace of a product of operators, it can be easily deduced,

$$G_+(t - i\beta, \vec{x}) = G_-(t, \vec{x}) \qquad (16)$$

which is known as the Kubo-Martin-Schwinger relation [7].

We can now compute the two-point Green function (12) for a free scalar field,

$$\phi(x) = \int \frac{d^3p}{(2\pi)^{3/2}(2\omega_p)^{1/2}} \left[a(p)e^{-ipx} + a^\dagger(p)e^{ipx}\right] \qquad (17)$$

where $\omega_p = \sqrt{\vec{p}^{\,2} + m^2}$, which satisfies the equation

$$\left[\partial^\mu \partial_\mu + m^2\right] G^{(C)}(x-y) = -i\delta_C(x-y) \equiv -i\delta_C(x^0 - y^0)\delta^{(3)}(\vec{x} - \vec{y}) \qquad (18)$$

Using the time derivative of (17), and the equal time commutation relation,

$$\left[\phi(t,\vec{x}),\dot{\phi}(t,\vec{y})\right] = i\delta^{(3)}(\vec{x}-\vec{y}) \qquad (19)$$

one easily obtains the commutation relation for creation and annihilation operators,

$$[a(p),a^\dagger(k)] = \delta^{(3)}(\vec{p}-\vec{k}) \qquad (20)$$

and defining the Hamiltonian of the field as,

$$H = \int \frac{d^3p}{(2\pi)^3}\omega_p a^\dagger(p)a(p) \qquad (21)$$

one can obtain, using (20) the thermodynamical averages,

$$\langle a^\dagger(p)a(k)\rangle = n_B(\omega_p)\delta^{(3)}(\vec{p}-\vec{k}) \qquad (22)$$
$$\langle a(p)a^\dagger(k)\rangle = [1+n_B(\omega_p)]\delta^{(3)}(\vec{p}-\vec{k})$$

where $n_B(\omega)$ is the Bose distribution function,

$$n_B(\omega) = \frac{1}{e^{\beta\omega}-1} \qquad (23)$$

We will give here a simplified derivation of expression (22). Consider the simpler example of a quantum mechanical state occupied by bosons of the **same** energy ω. There may be any number of bosons in that state and no interaction between the particles: we will denote that state by $|n\rangle$. The set $\{|n\rangle\}$ is complete. Creation and annihilation operators are denoted by $a^\dagger$ and a, respectively. They act on the states $|n\rangle$ as, $a^\dagger|n\rangle = \sqrt{n+1}|n+1\rangle$ and $a|n\rangle = \sqrt{n}|n-1\rangle$, and satisfy the commutation relation, $[a,a^\dagger] = 1$. The hamiltonian and number operators are defined as $H = \omega N$ and $N = a^\dagger a$, with eigenvalues ωn and n, respectively.

It is very easy to compute now $\langle a^\dagger a\rangle$ and $\langle aa^\dagger\rangle$ as in (22) using the completeness of $\{|n\rangle\}$. In particular,

$$Tr(e^{-\beta H}) = \sum_{n=0}^{\infty}\langle n|e^{-\beta H}|n\rangle = \sum_{n=0}^{\infty}e^{-\beta\omega n} = \frac{1}{1-e^{-\beta\omega}}$$

and

$$Tr(e^{-\beta H}a^\dagger a) = \sum_{n=0}^{\infty}n e^{-\beta\omega n} = \frac{e^{-\beta\omega}}{(1-e^{-\beta\omega})^2}$$

from where $\langle a^\dagger a\rangle = n_B(\omega)$, and $\langle aa^\dagger\rangle = 1+n_B(\omega)$, as we wanted to prove.

Using now (22) we can cast the two-point Green function as,

$$G^{(C)}(x-y) = \int \frac{d^4p}{(2\pi)^4}\rho(p)e^{-ip(x-y)}\left[\theta_C(x^0-y^0)+n_B(p^0)\right] \qquad (24)$$

147

where the function $\rho(p)$ is defined by $\rho(p) = 2\pi[\theta(p^0) - \theta(-p^0)]\delta(p^2 - m^2)$. Now the particular value of the Green function (24) depends on the chosen contour C. We will show later on two particular contours giving rise to the so-called imaginary and real time formalisms. Before coming to them we will describe how the previous formulae apply to the case of fermion fields.

Fermion fields

We will replace here (12) and (13) by,

$$S^{(C)}_{\alpha\beta}(x-y) \equiv \langle T_C \psi_\alpha(x)\overline{\psi}_\beta(y)\rangle = \theta_C(x^0-y^0)S^+_{\alpha\beta} - \theta_C(y^0-x^0)S^-_{\alpha\beta} \quad (25)$$

where α and β are spinor indices, and

$$S^+_{\alpha\beta}(x-y) = \langle \psi_\alpha(x)\overline{\psi}_\beta(y)\rangle \quad (26)$$

are the reduced Green function, which satisfy the Kubo-Martin-Schwinger relation,

$$S^+_{\alpha\beta}(t-i\beta,\vec{x}) = -S^-_{\alpha\beta}(t,\vec{x}) \quad (27)$$

The calculation of the two-Green function for a free fermion field, satisfying the equation

$$(i\gamma \cdot \partial - m)_{\alpha\sigma} S^{(C)}_{\sigma\beta}(x-y) = i\delta_C(x-y)\delta_{\alpha\beta} \quad (28)$$

follows lines similar to Eqs. (17) to (24). In particular, one can define a Green function $S^{(C)}$ as

$$S^{(C)}_{\alpha\beta}(x-y) \equiv (i\gamma \cdot \partial + m)_{\alpha\beta} S^{(C)}(x-y) \quad (29)$$

where $S^{(C)}(x-y)$ satisfies the Klein-Gordon propagator equation (18). One can obtain for $S^{(C)}$ the expression,

$$S^{(C)}(x-y) = \int \frac{d^4p}{(2\pi)^4} \rho(p) e^{-ip(x-y)} \left[\theta_C(x^0-y^0) - n_F(p^0)\right] \quad (30)$$

where $n_F(\omega)$ is the Fermi distribution function

$$n_F(\omega) = \frac{1}{e^{\beta\omega}+1}. \quad (31)$$

Eq. (31) can be derived similarly to (23) as the mean number of fermions for a Fermi gas. This time the Pauli exclusion principle forbids more than one fermion occupying a single state, so that only the states $|0\rangle$ and $|1\rangle$ exist. They are acted on by creation and annihilation operators $b^\dagger$ and b, respectively as: $b^\dagger|0\rangle = |1\rangle$, $b^\dagger|1\rangle = 0$, $b|0\rangle = 0$, $b|1\rangle = |0\rangle$, and satisfy anticommutation rules, $\{b,b^\dagger\} = 1$.

Defining the hamiltonian and number operators as $H = \omega N$ and $N = b^\dagger b$, we can compute now the statistical averages of $\langle b^\dagger b \rangle$ and $\langle bb^\dagger \rangle$ using the completeness of $\{|n\rangle\}$.

$$Tr(e^{-\beta H}) = \sum_{n=0}^{1} \langle n|e^{-\beta H}|n\rangle = \sum_{n=0}^{1} e^{-\beta \omega n} = 1 + e^{-\beta \omega}$$

and

$$Tr(e^{-\beta H} b^\dagger b) = \sum_{n=0}^{1} n e^{-\beta \omega n} = e^{-\beta \omega}$$

from where $\langle b^\dagger b \rangle = n_F(\omega)$, and $\langle bb^\dagger \rangle = 1 - n_F(\omega)$, as we wanted to prove.

Imaginary time formalism

The calculation of the propagators in the previous sections depends on the chosen path C going from an initial arbitrary time t to $t - i\beta$, provided by the Kubo-Martin-Schwinger periodicity properties (16) and (27) of Green functions. The simplest path is to take a straight line along the imaginary axis $t = -i\tau$. It is called Matsubara contour, since Matsubara [8] was the first to set up a perturbation theory based upon this contour. In that case $\delta_C(t) = i\delta(\tau)$.

The two-point Green functions for scalar (24) and fermion (30) fields can be written as,

$$G(\tau, \vec{x}) = \int \frac{d^4 p}{(2\pi)^4} \rho(p) e^{i\vec{p}\vec{x}} e^{-\tau p^0} \left[\theta(\tau) + \eta n(p^0)\right] \tag{32}$$

where the symbol η stands as: $\eta_B = 1$ ($\eta_F = -1$) for bosons (for fermions). Analogously, $n(p^0)$ stands either for $n_B(p^0)$, as given by (23) for bosons, or $n_F(p^0)$, as given by (31) for fermions. It can be defined as a function of η as,

$$n(\omega) = \frac{1}{e^{\beta\omega} - \eta} \tag{33}$$

The Green function (32) can be decomposed as in (12)

$$G(\tau, \vec{x}) = G_+(\tau, \vec{x})\theta(\tau) + G_-(\tau, \vec{x})\theta(-\tau) \tag{34}$$

Using now the Kubo-Martin-Schwinger relations, Eqs. (16) and (27), we can write $G(\tau + \beta) = \eta G(\tau)$ for $-\beta \leq \tau \leq 0$, $G(\tau - \beta) = \eta G(\tau)$ for $0 \leq \tau \leq \beta$, which means that the propagator for bosons (fermions) is periodic (antiperiodic) in the *time* variable τ, with period β.

It follows that the Fourier transform of (32)

$$\widetilde{G}(\omega_n, \vec{p}) = \int_{\alpha-\beta}^{\alpha} d\tau \int d^3 x \, e^{i\omega_n \tau - i\vec{x}\vec{p}} G(\tau, \vec{x}) \tag{35}$$

(where $0 \leq \alpha \leq \beta$) is independent of α and the discrete frequencies satisfy the relation $\eta e^{i\omega_n \beta} = 1$, i.e. $\omega_n = 2n\pi\beta^{-1}$ for bosons, and $\omega_n = (2n+1)\pi\beta^{-1}$ for fermions.

Inserting now (32) into (35) we can obtain the propagator in momentum space $\widetilde{G}$

$$\widetilde{G}(\omega_n, \vec{p}) = \frac{1}{\vec{p}^{\,2} + m^2 + \omega_n^2}. \tag{36}$$

We can now define the euclidean propagator, $\Delta(-i\tau, \vec{x})$, by

$$G(\tau, \vec{x}) = i\Delta(-i\tau, \vec{x}) \tag{37}$$

where $G(\tau, \vec{x})$ is the propagator defined in (32). Therefore, using (36), we can write the inverse Fourier transformation,

$$\Delta(x) = \frac{1}{\beta} \sum_{n=-\infty}^{\infty} \int \frac{d^3p}{(2\pi)^3} e^{-i\omega_n \tau + i\vec{p}\vec{x}} \frac{-i}{\vec{p}^{\,2} + m^2 + \omega_n^2} \tag{38}$$

where the Matsubara frequencies ω_n are either for bosons or for fermions.

From (38) one can deduce the Feynman rules for the different fields in the imaginary time formalism. We can summarize them in the following way:

$$\begin{aligned}
\text{Boson propagator} &: \frac{i}{p^2 - m^2}; \quad p^\mu = [2ni\pi\beta^{-1}, \vec{p}\,] \\
\text{Fermion propagator} &: \frac{i}{\gamma \cdot p - m}; \quad p^\mu = [(2n+1)i\pi\beta^{-1}, \vec{p}\,] \\
\text{Loop integral} &: \frac{i}{\beta} \sum_{n=-\infty}^{\infty} \int \frac{d^3p}{(2\pi)^3} \\
\text{Vertex function} &: -i\beta(2\pi)^3 \delta_{\sum \omega_i} \delta^{(3)}\Big(\sum_i \vec{p}_i\Big)
\end{aligned} \tag{39}$$

There is a standard trick to perform infinite summations as in (39). For the case of bosons we can have frequency sums as,

$$\frac{1}{\beta} \sum_{n=-\infty}^{\infty} f(p^0 = i\omega_n) \tag{40}$$

with $\omega_n = 2n\pi\beta^{-1}$. Since the function $\frac{1}{2}\beta \coth(\frac{1}{2}\beta z)$ has poles at $z = i\omega_n$ and is analytic and bounded everywhere else, we can write (40) as,

$$\frac{1}{2\pi i \beta} \int_\gamma dz f(z) \frac{\beta}{2} \coth(\frac{1}{2}\beta z)$$

where the contour γ encircles anticlockwise all the previous poles of the imaginary axis. We are assuming that $f(z)$ does not have singularities along the imaginary axis (otherwise the previous expression is obviously not correct). The contour γ can be deformed to a new contour consisting in two straight lines: the first one starting at $-i\infty + \epsilon$ and going to $i\infty + \epsilon$, and the second one starting at $i\infty - \epsilon$ and ending at $-i\infty - \epsilon$. Rearranging the exponentials in the hyperbolic cotangent one can write the previous expression as,

$$\frac{1}{2\pi i}\int_{-i\infty}^{i\infty} dz \frac{1}{2}[f(z) + f(-z)] + \frac{1}{2\pi i}\int_{-i\infty+\epsilon}^{i\infty+\epsilon} dz [f(z) + f(-z)]\frac{1}{e^{\beta z} - 1}$$

and the contour of the second integral can be deformed to a contour C which encircles clockwise all singularities of the functions $f(z)$ and $f(-z)$ in the right half plane. Therefore we can write (40) as

$$\frac{1}{\beta}\sum_{n=-\infty}^{\infty} f(p^0 = i\omega_n) = \int_{-i\infty}^{i\infty} \frac{dz}{4\pi i}[f(z) + f(-z)] + \int_C \frac{dz}{2\pi i} n_B(z)[f(z) + f(-z)] \quad (41)$$

where $n_B(z)$ is the Bose distribution function (23).

Eq. (41) can be generalized for both bosons and fermions as,

$$\frac{1}{\beta}\sum_{n=-\infty}^{\infty} f(p^0 = i\omega_n) = \int_{-i\infty}^{i\infty} \frac{dz}{4\pi i}[f(z) + f(-z)] + \eta \int_C \frac{dz}{2\pi i} n(z)[f(z) + f(-z)] \quad (42)$$

where the distribution functions $n(z)$ are defined in (33). Eq. (42) shows that the frequency sum naturally separates into a T independent piece, which should coincide with the similar quantity computed in the field theory at zero temperature, and a T dependent piece which vanishes in the limit $T \to 0$, i.e. $\beta \to \infty$.

Real time formalism

The obvious disadvantage of the imaginary time formalism is to compute Green functions along imaginary time, so that going to the real time has to be done through a process of analytic continuation. However, a direct evaluation of Green function in the real time is possible by a judicious choice of the contour C in (4). The family of such real time contours is $C = C_1 \bigcup C_2 \bigcup C_3 \bigcup C_4$ where C_1 goes from the initial time t_i to the final time t_f, C_3 from t_f to $t_f - i\sigma$, with $0 \leq \sigma \leq \beta$, C_2 from $t_f - i\sigma$ to $t_i - i\sigma$, and C_4 from $t_i - i\sigma$ to $t_i - i\beta$. Different choices of σ lead to an equivalence class of quantum field theories at finite temperature [9]. For instance the choice $\sigma = 0$ leads to the Keldysh perturbation expansion [10], while the choice $\sigma = \beta/2$ is the preferred one to compute Green functions.

Computing the Green function for scalar (24) and fermion (30) fields taking path C is a matter of calculation, as we did for the imaginary time formalism in (32)-(36). One can prove that the contribution from the contours C_3 and C_4 can be neglected [4,11]. Therefore, for the propagator between x^0 and y^0 there are four possibilities depending on whether they are on C_1 or C_2. Correspondingly, there are four propagators which are labeled by (11), (12), (21) and (22).

Making the choice $\sigma = \beta/2$, the propagators for scalar fields (24) can be written, in momentum space, as

$$G(p) \equiv \begin{pmatrix} G^{(11)}(p) & G^{(12)}(p) \\ G^{(21)}(p) & G^{(22)}(p) \end{pmatrix} = M_B(\beta,p) \begin{pmatrix} \Delta(p) & 0 \\ 0 & \Delta^*(p) \end{pmatrix} M_B(\beta,p) \quad (43)$$

where $\Delta(p)$ is the boson propagator at zero temperature, and the matrix $M_B(\beta,p)$ is given by,

$$M_B(\beta,p) = \begin{pmatrix} \cosh\theta(p) & \sinh\theta(p) \\ \sinh\theta(p) & \cosh\theta(p) \end{pmatrix} \quad (44)$$

where

$$\sinh\theta(p) = e^{-\beta\omega_p/2} \left(1 - e^{-\beta\omega_p}\right)^{-1/2}$$
$$\cosh\theta(p) = \left(1 - e^{-\beta\omega_p}\right)^{-1/2} \quad (45)$$

Using now (43), (44), (45), one can easily write the expression for the four bosonic propagators, as

$$G^{(11)}(p) = \Delta(p) + 2\pi n_B(\omega_p)\delta(p^2 - m^2)$$
$$G^{(22)}(p) = G^{(11)*}$$
$$G^{(12)} = 2\pi e^{\beta\omega_p/2} n_B(\omega_p)\delta(p^2 - m^2) \quad (46)$$
$$G^{(21)} = G^{(12)}$$

Similarly, the propagators for fermion fields can be written as

$$S(p)_{\alpha\beta} \equiv \begin{pmatrix} S^{(11)}_{\alpha\beta}(p) & S^{(12)}_{\alpha\beta}(p) \\ S^{(21)}_{\alpha\beta}(p) & S^{(22)}_{\alpha\beta}(p) \end{pmatrix}$$
$$= M_F(\beta,p) \begin{pmatrix} (\gamma\cdot p + m)_{\alpha\beta}\Delta(p) & 0 \\ 0 & (\gamma\cdot p + m)_{\alpha\beta}\Delta^*(p) \end{pmatrix} M_F(\beta,p) \quad (47)$$

where the matrix $M_F(\beta,p)$ is,

$$M_F(\beta,p) = \begin{pmatrix} \cos\theta(p) & \sin\theta(p) \\ \sin\theta(p) & \cos\theta(p) \end{pmatrix} \quad (48)$$

with

$$\sin\theta(p) = e^{-\beta\omega_p/2} \left(1 + e^{-\beta\omega_p}\right)^{-1/2}$$
$$\cos\theta(p) = [\theta(p^0) - \theta(-p^0)] \left(1 + e^{-\beta\omega_p}\right)^{-1/2} \quad (49)$$

In the same way, using now (47), (48), and (49) one can easily write the expression for the four fermionic propagators, as

$$S^{(11)}(p) = (\gamma \cdot p + m)\left(\Delta(p) - 2\pi n_F(\omega_p)\delta(p^2 - m^2)\right)$$
$$S^{(22)}(p) = S^{(11)*}$$
$$S^{(12)} = -2\pi(\gamma \cdot p + m)[\theta(p^0) - \theta(-p^0)]e^{\beta\omega_p/2}n_F(\omega_p)\delta(p^2 - m^2) \quad (50)$$
$$S^{(21)} = -S^{(12)}$$

As one can see from (46) and (50), the main feature of the real time formalism is that the propagators come in two terms: one which is the same as in the zero temperature field theory, and a second one where all the temperature dependence is contained. This is welcome. However the propagators (12), (21) and (22) are unphysical since one of their time arguments has an imaginary component. They are required for the consistency of the theory. The only physical propagator is the (11) component in (46) and (50).

Now the Feynman rules in the real time formalism are very similar to those in the zero temperature field theory. In fact all diagrams have the same topology as in the zero temperature field theory and the same symmetry factors. However, associated to every field there are two possible vertices, 1 and 2, and four possible propagators, (11), (12), (21) and (22) connecting them. All of them have to be considered for the consistency of the theory. In the Feynman rules, type 2 vertices are hermitian conjugate with respect to type 1 vertices. The golden rule is that: *physical legs must always be attached to type 1 vertices.* Apart from the previous prescription, one must sum over all the configurations of type 1 and type 2 vertices, and use the propagator $G^{(ab)}$ or $S^{(ab)}$ to connect vertex a with vertex b.

There is now a general agreement in the sense that the imaginary time formalism and the real time formalism should give the same physical answer [12]. Using one or the other is sometimes a matter of taste, though in some cases the choice is dictated by calculational simplicity depending on the physical problem one is dealing with.

THE EFFECTIVE POTENTIAL AT FINITE TEMPERATURE

In this section we will construct the (one-loop) effective potential at finite temperature, using all the tools provided in the previous sections. The usefulness of this construction is addressed to the theory of phase transitions at finite temperature. The latter being essential for the understanding of phenomena as: inflation, baryon asymmetry generation, quark-gluon plasma transition in QCD,... We will compare different methods leading to the same result, including the use of both the imaginary and the real time formalisms. This exercise can be useful mainly to face more complicated problems than those which will be developed in this course.

Scalar fields

We will consider here the simplest model of one self-interacting scalar field. We have to compute the 1PI diagrams with any number of external legs using the Feynman rules described in (39), for the imaginary time formalism, or in (46) for the real time formalism. We will write the result as,

$$V_{\text{eff}}^\beta(\phi_c) = V_0(\phi_c) + V_1^\beta(\phi_c) \tag{51}$$

where $V_0(\phi_c) = \frac{1}{2}\phi_c^2 + \frac{\lambda}{4!}\phi_c^4$ is the tree level potential.

Imaginary time formalism

There is a very simple way of computing the effective potential: it consists in *computing its derivative* **in the shifted theory** *and then integrating*! In fact the derivative of the effective potential $dV_1^\beta/d\phi_c$ is described diagrammatically by the tadpole diagram. Using the Feynman rules in (39) one can easily write ($m^2(\phi_c) = m^2 + \lambda\phi_c^2/2$)

$$\frac{dV_1^\beta}{dm^2(\phi_c)} = \frac{1}{2\beta}\sum_{n=-\infty}^{\infty}\int\frac{d^3p}{(2\pi)^3}\frac{1}{\omega_n^2 + \omega^2} \tag{52}$$

Now we can perform the infinite sum in (52) using the result in Eq. (41) with a function f defined as,

$$f(z) = \frac{1}{\omega^2 - z^2} \tag{53}$$

and obtain for the tadpole (52) the result

$$\frac{dV_1^\beta}{dm^2(\phi_c)} = \int\frac{d^3p}{(2\pi)^3}\left\{\frac{1}{2}\int_{-i\infty}^{i\infty}\frac{dz}{2\pi i}\frac{1}{\omega^2 - z^2} + \int_C \frac{dz}{2\pi i}\frac{1}{e^{\beta z} - 1}\frac{1}{\omega^2 - z^2}\right\} \tag{54}$$

The first term in (54) gives the β-independent part of the tadpole contribution as,

$$\frac{1}{2}\int_{-i\infty}^{i\infty}\frac{dz}{2\pi i}\frac{1}{\omega^2 - z^2} \tag{55}$$

We can now close the integration contour of (55) anticlockwise and pick the pole of (53) at $z = -\omega$ with a residue $1/2\omega$. The result of (55) is $1/4\omega$. The second term in (54) gives the β-dependent part of the tadpole contribution. Here the integration contour encircles the pole at $z = \omega$ with a residue $-1/[2\omega(e^{\beta\omega} - 1)]$. Adding up we obtain for the tadpole the final expression,

$$\frac{dV_1(\phi_c)}{dm^2(\phi_c)} = \frac{1}{2}\int \frac{d^3p}{(2\pi)^3}\left[\frac{1}{2\omega} + \frac{1}{\omega}\frac{1}{e^{\beta\omega}-1}\right] \tag{56}$$

Now, integration of (56) with respect to $m^2(\phi_c)$ leads to the expression

$$V_1^\beta(\phi_c) = \int \frac{d^3p}{(2\pi)^3}\left[\frac{\omega}{2} + \frac{1}{\beta}\log\left(1-e^{-\beta\omega}\right)\right] \tag{57}$$

One can easily prove that the first integral in (57) is the one-loop effective potential at zero temperature. For that we have to prove the identity,

$$-\frac{i}{2}\int_{-\infty}^{\infty} \frac{dx}{2\pi}\log(-x^2+\omega^2-i\epsilon) = \frac{\omega}{2} + \text{constant} \tag{58}$$

i.e.

$$\omega\int_{-\infty}^{\infty} \frac{dx}{2\pi i}\frac{1}{-x^2+\omega^2-i\epsilon} = \frac{1}{2} \tag{59}$$

Integral (59) can be performed closing the integration interval $(-\infty,\infty)$ in the complex x plane along a contour going anticlockwise and picking the pole of the integrand at $x = -\sqrt{\omega^2-i\epsilon}$ with a residue $1/2\omega$. Using the residues theorem Eq. (59) can be easily checked. Now we can use identity (58) to write the temperature independent part of (57) and, after making the Wick rotation $p^0 = ip_E$ we obtain,

$$\frac{1}{2}\int \frac{d^3p}{(2\pi)^3}\omega = \frac{1}{2}\int \frac{d^4p}{(2\pi)^4}\log[p^2+m^2(\phi_c)] \tag{60}$$

which is the same result obtained in the zero temperature field theory.

Now the temperature dependent part in (57) can be easily written as,

$$\frac{1}{\beta}\int \frac{d^3p}{(2\pi)^3}\log\left(1-e^{-\beta\omega}\right) = \frac{1}{2\pi^2\beta^4}J_B[m^2(\phi_c)\beta^2] \tag{61}$$

where the thermal bosonic function J_B is defined as,

$$J_B[m^2\beta^2] = \int_0^\infty dx\, x^2 \log\left[1-e^{-\sqrt{x^2+\beta^2 m^2}}\right] \tag{62}$$

The integral (62) and therefore the thermal bosonic effective potential admits a high-temperature expansion which will be very useful for practical applications. It is given by

$$J_B(m^2/T^2) = -\frac{\pi^4}{45} + \frac{\pi^2}{12}\frac{m^2}{T^2} - \frac{\pi}{6}\left(\frac{m^2}{T^2}\right)^{3/2} - \frac{1}{32}\frac{m^4}{T^4}\log\frac{m^2}{a_b T^2} \tag{63}$$
$$-2\pi^{7/2}\sum_{\ell=1}^{\infty}(-1)^\ell \frac{\zeta(2\ell+1)}{(\ell+1)!}\Gamma\left(\ell+\frac{1}{2}\right)\left(\frac{m^2}{4\pi^2 T^2}\right)^{\ell+2}$$

where $a_b = 16\pi^2 \exp(3/2-2\gamma_E)$ ($\log a_b = 5.4076$) and ζ is the Riemann ζ-function.

Real time formalism

As we will see in this section, the final result for the effective potential (57) can be also obtained using the real time formalism. Let us compute the tadpole diagram. Since physical legs must be attached to type 1 vertices, the vertex must be considered of type 1, and the propagator circulating around the loop has to be considered as a (11) propagator. Application of the Feynman rules (46) to the tadpole diagram leads to the expression

$$\frac{dV_1^\beta}{dm^2(\phi_c)} = \frac{1}{2}\int \frac{d^4p}{(2\pi)^4}\left[\frac{-i}{-p^2+m^2(\phi_c)-i\epsilon} + 2\pi n_B(\omega)\delta(p^2-m^2(\phi_c))\right] \quad (64)$$

Now the β-independent part of (64), after integration on $m^2(\phi_c)$ contributes to the effective potential as

$$-\frac{i}{2}\int \frac{d^4p}{(2\pi)^4}\log(-p^2+m^2(\phi_c)-i\epsilon) \quad (65)$$

Finally using Eq. (58) to perform the p^0 integral, we can cast Eq. (65) as $\int \frac{d^3p}{(2\pi)^3}\frac{\omega}{2}$ which coincides with the first term in (57).

Integration over p^0 in the β-dependent part of (64) can be easily performed leading to, $\int \frac{d^3p}{(2\pi)^3}\frac{1}{2\omega}n_B(\omega)$ which, upon integration over $m^2(\phi_c)$ leads to the second term of Eq. (57).

We have checked that trivially the real time and imaginary time formalisms lead to the same expression of the thermal effective potential, in the one loop approximation.

Fermion fields

We will consider here a theory with fermion fields. As in the scalar case, we have to compute the corresponding diagrams, using the Feynman rules either for the imaginary or for the real time formalism, and decompose the thermal effective potential as in (51).

Imaginary time formalism

As we did in the case of the scalar field, there is a very simple way of obtaining the effective potential, computing the tadpole in the shifted theory, and integrating over ϕ_c. Using for the fermion propagator (39) $[M_f(\phi_c) = \Gamma \phi_c]$

$$i\frac{\gamma \cdot p + M_f}{p^2 - M_f^2}$$

we can write for the tadpole the expression,

$$\frac{dV_1^\beta}{dM_f^2(\phi_c)} = -2\lambda \frac{1}{2\beta} \sum_{n=-\infty}^{\infty} \int \frac{d^3p}{(2\pi)^3} \frac{1}{\omega_n^2 + \omega^2} \tag{66}$$

where $\lambda = 4$ (2) for Dirac (Weyl) fermions.

Now the infinite sum in (66) can be done with the help of (42), with $f(z)$ given by (53), as

$$\frac{dV_1^\beta}{dM_f^2(\phi_c)} = -2\lambda \int \frac{d^3p}{(2\pi)^3} \left\{ \frac{1}{2} \int_{-i\infty}^{i\infty} \frac{dz}{2\pi i} \frac{1}{\omega^2 - z^2} - \int_C \frac{dz}{2\pi i} \frac{1}{e^{\beta z} + 1} \frac{1}{\omega^2 - z^2} \right\} \tag{67}$$

The first term of (67) reproduces the zero temperature result, after M_f^2 integration, by closing the integration contour of (55) anticlockwise and picking the pole at $z = -\omega$ with a residue $1/2\omega$. The second term in (67) gives the β-dependent part of the tadpole contribution. Here the integration contour C encircles the pole at $z = \omega$ with a residue $(-2\lambda)\frac{1}{2\omega}\frac{1}{e^{\beta\omega}+1}$. Adding all of them together, we obtain for the tadpole the final expression

$$\frac{dV_1(\phi_c)}{dM_f^2(\phi_c)} = -\lambda \int \frac{d^3p}{(2\pi)^3} \left[\frac{1}{2\omega} - \frac{1}{\omega} \frac{1}{e^{\beta\omega}+1} \right] \tag{68}$$

and, upon integration with respect to M_f^2

$$V_1^\beta(\phi_c) = -2\lambda \int \frac{d^3p}{(2\pi)^3} \left[\frac{\omega}{2} + \frac{1}{\beta} \log\left(1 + e^{-\beta\omega}\right) \right] \tag{69}$$

The first integral in (69) can be proven, as in (58)-(60), to lead to the one-loop effective potential at zero temperature. The second integral, which contains all the temperature dependent part, can be written as,

$$-2\lambda \frac{1}{\beta} \int \frac{d^3p}{(2\pi)^3} \log\left(1 + e^{-\beta\omega}\right) = -2\lambda \frac{1}{2\pi^2 \beta^4} J_F[M_f^2(\phi_c)\beta^2] \tag{70}$$

where the thermal fermionic function J_F is defined as,

$$J_F[m^2\beta^2] = \int_0^\infty dx \, x^2 \log\left[1 + e^{-\sqrt{x^2 + \beta^2 m^2}}\right] \tag{71}$$

As in the scalar field, the integral (71) and therefore the thermal fermionic effective potential admits a high-temperature expansion which will be very useful for practical applications. It is given by

$$J_F(m^2/T^2) = \frac{7\pi^4}{360} - \frac{\pi^2}{24}\frac{m^2}{T^2} - \frac{1}{32}\frac{m^4}{T^4} \log \frac{m^2}{a_f T^2} \tag{72}$$

$$- \frac{\pi^{7/2}}{4} \sum_{\ell=1}^{\infty} (-1)^\ell \frac{\zeta(2\ell+1)}{(\ell+1)!} \left(1 - 2^{-2\ell-1}\right) \Gamma\left(\ell + \frac{1}{2}\right) \left(\frac{m^2}{\pi^2 T^2}\right)^{\ell+2}$$

where $a_f = \pi^2 \exp(3/2 - 2\gamma_E)$ ($\log a_f = 2.6351$) and ζ is the Riemann ζ-function.

Real time formalism

As for the case of scalar fields, the thermal effective potential for fermions (69) can also be very easily obtained using the real time formalism. We compute again the tadpole diagram, where the vertex between the two fermions and the scalar is of type 1 and the fermion propagator circulating along the loop is a (11) propagator. Application of the Feynman rules (50) leads to the expression

$$\frac{dV_1^\beta}{dM_f^2(\phi_c)} = -\frac{Tr\mathbf{1}}{2} \int \frac{d^4p}{(2\pi)^4} \left[\frac{-i}{-p^2 + M_f^2 - i\epsilon} - 2\pi n_F(\omega)\delta(p^2 - M_f^2) \right] \quad (73)$$

Now the β-independent part of (73), after integration on M_f^2, contributes to the effective potential, $-Tr\mathbf{1} \int \frac{d^3p}{(2\pi)^3} \frac{\omega}{2}$, which coincides with the first term in (69). Integration over p^0 in the β-dependent part of (73) can be easily performed leading to $\int \frac{d^3p}{(2\pi)^3} \frac{1}{2\omega}[-n_F(\omega)]$ which, upon integration over M_f^2 leads to the second term of Eq. (69).

Gauge bosons

The thermal effective potential for gauge bosons is computed in the same way as for previous fields. The simplest thing is to compute the tadpole diagram using the shifted mass for the gauge boson. In the Landau gauge, the gauge boson propagator reads as,

$$\Pi^\mu_{\ \nu}(p)^{\alpha\beta} = \Delta^\mu_{\ \nu} G^{\alpha\beta}(p) \quad (74)$$

where Δ is the projector with a trace equal to 3. Therefore the final expression for the thermal effective potential is computed as,

$$V_1^\beta(\phi_c) = Tr(\Delta) \left\{ \frac{1}{2} \int \frac{d^4p}{(2\pi)^4} \log[p^2 + M_{gb}^2(\phi_c)] + \frac{1}{2\pi^2\beta^4} J_B[M_{gb}^2(\phi_c)\beta^2] \right\}. \quad (75)$$

The first term of (75) agrees with the zero temperature effective potential and the second one just counts that of a scalar field theory a number of times equal to the number of degrees of freedom (3) of the gauge boson.

COSMOLOGICAL PHASE TRANSITIONS

All cosmological applications of field theories are based on the theory of phase transitions at finite temperature, that we will briefly describe throughout this section. The main point here is that at finite temperature, the equilibrium value of the scalar field ϕ, $\langle \phi(T) \rangle$, does not correspond to the minimum of the effective potential $V_{\text{eff}}^{T=0}(\phi)$, but to the minimum of the finite temperature effective potential $V_{\text{eff}}^{\beta}(\phi)$, as given by (51). Thus, even if the minimum of $V_{\text{eff}}^{T=0}(\phi)$ occurs at $\langle \phi \rangle = \sigma \neq 0$, very often, for sufficiently large temperatures, the minimum of $V_{\text{eff}}^{\beta}(\phi)$ occurs at $\langle \phi(T) \rangle = 0$: this phenomenon is known as **symmetry restoration** at high temperature, and gives rise to the phase transition from $\phi(T) = 0$ to $\phi = \sigma$. It was discovered by Kirzhnits [13] in the context of the electroweak theory (symmetry breaking between weak and electromagnetic interactions occurs when the universe cools down to a critical temperature $T_c \sim 10^2\ GeV$) and subsequently confirmed and developed by other authors [14,1,2,15].

The cosmological scenario can be drawn as follows: In the theory of the hot big bang, the universe is initially at very high temperature and, depending on the function $V_{\text{eff}}^{\beta}(\phi)$, it can be in the **symmetric phase** $\langle \phi(T) \rangle = 0$, *i.e.* $\phi = 0$ can be the stable absolute minimum. At some critical temperature T_c the minimum at $\phi = 0$ becomes metastable and the phase transition may proceed. The phase transition may be **first** or **second** order. First-order phase transitions have supercooled (out of equilibrium) symmetric states when the temperature decreases and are of use for baryogenesis purposes. Second-order phase transitions are used in the so-called new inflationary models [16]. We will illustrate these kinds of phase transitions with very simple examples.

First and second order phase transitions

We will illustrate the difference between **first** and **second** order phase transitions by considering first the simple example of a potential [2] described by the function,

$$V(\phi, T) = D(T^2 - T_o^2)\phi^2 + \frac{\lambda(T)}{4}\phi^4 \qquad (76)$$

where D and T_o^2 are constant terms and λ is a slowly varying function of T [3]. A quick glance at (63) and (72) shows that the potential (76) can be part of the one-loop finite temperature effective potential in field theories.

At zero temperature, the potential has a negative mass-squared term, which indicates that the state $\phi = 0$ is unstable, and the energetically favored state corresponds to the minimum at $\phi(0) = \pm\sqrt{\frac{2D}{\lambda}}T_o$, where the symmetry $\phi \leftrightarrow -\phi$ of the original theory is spontaneously broken.

The curvature of the finite temperature potential (76) is now T-dependent,

$$m^2(\phi, T) = 3\lambda\phi^2 + 2D(T^2 - T_o^2) \qquad (77)$$

and its stationary points, *i.e.* solutions to $dV(\phi, T)/d\phi = 0$, given by,

---

[2] The ϕ independent terms in (76), *i.e.* $V(0, T)$, are not explicitly considered.
[3] The T dependence of λ will often be neglected in this section.

$$\phi(T) = 0$$
and (78)
$$\phi(T) = \sqrt{\frac{2D(T_o^2 - T^2)}{\lambda(T)}}$$

Therefore the critical temperature is given by T_o. At $T > T_o$, $m^2(0,T) > 0$ and the origin $\phi = 0$ is a minimum. At the same time only the solution $\phi = 0$ in (78) does exist. At $T = T_o$, $m^2(0,T_o) = 0$ and both solutions in (78) collapse at $\phi = 0$. The potential (76) becomes,

$$V(\phi, T_o) = \frac{\lambda(T_o)}{4}\phi^4 \qquad (79)$$

At $T < T_o$, $m^2(0,T) < 0$ and the origin becomes a maximum. Simultaneously, the solution $\phi(T) \neq 0$ does appear in (78). This phase transition is called of **second order**, because there is no barrier between the symmetric and broken phases. Actually, when the broken phase is formed, the origin (symmetric phase) becomes a maximum. The phase transition may be achieved by a thermal fluctuation for a field located at the origin.

However, in many interesting theories there is a **barrier** between the symmetric and broken phases. This is characteristic of **first order** phase transitions. A typical example is provided by the potential [4],

$$V(\phi, T) = D(T^2 - T_o^2)\phi^2 - ET\phi^3 + \frac{\lambda(T)}{4}\phi^4 \qquad (80)$$

where, as before, D, T_0 and E are T independent coefficients, and λ is a slowly varying T-dependent function. Notice that the difference between (80) and (76) is the cubic term with coefficient E. This term can be provided by the contribution to the effective potential of bosonic fields (63). The behaviour of (80) for the different temperatures is reviewed in Refs. [17,18]. At $T > T_1$ the only minimum is at $\phi = 0$. At $T = T_1$

$$T_1^2 = \frac{8\lambda(T_1)DT_o^2}{8\lambda(T_1)D - 9E^2} \qquad (81)$$

a local minimum at $\phi(T) \neq 0$ appears as an inflection point. The value of the field ϕ at $T = T_1$ is,

$$\langle\phi(T_1)\rangle = \frac{3ET_1}{2\lambda(T_1)} \qquad (82)$$

A barrier between the latter and the minimum at $\phi = 0$ starts to develop at lower temperatures. Then the point (82) splits into a maximum

---

[4] See, e.g. the one-loop effective potential for the Standard Model, Eq. (138).

$$\phi_M(T) = \frac{3ET}{2\lambda(T)} - \frac{1}{2\lambda(T)}\sqrt{9E^2T^2 - 8\lambda(T)D(T^2 - T_o^2)} \qquad (83)$$

and a local minimum

$$\phi_m(T) = \frac{3ET}{2\lambda(T)} + \frac{1}{2\lambda(T)}\sqrt{9E^2T^2 - 8\lambda(T)D(T^2 - T_o^2)} \qquad (84)$$

At a given temperature $T = T_c$

$$T_c^2 = \frac{\lambda(T_c)DT_o^2}{\lambda(T_c)D - E^2} \qquad (85)$$

the origin and the minimum (84) become degenerate. From (83) and (84) we find that

$$\phi_M(T_c) = \frac{ET_c}{\lambda(T_c)} \qquad (86)$$

and

$$\phi_m(T_c) = \frac{2ET_c}{\lambda(T_c)} \qquad (87)$$

For $T < T_c$ the minimum at $\phi = 0$ becomes metastable and the minimum at $\phi_m(T) \neq 0$ becomes the global one. At $T = T_o$ the barrier disappears, the origin becomes a maximum

$$\phi_M(T_o) = 0 \qquad (88)$$

and the second minimum becomes equal to

$$\phi_m(T_o) = \frac{3ET_o}{\lambda(T_o)} \qquad (89)$$

The phase transition starts at $T = T_c$ by tunneling. However, if the barrier is high enough the tunneling effect is very small and the phase transition does effectively start at a temperature $T_c > T_t > T_o$. In some models T_o can be equal to zero. The details of the phase transition depend therefore on the process of tunneling from the false to the global minimum. These details will be studied in the rest of this section.

Thermal tunneling

The transition from the false to the true vacuum proceeds via **thermal tunneling** at finite temperature. It can be understood in terms of formation of bubbles of the broken phase in the sea of the symmetric phase. Once this has happened, the bubble spreads throughout the universe converting false vacuum into true one.

The tunneling rate [19–21] is computed by using the rules of field theory at finite temperature [22]. In the previous section we defined the critical temperature T_c as the temperature at which the two minima of the potential $V(\phi, T)$ have the same depth. However, tunneling with formation of bubbles of the field ϕ corresponding to the second minimum starts somewhat later, and goes sufficiently fast to fill the universe with bubbles of the new phase only at some lower temperature T_t when the corresponding euclidean action $S_E = S_3/T$ suppressing the tunneling becomes $\mathcal{O}(130 - 140)$ [23,24,17], as we will see in the next section.

We will use as prototype the potential of Eq. (80) which can trigger, as we showed in this section, a first order phase transition. In this case the false minimum is $\phi = 0$, and the value of the potential at the origin is zero, $V(0, T) = 0$. The tunneling probability per unit time per unit volume is given by [22]

$$\frac{\Gamma}{\nu} \sim A(T) e^{-S_3/T} \tag{90}$$

In (90) the prefactor $A(T)$ is roughly of $\mathcal{O}(T^4)$ while S_3 is the three-dimensional euclidean action defined as

$$S_3 = \int d^3x \left[\frac{1}{2} \left(\vec{\nabla} \phi \right)^2 + V(\phi, T) \right]. \tag{91}$$

At very high temperature the **bounce** solution has $O(3)$ symmetry [22] and the euclidean action is then simplified to,

$$S_3 = 4\pi \int_0^\infty r^2 dr \left[\frac{1}{2} \left(\frac{d\phi}{dr} \right)^2 + V(\phi(r), T) \right] \tag{92}$$

where $r^2 = \vec{x}^2$, and the euclidean equation of motion yields,

$$\frac{d^2\phi}{dr^2} + \frac{2}{r} \frac{d\phi}{dr} = V'(\phi, T) \tag{93}$$

with the boundary conditions

$$\lim_{r \to \infty} \phi(r) = 0 \tag{94}$$

$$\left. \frac{d\phi}{dr} \right|_{r=0} = 0 \tag{95}$$

From here on we will follow the discussion in Ref. [17]. Let us take $\phi = 0$ outside a bubble. Then (92), which is also the surplus free energy of a true vacuum bubble, can be written as

$$S_3 = 4\pi \int_0^R r^2 dr \left[\frac{1}{2}\left(\frac{d\phi}{dr}\right)^2 + V(\phi(r),T) \right] \tag{96}$$

where R is the bubble radius. There are two contributions to (96): a surface term F_S, coming from the derivative term in (96), and a volume term F_V, coming from the second term in (96). They scale like,

$$S_3 \sim 2\pi R^2 \left(\frac{\delta\phi}{\delta R}\right)^2 \delta R + \frac{4\pi R^3 \langle V \rangle}{3} \tag{97}$$

where δR is the thickness of the bubble wall, $\delta\phi = \phi_m$ and $\langle V \rangle$ is the average of the potential inside the bubble.

For temperatures just below T_c, the height of the barrier $V(\phi_M, T)$ is large compared to the depth of the potential at the minimum, $-V(\phi_m, T)$. In that case, the solution of minimal action corresponds to minimizing the contribution to F_V coming from the region $\phi = \phi_M$. This amounts to a very small bubble wall $\delta R/R \ll 1$ and so a very quick change of the field from $\phi = 0$ outside the bubble to $\phi = \phi_m$ inside the bubble. Therefore, the first formed bubbles after T_c are **thin wall** bubbles.

Subsequently, when the temperature drops towards T_o the height of the barrier $V(\phi_M, T)$ becomes small as compared with the depth of the potential at the minimum $-V(\phi_m, T)$. In that case the contribution to F_V from the region $\phi = \phi_M$ is negligible, and the minimal action corresponds to minimizing the surface term F_S. This amounts to a configuration where δR is as large as possible, i.e. $\delta R/R = \mathcal{O}(1)$: **thick wall** bubbles. So whether the phase transition proceeds through thin or thick wall bubbles depends on how large the bubble nucleation rate (90) is, or how small S_3 is, before thick bubbles are energetically favoured.

For the potential (80) an analytic formula has been obtained in Ref. [18] without assuming the thin wall approximation. It is given by,

$$\frac{S_3}{T} = \frac{13.72}{E^2}\left[D\left(1 - \frac{T_o^2}{T^2}\right)\right]^{3/2} f\left[\frac{\lambda(T)D}{E^2}\left(1 - \frac{T_o^2}{T^2}\right)\right] \tag{98}$$

$$f(x) = 1 + \frac{x}{4}\left[1 + \frac{2.4}{1-x} + \frac{0.26}{(1-x)^2}\right] \tag{99}$$

The case of two fields is extremely more complicated. In particular the two-Higgs situation in the supersymmetric standard theory has been recently solved in Ref. [25]. The connection between zero temperature and finite temperature tunneling is manifest. In particular at temperatures much less than the inverse radius the $(T=0)$ $O(4)$ solution has the least action. This can happen for theories with a supercooled symmetric phase: for instance in the presence of a barrier that does not disappear when the temperature drops to zero. At temperatures much larger than the inverse radius, the $O(3)$ solution has the least action.

Bubble nucleation

In the previous subsection we have established the free energy and the critical radius of a bubble large enough to grow after formation. The subsequent progress of the phase transition depends on the ratio of the rate of production of bubbles of true vacuum, as given by (90), over the expansion rate of the universe. For example if the former remains always smaller than the latter, then the state will be trapped in the supercooled false vacuum. Otherwise the phase transition will start at some temperature T_t by **bubble nucleation**. The probability of bubble formation per unit time per unit volume is given by (90) where $B(T) = S_3(T)/T$, $A(T) = \omega T^4$, where the parameter ω will be taken of $\mathcal{O}(1)$.

Since the progress of the phase transition should depend on the expansion rate of the universe, we have to describe the universe at temperatures close to the electroweak phase transition. A homogeneous and isotropic (flat) universe is described by a Robertson-Walker metric which, in comoving coordinates, is given by $ds^2 = dt^2 - a(t)^2 (dr^2 + r^2 d\Omega^2)$, where $a(t)$ is the scale factor of the universe. The universe expansion is governed by the equation

$$\left(\frac{\dot{a}}{a}\right)^2 = \frac{8\pi}{3M_{P\ell}^2}\rho \qquad (100)$$

where $M_{P\ell}$ is the Planck mass, and ρ is the energy density. For temperatures $T \sim 10^2 \; GeV$ the universe is radiation dominated, and its energy density is given by,

$$\rho = \frac{\pi^2}{30}g(T)T^4 \qquad (101)$$

where $g(T) = g_B(T) + \frac{7}{8}g_F(T)$, and $g_B(T)$ ($g_F(T)$) is the effective number of bosonic (fermionic) degrees of freedom at the temperature T. For the standard model we have $g^{SM} = 106.75$ which can be considered as temperature independent.

The equation of motion (100) can be solved, and assuming an adiabatic expansion of the universe, $a(T_1)T_1 = a(T_2)T_2$, one obtains the following relationship,

$$t = \zeta \frac{M_{P\ell}}{T^2} \qquad (102)$$

where $\zeta = \frac{1}{4\pi}\sqrt{\frac{45}{\pi g}} \sim 3 \times 10^{-2}$. Using (102) the horizon volume is given by

$$V_H(t) = 8\zeta^3 \frac{M_{P\ell}^3}{T^6} \qquad (103)$$

The onset of nucleation happens at a temperature T_t such that **the probability for a single bubble to be nucleated within one horizon volume is** ~ 1, i.e. [25]

$$\int_{T_t}^{\infty} \frac{dT}{T} \left(\frac{2\zeta M_{\text{Pl}}}{T}\right)^4 \exp\{-S_3(T)/T\} = \mathcal{O}(1) \ . \tag{104}$$

which implies numerically,

$$B(T_t) \sim 137 + \log \frac{10^2 E^2}{\lambda D} + 4 \log \frac{100 \ GeV}{T_t} \tag{105}$$

where we have normalized $T_t \sim 100 \ GeV$ and $E^2/(\lambda D) \sim 10^{-2}$ which are typical values which will be obtained in the standard model of electroweak interactions.

BARYOGENESIS AT PHASE TRANSITIONS

There are two essential problems to be understood related with the baryon number of the universe:

i) There is no evidence of antimatter in the universe. In fact, there is no antimatter in the solar system, and only $\bar{p}$ in cosmic rays. However antiprotons can be produced as secondaries in collisions ($pp \to 3p + \bar{p}$) at a rate similar to the observed one. Numerically, $\frac{n_{\bar{p}}}{n_p} \sim 3 \times 10^{-4}$, and $\frac{n_{4He}}{n_{4\overline{He}}} \sim 10^{-5}$. We can conclude that $n_B \gg n_{\bar{B}}$, so $n_{\Delta B} \equiv n_B - n_{\bar{B}} \sim n_B$.

ii) The second problem is to understand the origin of

$$\eta \equiv \frac{n_B}{n_\gamma} \sim (0.3 - 1.0) \times 10^{-9} \tag{106}$$

today. This parameter is essential for primordial nucleosynthesis [26]. η may not have changed since nucleosynthesis. At these energy scales (~ 1 MeV) baryon number is conserved if there are no processes which would have produced entropy to change the photon number. We can easily estimate from η the baryon to entropy ratio by using

$$s = \frac{\pi^4}{45\zeta(3)} 3.91 \ n_\gamma = 7.04 \ n_\gamma \tag{107}$$

and the range (106).

In the standard cosmological model there is no explanation for the smallness of the ratio (106) if we start from $n_{\Delta B} = 0$. An initial asymmetry has to be imposed by hand as an initial condition (which violates any naturalness principle) or has to be dynamically generated at phase transitions, which is the way we will explore all along this section.

Conditions for baryogenesis

As we have seen in the previous subsection the universe was initially baryon symmetric ($n_B \simeq n_{\bar{B}}$) although the matter-antimatter asymmetry appears to be large today ($n_{\Delta B} \simeq n_B \gg n_{\bar{B}}$). In the standard cosmological model there is no explanation for the value of η consistent with nucleosynthesis, Eq. (106), and it has to be imposed by hand as an initial condition. However, it was suggested by Sakharov long ago [27] that a tiny $n_{\Delta B}$ might have been produced in the early universe leading, after $p\bar{p}$ annihilations, to (106). The three ingredients necessary for baryogenesis are:

B-nonconserving interactions

This condition is obvious since we want to start with a baryon symmetric universe ($\Delta B = 0$) and evolve it to a universe where $\Delta B \neq 0$. B-nonconserving interactions might mediate proton decay; in that case the phenomenological constraints are provided by the proton lifetime measurements [28] $\tau_p \gtrsim 10^{32} yr$.

C and CP violation

The action of C (charge conjugation) and CP (combined action of charge conjugation and parity) interchanges particles with antiparticles, changing therefore the sign of B. For instance if we describe spin-$\frac{1}{2}$ fermions by two-component fields of definite chirality (left-handed fields ψ_L and right-handed fields ψ_R) the action of C and CP over them is given by

$$P : \psi_L \longrightarrow \psi_R, \quad \psi_R \longrightarrow \psi_L \qquad (108)$$
$$C : \psi_L \longrightarrow \psi_L^C \equiv \sigma_2 \psi_R^*, \quad \psi_R \longrightarrow \psi_R^C \equiv -\sigma_2 \psi_L^*$$
$$CP : \psi_L \longrightarrow \psi_R^C, \quad \psi_R \longrightarrow \psi_L^C$$

If the universe is initially matter-antimatter symmetric, and without a preferred direction of time as in the standard cosmological model, it is represented by a C and CP invariant state, $|\phi_o\rangle$, with $B = 0$. If C and CP were conserved, i.e. $[C, H] = [CP, H] = 0$, H being the hamiltonian, then the state of the universe at a later time t, $|\phi(t)\rangle = e^{iHt}|\phi_o\rangle$ would be C and CP invariant and, therefore, baryon number conserving, $\Delta B = 0$. The only way to generate a net $\Delta B \neq 0$ is to have C and CP violating interactions.

Departure from thermal equilibrium

If all particles in the universe remained in thermal equilibrium, then no direction for time would be defined and CPT invariance would prevent the appearance of any baryon excess, rendering CP violating interactions irrelevant [29].

A particle species is in thermal equilibrium if all its reaction rates, Γ, are much faster than the expansion rate of the universe, H. On the other hand a departure from thermal equilibrium is expected whenever a rate crucial for maintaining it is less than the expansion rate ($\Gamma < H$). Deviation from thermal equilibrium cannot occur in a homogeneous isotropic universe containing only massless species: massive species are needed in general for such deviations to occur.

Baryogenesis at the electroweak phase transitions

It has been recently realized [30,31] that the three Sakharov's conditions for baryogenesis can be fulfilled at the electroweak phase transition:

- Baryonic charge non-conservation was discovered by 't Hooft [32]. In fact baryon and lepton number are conserved anomalous global symmetries in the Standard Model. They are violated by non-perturbative effects.

- CP violation can be generated in the Standard Model from phases in the fermion mass matrix, Cabibbo, Kobayashi, Maskawa (CKM) phases [33]. This effect is much too small to explain the observed baryon to entropy ratio. However, in extensions of the Standard Model as the minimal supersymmetric standard model (MSSM), a sizeable CP violation can happen through an extended Higgs sector.

- The out of equilibrium condition can be achieved, if the phase transition is strong enough first order, in the bubble walls. In that case the B-violating interactions are out of equilibrium in the bubble walls and a net B-number can be generated during the phase transition [31].

Baryon and lepton number violation in the electroweak theory

Violation of baryon and lepton number in the electroweak theory is a very striking phenomenon. Classically, baryonic and leptonic currents are conserved in the electroweak theory. However, that conservation is spoiled by quantum corrections through the chiral anomaly associated with triangle fermionic loop in external gauge fields. The calculation gives,

$$\partial_\mu j_B^\mu = \partial_\mu j_L^\mu = N_f \left(\frac{g^2}{32\pi^2} W\widetilde{W} - \frac{g'^2}{32\pi^2} Y\widetilde{Y} \right) \tag{109}$$

where N_f is the number of fermion generations, $W_{\mu\nu}$ and $Y_{\mu\nu}$ are the gauge field strength tensors for $SU(2)$ and $U(1)_Y$, respectively, and the tilde means the dual tensor.

A very important feature of (109) is that the difference $B-L$ is strictly conserved, and so only the sum $B+L$ is anomalous and can be violated. Another feature is that fluctuations of the gauge field strengths can lead to fluctuations of the corresponding value of $B + L$. The product of gauge field strengths on the right hand side of Eq. (109) can be written as four-divergences, $W\widetilde{W} = \partial_\mu k_W^\mu$, $Y\widetilde{Y} = \partial_\mu k_Y^\mu$, where

$$k_Y^\mu = \epsilon^{\mu\nu\alpha\beta} Y_{\nu\alpha} Y_\beta \qquad (110)$$

$$k_W^\mu = \epsilon^{\mu\nu\alpha\beta} \left(W_{\nu\alpha}^a W_\beta^a - \frac{g}{3}\epsilon_{abc} W_\nu^a W_\alpha^b W_\beta^c \right)$$

and W_μ, Y_μ are the gauge fields of $SU(2)$ and $U(1)_Y$, respectively. In general total derivatives are unobservable because they can be integrated by parts and drop from the integrals. This is true for the terms in the four-vectors (110) proportional to the field strengths $W_{\mu\nu}$ and $Y_{\mu\nu}$. This means that for the abelian subgroup $U(1)_Y$ the current non conservation induced by quantum effects becomes non observable. However this is not mandatory for gauge fields, for which the integral can be nonzero. Hence only for non-abelian groups can the current non conservation induced by quantum effects become observable. In particular one can write $\Delta B = \Delta L = N_f \Delta N_{CS}$, where N_{CS} is the so-called Chern-Simons number characterizing the topology of the gauge field configuration,

$$N_{CS} = \frac{g^2}{32\pi^2} \int d^3x \epsilon^{ijk} \left(W_{ij}^a W_k^a - \frac{g}{3}\epsilon_{abc} W_i^a W_j^b W_k^c \right) \qquad (111)$$

Note that though N_{CS} is not gauge invariant, its variation ΔN_{CS} is.

We want to compute now ΔB between an initial and a final configuration of gauge fields. We are considering vacuum field strength tensors $W_{\mu\nu}$ which vanish. The corresponding potentials are not necessarily zero but can be represented by purely gauge fields,

$$W_\mu = -\frac{i}{g} U(x) \partial_\mu U^{-1}(x) \qquad (112)$$

There are two classes of gauge transformations keeping $W_{\mu\nu} = 0$:

- Continuous transformations of the potentials yielding $\Delta N_{CS} = 0$.

- If one tries to generate $\Delta N_{CS} \neq 0$ by a continuous variation of the potentials, then one has to enter a region where $W_{\mu\nu} \neq 0$. This means that vacuum states with different topological charges are separated by potential barriers.

The probability of barrier penetration can be calculated using the quasi-classical approximation [19]. In euclidean space time, the trajectory in field space configuration which connects two vacua differing by a unit of topological charge is called instanton. The euclidean action evaluated at this trajectory gives the probability for barrier penetration as $\Gamma \sim \exp\left(-\frac{4\pi}{\alpha_W}\right) \sim 10^{-162}$, where $\alpha_W = g^2/4\pi$. This number is so small that the calculation of the pre-exponential is unnecessary and the probability for barrier penetration is essentially zero.

Baryon violation at finite temperature: sphalerons

However, in a system with non zero temperature a particle may classically go over the barrier with a probability determined by the Boltzmann exponent, as we have seen.

What we have is a potential which depends on the gauge field configuration W_μ. This potential has an infinite number of degenerate minima, labeled as Ω_n. These minima are characterized by different values of the Chern-Simons number. The minimum Ω_0 corresponds to the configuration $W_\mu = 0$ and we can take conventionally the value of the potential at this point to be zero. Other minima have gauge fields given by (112). In the temporal gauge $W_0 = 0$, the gauge transformation U must be time independent (since we are considering gauge configurations with $W_{\mu\nu} = 0$), i.e. $U = U(\vec{x})$, and so functions U define maps,

$$U : S^3 \longrightarrow SU(2)$$

All the minima with $W_{\mu\nu} = 0$ have equally zero potential energy, but those defined by a map $U(\vec{x})$ with nonzero Chern-Simons number

$$n[U] = \frac{1}{24\pi^2} \int d^3x \epsilon^{ijk} Tr(U\partial_i U^{-1} U\partial_j U^{-1} U\partial_k U^{-1}) \tag{113}$$

correspond to degenerate minima in the configuration space with non-zero baryon and lepton number.

Degenerate minima are separated by a potential barrier. The field configuration at the top of the barrier is called **sphaleron**, which is a *static unstable* solution to the classic equations of motion [34]. The sphaleron solution has been explicitly computed in Ref. [34] for the case of zero Weinberg angle, (*i.e.* neglecting terms $\mathcal{O}(g')$), and for an arbitrary value of $\sin^2 \theta_W$ in Ref. [35].

An ansatz for the sphaleron solution for the case of zero Weinberg angle was given (for the zero temperature potential) in Ref. [34], for the Standard Model with a single Higgs doublet, as,

$$W_i^a \sigma^a dx^i = -\frac{2i}{g} f(\xi) dU \, U^{-1} \tag{114}$$

for the gauge field, and

$$\Phi = \frac{v}{\sqrt{2}} h(\xi) U \begin{pmatrix} 0 \\ 1 \end{pmatrix} \tag{115}$$

for the Higgs field, where the gauge transformation U is taken to be,

$$U = \frac{1}{r} \begin{pmatrix} z & x+iy \\ -x+iy & z \end{pmatrix} \tag{116}$$

and we have introduced the dimensionless radial distance $\xi = gvr$.

Using the ansatz (114), (115) and (116) the field equations reduce to,

$$\xi^2 \frac{d^2 f}{d\xi^2} = 2f(1-f)(1-2f) - \frac{\xi^2}{4}h^2(1-f) \tag{117}$$

$$\frac{d}{d\xi}\left(\xi^2 \frac{dh}{d\xi}\right) = 2h(1-f)^2 + \frac{\lambda}{g^2}\xi^2(h^2-1)h$$

with the boundary conditions, $f(0) = h(0) = 0$ and $f(\infty) = h(\infty) = 1$. The energy functional becomes then,

$$E = \frac{4\pi v}{g}\int_0^\infty \left\{ 4\left(\frac{df}{d\xi}\right)^2 + \frac{8}{\xi^2}[f(1-f)]^2 + \frac{1}{2}\xi^2\left(\frac{dh}{d\xi}\right)^2 \right.$$
$$\left. + [h(1-f)]^2 + \frac{1}{4}\left(\frac{\lambda}{g^2}\right)\xi^2(h^2-1)^2 \right\} d\xi \tag{118}$$

The solution to Eqs. (117) has to be found numerically. The solutions depend on the gauge and quartic couplings, g and λ. Once replaced into the energy functional (118) they give the sphaleron energy which is the height of the barrier between different degenerate minima. It is customary to write the solution as,

$$E_{\text{sph}} = \frac{2m_W}{\alpha_W}B(\lambda/g^2) \tag{119}$$

where B is the constant which requires numerical evaluation. For the standard model with a single Higgs doublet this parameter ranges from $B(0) = 1.5$ to $B(\infty) = 2.7$. A fit valid for values of the Higgs mass $25\ GeV \leq m_h \leq 250\ GeV$ can be written as,

$$B(x) = 1.58 + 0.32x - 0.05x^2 \tag{120}$$

where $x = m_h/m_W$.

The previous calculation of the sphaleron energy was performed at zero temperature. The sphaleron at finite temperature was computed in Ref. [36,37]. Its energy follows the approximate scaling law, $E_{\text{sph}}(T) = E_{\text{sph}}\langle\phi(T)\rangle/\langle\phi(0)\rangle$ which, using (119), can be written as,

$$E_{\text{sph}}(T) = \frac{2m_W(T)}{\alpha_W}B(\lambda/g^2) \tag{121}$$

where $m_W(T) = \frac{1}{2}g\langle\phi(T)\rangle$

Baryon violation rate at $T > T_c$

The calculation of the baryon violation rate at $T > T_c$, i.e. in the symmetric phase, is very different from that in the broken phase, that will be reviewed in the next section. In the symmetric phase, at $\phi = 0$, the perturbation theory is spoiled by infrared divergences, and so we cannot rely upon perturbative calculations to compute the baryon violation rate in this phase. In fact, the infrared divergences are cut off by the non-perturbative generation of a **magnetic mass**, $m_M \sim \alpha_W T$, i.e. a **magnetic screening length** $\xi_M \sim (\alpha_W T)^{-1}$. The rate of baryon violation per unit time and unit volume Γ does not contain any exponential Boltzmann factor [5]. The pre-exponential can be computed from dimensional grounds, [38] as

$$\Gamma = k(\alpha_W T)^4 \qquad (122)$$

where the coefficient k has been evaluated numerically in Ref. [39] with the result $0.1 \lesssim k \lesssim 1.0$.

Baryon violation rate at $T < T_c$

After the phase transition, the calculation of baryon violation rate can be done using the semiclassical approximations given by Eq. (90). The rate per unit time and unit volume for fluctuations between neighboring minima contains a Boltzmann suppression factor $\exp(-E_{\text{sph}}(T)/T)$, where $E_{\text{sph}}(T)$ is given by (121), and a prefactor containing the determinant of all zero and non-zero modes. The prefactor was computed in Ref. [40] as

$$\Gamma \sim 2.8 \times 10^5 T^4 \left(\frac{\alpha_W}{4\pi}\right)^4 \kappa \frac{\zeta^7}{B^7} e^{-\zeta} \qquad (123)$$

where we have defined $\zeta(T) = E_{\text{sph}}(T)/T$, the coefficient B is the function of λ/g^2 defined in (120) and κ is the functional determinant associated with fluctuations about the sphaleron. It has been estimated [24] to be in the range, $10^{-4} \lesssim \kappa \lesssim 10^{-1}$.

The equation describing the dilution S of the baryon asymmetry in the anomalous electroweak processes reads [41]

$$\frac{\partial S}{\partial t} = -V_B(t) S \qquad (124)$$

where $V_B(t)$ is the rate of the baryon number non-conserving processes. Assuming T is constant during the phase transition the integration of (124) yields $S = e^{-X}$ and $X = \frac{13}{2} N_f \frac{\Gamma}{T^3} t$. Using now (123) and (102) we can write the exponent X as, $X \sim 10^{10} \kappa \zeta^7 e^{-\zeta}$, where we have taken the values of the parameters, $B = 1.87$, $\alpha_W = 0.0336$, $N_f = 3$, $T_c \sim 10^2$ GeV. Imposing now the condition $S \gtrsim 10^{-5}$, or $X \lesssim 10$, leads to the condition on $\zeta(T_c)$,

$$\zeta(T_c) \gtrsim 7 \log \zeta(T_c) + 9 \log 10 + \log \kappa \qquad (125)$$

---

[5] It would disappear from (90) in the limit $T \to \infty$.

Now, taking κ at its upper bound, $\kappa = 10^{-1}$, we obtain from (125) the bound [42]

$$\frac{E_{\text{sph}}(T_c)}{T_c} \gtrsim 45, \qquad (126)$$

and using the lower bound, $\kappa = 10^{-4}$ we obtain,

$$\frac{E_{\text{sph}}(T_c)}{T_c} \gtrsim 37, \qquad (127)$$

Eq. (126) is the usual bound used to test different theories while Eq. (127) gives an idea on how much can one move away from the bound (126), *i.e.* the uncertainty on the bound (126).

The bounds (126) and (127) can be translated into bounds on $\phi(T_c)/T_c$. Using the relation (121) we can write

$$\frac{\phi(T_c)}{T_c} = \frac{g}{4\pi B} \frac{E_{\text{sph}}(T_c)}{T_c} \sim \frac{1}{36} \frac{E_{\text{sph}}(T_c)}{T_c} \qquad (128)$$

where we have used the previous values of the parameters. The bound (126) translates into

$$\frac{\phi(T_c)}{T_c} \gtrsim 1.3 \qquad (129)$$

while the bound (127) translates into,

$$\frac{\phi(T_c)}{T_c} \gtrsim 1.0 \qquad (130)$$

These bounds, Eqs. (129) and (126), require that the phase transition is strong enough first order. In fact for a second order phase transition, $\phi(T_c) \simeq 0$ and any previously generated baryon asymmetry would be washed out during the phase transition. For the case of the Standard Model the previous bounds translate into a bound on the Higgs mass, as we will see.

ON THE VALIDITY OF THE PERTURBATIVE EXPANSION

The approach of Ref. [2] to the finite temperature effective potential relied on the observation that **symmetry restoration implies that ordinary perturbation theory must break down at high temperature**. In fact, otherwise perturbation theory should hold and, since the tree level potential is temperature independent, radiative corrections (which are temperature dependent) should be unable to restore the symmetry. We will see that the failure of perturbative expansion is intimately linked to the appearance of infrared divergences for the zero Matsubara modes of bosonic degrees of freedom. This just means that the usual perturbative expansion in powers of the coupling constant fails at temperatures beyond the critical temperature. It has to be replaced by an improved perturbative expansion where an infinite number of diagrams are resummed at each order in the new expansion. We will review the actual situation in this section.

The breakdown of perturbative expansion

We will examine the simplest model of one self-interacting real scalar field, described by a lagrangian with a squared-mass term, m^2 and a quartic coupling λ. We will use now power counting arguments to investigate the high temperature behaviour of higher loop diagrams contributing to the effective potential [2,43,44]. After rescaling all loop momenta and energies by T, a loop amplitude with superficial divergence D takes the form, $T^D f(m/T)$. If there are no infrared divergences when $m/T \to 0$, then the loop goes like T^D. For instance the diagram contributing to the self-energy of Fig. 1 is quadratically divergent ($D = 2$), and so behaves like

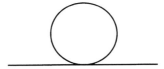

FIGURE 1. One-loop contribution to the self-energy for the scalar theory

λT^2. For $D \leq 0$, there are infrared divergences associated to the zero modes of bosonic propagators in the imaginary time formalism [$n = 0$ in (39)] and the only T dependence comes from the T in front of the loop integral in (39). Then every logarithmically divergent or convergent loop contributes a factor of T. For instance the diagram contributing to the self-energy in Fig. 2 contains two logarithmically

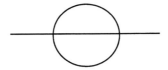

FIGURE 2. Two-loop contribution to the self-energy for the scalar theory

divergent loops and so behaves like, $\lambda^2 T^2 = \lambda(\lambda T^2)$. It is clear that to a fixed order in the loop expansion the largest graphs are those with the maximum number of quadratically divergent loops. These diagrams are obtained from the diagram in Fig. 1 by adding n quadratically divergent loops on top of it, as shown in Fig. 3. They behave as,

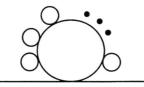

FIGURE 3. Daisy (n+1)-loop contribution to the self-energy for the scalar theory

$$\lambda^{n+1}\frac{T^{2n+1}}{M^{2n-1}} = \lambda^2 \frac{T^3}{M} \left(\frac{\lambda T^2}{M^2}\right)^{n-1} \tag{131}$$

where M is the mass scale of the theory, and has been introduced to rescale the powers of the temperature [6]. As was clear from Eq. (131), adding a quadratically divergent bubble to a propagator which is part of a logarithmically divergent or finite loop amounts to multiplying the diagram by

$$\alpha \equiv \lambda \frac{T^2}{M^2} \tag{132}$$

This means that for the one-loop approximation to be valid it is required that

$$\lambda \frac{T^2}{M^2} \ll 1$$

along with the usual requirement for the ordinary perturbation expansion

$$\lambda \ll 1$$

However at the critical temperature we have that $T_c \sim M/\sqrt{\lambda}$ [see *e.g.* eqs. (76)-(77)]. Therefore **at the critical temperature the one-loop approximation is not valid** and higher loop diagrams where multiple quadratically divergent bubbles are inserted cannot be neglected. Daisy resummation [45] consists precisely to resum all powers of α and provides a theory where $m^2(\phi_c) \to m^2_{\text{eff}} \equiv m^2(\phi_c) + \Pi$, where Π is the self-energy corresponding to the one-loop resummed diagrams to leading order in powers of the temperature T (e.g. $\sim T^2$). This method was pursued systematically by Parwani [46] and applied to the Standard Model by Arnold and Espinosa [47].

What about the diagrams which are not considered in the **improved** expansion? The two-loop diagram of Fig. 2 is suppressed with respect to the diagram of Fig. 1 by λ. On the other hand the multiple loop diagram obtained from that of Fig. 2 by adding n quadratically divergent loops on top of it, see Fig. 4, behaves as

---

[6] In fact, the mass M has a different meaning for the improved and the unimproved theories, as we shall see. For the unimproved theory, M is the mass in the shifted lagrangian, $M^2 = m^2(\phi)$, while in the improved theory, M is given by the thermal mass.

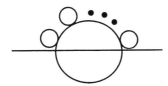

FIGURE 4. Non-daisy (n+2)-loop contribution to the self-energy for the scalar theory

$$\lambda^{n+2} \frac{T^{2n+2}}{M^{2n}} = \lambda^{n+1} \frac{T^{2n+1}}{M^{2n-1}} \left(\lambda \frac{T}{M} \right) \qquad (133)$$

and it is suppressed with respect to the multiple loop diagram of Eq. (131) by $\lambda T/M$ [7]. Therefore the validity of the improved expansion is guaranteed provided that,

$$\lambda \ll 1 \qquad (134)$$
$$\beta \equiv \lambda \frac{T}{M} \ll 1$$

This simplified discussion can be done of course in more complicated field theories, as e.g. the Standard Model, with similar arguments [47]. We will now apply these results to two different models.

The scalar theory

In the case just considered of the scalar theory, the effective potential can be written in the one-loop daisy resummed approximation as

$$V(\phi, T) = \frac{1}{2} m_{\text{eff}}^2 \phi^2 - \frac{1}{12\pi} \left(m_{\text{eff}}^2 + \frac{1}{2} \lambda \phi^2 \right)^{3/2} + \frac{\lambda}{4!} \phi^4 + \cdots \qquad (135)$$

where $m_{\text{eff}}^2 \equiv m^2 + \frac{\lambda'}{24} T^2$.

The potential (135) apparently yields a first order phase transition [48]. However, the symmetry breaking minimum occurs when the three terms in the potential are similar. Then, at $T \sim T_c$, at the minimum

$$\phi \sim \sqrt{\lambda} T, \quad m_{\text{eff}}^2 \sim \lambda^2 T^2 \qquad (136)$$

and then,

$$\beta \sim \lambda \frac{T}{m_{\text{eff}}} \sim \lambda \frac{T}{\lambda T} = \mathcal{O}(1) \qquad (137)$$

which shows that the result of perturbation theory is fake [47].

---

[7] Non-daisy contributions to the self-energy are suppressed, with respect to daisy contributions, by $\mathcal{O}(\beta)$, where β is defined in (134). The corresponding contributions to the vacuum diagrams (i.e. effective potential) are suppressed by $\mathcal{O}(\beta^2)$ [44].

The Standard Model

In cases where there are gauge and/or Yukawa interactions, as the Standard Model case, the situation is completely different. The effective potential in the one-loop improved approximation can be written as

$$V(\phi, T) = \frac{1}{2} m_{\text{eff}}^2 \phi^2 - bg^3 \phi^3 T + \frac{\lambda}{4!} \phi^4 + \cdots \quad (138)$$

where

$$bg^3 \phi^3 T \equiv \left[4 M_W^3(\phi) + 2 M_Z^3(\phi)\right] \frac{T}{12\pi} \equiv \left[\frac{1}{2} g^3 + (g^2 + g'^2)^{3/2}\right] \frac{T}{12\pi} \phi^3$$

and $m_{\text{eff}}^2 = m^2 + ag^2 T^2$, ag^2 denoting the contribution of gauge and Yukawa couplings to the one loop self-energy. We have only considered the contribution of transverse gauge bosons to the phase transition strength, and neglected that of the (screened) longitudinal gauge bosons.

Now again, symmetry breaking occurs when all terms are similar,

$$\phi \sim \frac{g^3}{\lambda} T, \quad m_{\text{eff}}^2 \sim \frac{g^6}{\lambda} T^2 \quad (139)$$

and the expansion parameter β,

$$\beta_{\text{SM}} \sim g^2 \frac{T}{M_W(\phi)} \sim g \frac{T}{\phi} \sim \frac{\lambda}{g^2} \sim \frac{m_H^2}{m_W^2} \quad (140)$$

Therefore the validity of perturbation theory implies that $m_H \lesssim m_W$. This behaviour is supported by non-perturbative calculations [49].

STANDARD MODEL RESULTS

The effective potential for the Standard Model was analyzed in Eq. (138) in the one-loop approximation, including leading order plasma effects. In this approximation, the longitudinal components of the gauge bosons are screened by plasma effects while the transverse components remain unscreened. In this way a good approximation to the effective potential including these plasma effects is provided by Eq. (80), where the coefficient E is given by

$$E_{\text{SM}} = \frac{2}{3} \frac{2 m_W^3 + m_Z^3}{4\pi v^3} \sim 9.5 \times 10^{-3} \quad (141)$$

Now we can use Eq. (87) and $m_h^2 = 2\lambda v^2$ to write,

$$\frac{\phi(T_c)}{T_c} \sim \frac{4 E v^2}{m_h^2} \quad (142)$$

In this way the bound (129) translates into the bound on the Higgs mass,

$$m_h \lesssim \sqrt{\frac{4E}{1.3}} \sim 42 \ GeV. \tag{143}$$

The bound (143) is excluded by LEP measurements [28], and so the Standard Model is unable to keep any previously generated baryon asymmetry. Including two-loop effects the bound is slightly increased to ~ 45 GeV [47]. Is it possible, in extensions of the Standard Model, to overcome this difficulty? We will see in the next sections one typical example where the Standard Model is extended: the well motivated supersymmetric extension of the Standard Model.

MSSM RESULTS

Among the extensions of the Standard Model, the physically most motivated and phenomenologically most acceptable one is the Minimal Supersymmetric Standard Model. This model allows for extra CP-violating phases besides the Kobayashi-Maskawa one, which could help in generating the observed baryon asymmetry [50]. It is then interesting to study whether in the MSSM the nature of the phase transition can be significantly modified with respect to the Standard Model.

In this section we extend the considerations of the previous section to the full parameter space, characterizing the Higgs sector of the MSSM [51-53]. The main tool for our study is the one-loop, daisy-improved finite-temperature effective potential of the MSSM, $V_{\text{eff}}(\phi, T)$. We are actually interested in the dependence of the potential on $\phi_1 \equiv \text{Re } H_1^0$ and $\phi_2 \equiv \text{Re } H_2^0$ only, where H_1^0 and H_2^0 are the neutral components of the Higgs doublets H_1 and H_2, thus ϕ will stand for (ϕ_1, ϕ_2). Working in the 't Hooft-Landau gauge and in the $\overline{DR}$-scheme, we can write

$$V_{\text{eff}}(\phi, T) = V_0(\phi) + V_1(\phi, 0) + \Delta V_1(\phi, T) + \Delta V_{\text{daisy}}(\phi, T) + V_2(\phi, T), \tag{144}$$

where

$$V_0(\phi) = m_1^2 \phi_1^2 + m_2^2 \phi_2^2 + 2m_3^2 \phi_1 \phi_2 + \frac{g^2 + g'^2}{8}(\phi_1^2 - \phi_2^2)^2, \tag{145}$$

$$V_1(\phi, 0) = \sum_i \frac{n_i}{64\pi^2} m_i^4(\phi) \left[\log \frac{m_i^2(\phi)}{Q^2} - \frac{3}{2}\right], \tag{146}$$

$$\Delta V_1(\phi, T) = \frac{T^4}{2\pi^2} \left\{ \sum_i n_i J_i \left[\frac{m_i^2(\phi)}{T^2}\right] \right\}, \tag{147}$$

$$\Delta V_{\text{daisy}}(\phi, T) = -\frac{T}{12\pi} \sum_{i=\text{bosons}} n_i \left[\mathcal{M}_i^3(\phi, T) - m_i^3(\phi)\right]. \tag{148}$$

The masses $\mathcal{M}_i^2(\phi, T)$ are obtained from the $m_i^2(\phi)$ by adding the leading T-dependent self-energy contributions, which are proportional to T^2. We recall that, in the gauge boson sector, only the longitudinal components (W_L, Z_L, γ_L) receive such contributions.

The relevant degrees of freedom for our calculation are: $n_t = -12$, $n_{\tilde{t}_1} = n_{\tilde{t}_2} = 6$, $n_W = 6$, $n_Z = 3$, $n_{W_L} = 2$, $n_{Z_L} = n_{\gamma_L} = 1$. The field-dependent top mass is $m_t^2(\phi) = h_t^2 \phi_2^2$. The entries of the field-dependent stop mass matrix are

$$m_{\tilde{t}_L}^2(\phi) = m_{Q_3}^2 + m_t^2(\phi) + D_{\tilde{t}_L}^2(\phi), \tag{149}$$

$$m_{\tilde{t}_R}^2(\phi) = m_{U_3}^2 + m_t^2(\phi) + D_{\tilde{t}_R}^2(\phi), \tag{150}$$

$$m_X^2(\phi) = h_t(A_t \phi_2 - \mu \phi_1), \tag{151}$$

where m_Q, m_U and A_t are soft supersymmetry-breaking mass parameters, μ is a superpotential Higgs mass term, and

$$D_{\tilde{t}_L}^2(\phi) = \left(\frac{1}{2} - \frac{2}{3}\sin^2\theta_W\right) \frac{g^2 + g'^2}{2} (\phi_1^2 - \phi_2^2), \tag{152}$$

$$D_{\tilde{t}_R}^2(\phi) = \left(\frac{2}{3}\sin^2\theta_W\right) \frac{g^2 + g'^2}{2} (\phi_1^2 - \phi_2^2) \tag{153}$$

are the D-term contributions. The field-dependent stop masses are then

$$m_{\tilde{t}_{1,2}}^2(\phi) = \frac{m_{\tilde{t}_L}^2(\phi) + m_{\tilde{t}_R}^2(\phi)}{2} \pm \sqrt{\left[\frac{m_{\tilde{t}_L}^2(\phi) - m_{\tilde{t}_R}^2(\phi)}{2}\right]^2 + [m_X^2(\phi)]^2}. \tag{154}$$

The corresponding effective T-dependent masses, $\mathcal{M}_{\tilde{t}_{1,2}}^2(\phi, T)$, are given by expressions identical to (154), apart from the replacement

$$m_{\tilde{t}_{L,R}}^2(\phi) \to \mathcal{M}_{\tilde{t}_{L,R}}^2(\phi, T) \equiv m_{\tilde{t}_{L,R}}^2(\phi) + \Pi_{\tilde{t}_{L,R}}(T). \tag{155}$$

The $\Pi_{\tilde{t}_{L,R}}(T)$ are the leading parts of the T-dependent self-energies of $\tilde{t}_{L,R}$,

$$\Pi_{\tilde{t}_L}(T) = \frac{4}{9} g_s^2 T^2 + \frac{1}{4} g^2 T^2 + \frac{1}{108} g'^2 T^2 + \frac{1}{6} h_t^2 T^2, \tag{156}$$

$$\Pi_{\tilde{t}_R}(T) = \frac{4}{9} g_s^2 T^2 + \frac{4}{27} g'^2 T^2 + \frac{1}{3} h_t^2 T^2, \tag{157}$$

where g_s is the strong gauge coupling constant. Only loops of gauge bosons, Higgs bosons and third generation squarks have been included, implicitly assuming that all remaining supersymmetric particles are heavy and decouple.

We shall work in the limit in which the left handed stop is heavy, $m_Q \gtrsim 500$ GeV. In this limit, the supersymmetric corrections to the precision electroweak parameter $\Delta\rho$ become small and hence, this allows a good fit to the electroweak precision data coming from LEP and SLD. Lower values of m_Q make the phase transition stronger and we are hence taking a conservative assumption from the point of view of defining the region consistent with electroweak baryogenesis. The left handed stop decouples at finite temperature, but, at zero temperature, it sets the scale of the Higgs masses as a function of $\tan\beta$. For right-handed stop masses below, or of order of, the top quark mass, and for large values of the CP-odd Higgs mass, $m_A \gg M_Z$, the one-loop improved Higgs effective potential admits a high temperature expansion,

$$V_0 + V_1 = -\frac{m^2(T)}{2}\phi^2 - T\left[E_{\rm SM}\phi^3 + (2N_c)\frac{\left(m_{\tilde{t}}^2 + \Pi_{\tilde{t}_R}(T)\right)^{3/2}}{12\pi}\right] + \frac{\lambda(T)}{8}\phi^4 + \cdots \tag{158}$$

where $N_c = 3$ is the number of colours and $E_{\rm SM}$ is the cubic term coefficient in the Standard Model case.

Within our approximation, the lightest stop mass is approximately given by

$$m_{\tilde{t}}^2 \simeq m_U^2 + 0.15 M_Z^2 \cos 2\beta + m_t^2\left(1 - \frac{\widetilde{A}_t^2}{m_Q^2}\right) \tag{159}$$

where $\widetilde{A}_t = A_t - \mu/\tan\beta$ is the stop mixing parameter. As was observed in Ref. [54], the phase transition strength is maximized for values of the soft breaking parameter $m_U^2 = -\Pi_{\tilde{t}_R}(T)$, for which the coefficient of the cubic term in the effective potential,

$$E \simeq E_{\rm SM} + \frac{h_t^3 \sin^3\beta \left(1 - \widetilde{A}_t^2/m_Q^2\right)^{3/2}}{4\sqrt{2}\pi}, \tag{160}$$

can be one order of magnitude larger than $E_{\rm SM}$ [54]. In principle, the above would allow a sufficiently strong first order phase transition for Higgs masses as large as 100 GeV. However, it was also noticed that such large negative values of m_U^2 may induce the presence of color breaking minima at zero or finite temperature [54,55]. Demanding the absence of such dangerous minima, the one loop analysis leads to an upper bound on the lightest CP-even Higgs mass of order 80 GeV. This bound was obtained for values of $\widetilde{m}_U^2 = -m_U^2$ of order $(80 \text{ GeV})^2$.

The most important two loop corrections are of the form $\phi^2 \log(\phi)$ and, as said above, are induced by the Standard Model weak gauge bosons as well as by the stop and gluon loops [56]. It was recently noticed that the coefficient of these terms can be efficiently obtained by the study of the three dimensional running mass of the scalar top and Higgs fields in the dimensionally reduced theory at high temperatures [57]. Equivalently, in a four dimensional computation of the MSSM Higgs effective potential with a heavy left-handed stop, we obtain [54]

$$V_2 \simeq \frac{\phi^2 T^2}{32\pi^2}\left[\frac{51}{16}g^2 - 3\left[h_t^2 \sin\beta^2\left(1 - \frac{\widetilde{A}_t^2}{m_Q^2}\right)\right]^2 + 8g_s^2 h_t^2 \sin^2\beta\left(1 - \frac{\widetilde{A}_t^2}{m_Q^2}\right)\right]\log\frac{\Lambda_H}{\phi} \tag{161}$$

where the first term comes from the Standard Model gauge boson-loop contributions, while the second and third terms come from the light supersymmetric particle loop contributions. The scale Λ_H depends on the finite corrections, which may be obtained by the expressions given in [58]. As mentioned above, the two-loop corrections are very important and, as has been shown in Ref. [56], they can make the phase transition strongly first order even for $m_U \simeq 0$ [58]. Concerning the validity of the perturbative expansion, the β-parameter, similarly to β_{SM}, (140), can be shown to be given by

$$\beta_{\text{MSSM}} \sim \frac{m_h^2}{m_t^2}, \qquad (162)$$

which leads to a reliable perturbative expansion for a value of the Higgs mass enhanced, with respect to its Standard Model value, by a factor $\sim (m_t/m_h)^2$.

An analogous situation occurs in the U-direction ($U \equiv \widetilde{t}_R$). The one-loop expression is approximately given by

$$V_0(U) + V_1(U,T) = \left(-\widetilde{m}_U^2 + \gamma_U T^2\right) U^2 - T E_U U^3 + \frac{\lambda_U}{2} U^4, \qquad (163)$$

where γ_U and E_U were given in [58].

Analogous to the case of the field ϕ, the two loop corrections to the U-potential are dominated by gluon and stop loops and are approximately given by

$$V_2(U,T) = \frac{U^2 T^2}{16\pi^2} \left[\frac{100}{9} g_s^4 - 2 h_t^2 \sin^2\beta \left(1 - \frac{\widetilde{A}_t^2}{m_Q^2}\right) \right] \log\left(\frac{\Lambda_U}{U}\right) \qquad (164)$$

where, as in the Higgs case, the scale Λ_U may only be obtained after the finite corrections to the effective potential are computed [58].

Once the effective potential in the ϕ and U directions are computed, one can study the strength of the electroweak phase transition, as well as the presence of potential color breaking minima. At one-loop, it was observed that requiring the stability of the physical vacuum at zero temperature was enough to assure the absolute stability of the potential at finite temperature. As has been first noticed in Ref. [57], once two loop corrections are included, the situation is more complicated [58].

At zero temperature the minimization of the effective potential for the fields ϕ and U shows that the true minima are located for vanishing values of one of the two fields. The two set of minima are connected through a family of saddle points for which both fields acquire non-vanishing values. Due to the nature of the high temperature corrections, we do not expect a modification of this conclusion at finite temperature.

Two parameters control the presence of color breaking minima: $\widetilde{m}_U^c$, defined as the smallest value of $\widetilde{m}_U$ for which a color breaking minimum deeper than the electroweak breaking minimum is present at $T = 0$, and T_c^U, the critical temperature for the transition into a color breaking minimum in the U-direction. The value of $\widetilde{m}_U^c$ may be obtained by analysing the effective potential for the field U at zero temperature, and it is approximately given by [54]

$$\widetilde{m}_U^c \simeq \left(\frac{m_H^2 \, v^2 \, g_s^2}{12}\right)^{1/4}. \tag{165}$$

Defining the critical temperature as that at which the potential at the symmetry preserving and broken minima are degenerate, four situations can happen in the comparison of the critical temperatures along the ϕ (T_c) and U (T_c^U) transitions: **a)** $T_c^U < T_c$; $\widetilde{m}_U < \widetilde{m}_U^c$; **b)** $T_c^U < T_c$; $\widetilde{m}_U > \widetilde{m}_U^c$; **c)** $T_c^U > T_c$; $\widetilde{m}_U < \widetilde{m}_U^c$; **d)** $T_c^U > T_c$; $\widetilde{m}_U > \widetilde{m}_U^c$.

In case a), as the universe cools down, a phase transition into a color preserving minimum occurs, which remains stable until $T = 0$. This situation, of absolute stability of the physical vacuum, is the most conservative requirement to obtain electroweak baryogenesis. In case b), at $T = 0$ the color breaking minimum is deeper than the physical one implying that the color preserving minimum becomes unstable for finite values of the temperature, with $T < T_c$. A physically acceptable situation may only occur if the lifetime of the physical vacuum is smaller than the age of the universe. We shall denote this situation as "metastability". In case c), as the universe cools down, a color breaking minimum develops which, however, becomes metastable as the temperature approaches zero. A physically acceptable situation can only take place if a two step phase transition occurs, that is if the color breaking minimum has a lifetime lower than the age of the universe at some temperature $T < T_c$ [57]. Finally, in case d) the color breaking minimum is absolutely stable and hence, the situation becomes physically unacceptable.

Figure 5 shows the region of parameter space consistent with a sufficiently strong phase transition for $\widetilde{A}_t = 0, 200, 300$ GeV. For low values of the mixing, $\widetilde{A}_t \lesssim 200$ GeV, case a) or c) may occur but, contrary to what happens at one-loop, case b) is not realized. For the case of no mixing, this result is in agreement with the analysis of [57]. The region of absolute stability of the physical vacuum for $\widetilde{A}_t \simeq 0$ is bounded to values of the Higgs mass of order 95 GeV. There is a small region at the right of the solid line, in which a two-step phase transition may take place, for values of the parameters which would lead to $v/T < 1$ for $T = T_c$, but may evolve to larger values at some $T < T_c$ at which the second of the two step phase transition into the physical vacuum takes place. This region disappears for larger values of the stop mixing mass parameter. For values of the mixing parameter $\widetilde{A}_t$ between 200 GeV and 300 GeV, both situations, cases b) and c) may occur, depending on the value of $\tan\beta$. For large values of the stop mixing, $\widetilde{A}_t > 300$ GeV, a two-step phase transition does not take place.

All together, and even demanding absolute stability of the physical vacuum, electroweak baryogenesis seems to work for a wide region of Higgs and stop mass values. Higgs masses between the present experimental limit, of about 85 GeV [59], and around 105 GeV are consistent with this scenario. Similarly, the running stop

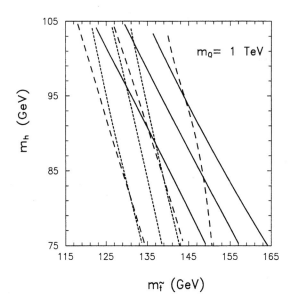

FIGURE 5. Values of m_h, $m_{\tilde{t}}$ for which $v(T_c)/T_c = 1$ (solid line), $T_c^U = T_c$ (dashed line), $\tilde{m}_U = \tilde{m}_U^c$ (short-dashed line), for $m_Q = 1$ TeV and $\widetilde{A}_t = 0, 200, 300$ GeV. The region on the left of the solid line is consistent with a strongly first order phase transition. A two step phase transition may occur in the regions on the left of the dashed line, while on the left of the short-dashed line, the physical vacuum at $T = 0$ becomes metastable. The region on the left of both the dashed and short-dashed lines leads to a stable color breaking vacuum state at zero temperature and is hence physically unacceptable.

mass may vary from values of order 165 GeV (of the same order as the top quark mass one) and 100 GeV. Observe that, due to the influence of the D-terms, values of $m_{\tilde{t}} \simeq 165$ GeV, $\widetilde{A}_t \simeq 0$ and $m_h \simeq 75$ GeV, are achieved for small positive values of m_U. Also observe that for lower values of m_Q the phase transition may become more strongly first order and slightly larger values of the stop masses may be obtained. Remind that these results are based on the two-loop improved effective potential. Recent non-perturbative calculations [60] confirm the validity of these perturbative results.

REFERENCES

1. L. Dolan and R. Jackiw, *Phys. Rev.* **D9** (1974) 3320.
2. S. Weinberg, *Phys. Rev.* **D9** (1974) 3357.
3. R.H. Brandenberger, *Rev. Mod. Phys.* **57** (1985) 1.
4. N.P. Landsman and Ch.G. van Weert, *Phys. Rep.* **145** (1987) 141.
5. M. Quirós, *Helv. Phys. Acta* **67** (1994) 451
6. J.I. Kapusta, *Finite-temperature field theory* (Cambridge University Press, 1989).

7. R. Kubo, *J. Phys. Soc. Japan* **12** (1957) 570; P.C. Martin and J. Schwinger, *Phys. Rev.* **115** (1959) 1342.
8. T. Matsubara, *Prog. Theor. Phys.* **14** (1955) 351.
9. H. Matsumoto, Y. Nakano, H. Umezawa, F. Mancini and M. Marinaro, *Prog. Theor. Phys.* **70** (1983) 599; H. Matsumoto, Y. Nakano and H. Umezawa, *J. Math. Phys.* **25** (1984) 3076.
10. L.V. Keldish, *Sov. Phys. JETP* **20** (1964) 1018.
11. T.S. Evans, *Phys. Rev.* **D47** (1993) R4196; T. Altherr, CERN preprint, CERN-TH.6942/93.
12. R. Kobes, *Phys. Rev.* **D42** (1990) 562 and *Phys. Rev. Lett.* **67** (1991) 1384; T.S. Evans, *Phys. Lett.* **B249** (1990) 286, *Phys. Lett.* **B252** (1990) 108, *Nucl. Phys.* **B371** (1992) 340; P. Aurenche and T. Becherrawy, *Nucl. Phys.* **B379** (1992) 259; M.A. van Eijck and Ch. G. van Weert, *Phys. Lett.* **B278** (1992) 305.
13. D.A. Kirzhnits, *JETP Lett.* **15** (1972) 529.
14. D.A. Kirzhnits and A.D. Linde, *Phys. Lett.* **42B** (1972) 471; D.A. Kirzhnits and A.D. Linde, *JETP* **40** (1974) 628; D.A. Kirzhnits and A.D. Linde, *Ann. Phys.* **101** (1976) 195.
15. A.D. Linde, *Rep. Prog. Phys.* **42** (1979) 389; A.D. Linde, *Phys. Lett.* **99B** (1981) 391; A.D. Linde, *Phys. Lett.* **99B** (1981) 391; A.D. Linde, *Particle Physics and Inflationary Cosmology* (Harwood, Chur, Switzerland, 1990).
16. A.D. Linde, *Phys. Lett.* **B108** (1982) 389; A. Albrecht and P.J. Steinhardt, *Phys. Rev. Lett.* **48** (1082) 1226.
17. G.W. Anderson and L.J. Hall, *Phys. Rev.* **D45** (1992) 2685.
18. M. Dine, R.G. Leigh, P. Huet, A. Linde and D. Linde, *Phys. Lett.* **B283** (1992) 319; *Phys. Rev.* **D46** (1992) 550.
19. S. Coleman, *Phys. Rev.* **D15** (1977) 2929.
20. C.G. Callan and S. Coleman, *Phys. Rev.* **D16** (1977) 1762.
21. S. Coleman and F. De Luccia, *Phys. Rev.* **D21** (1980) 3305.
22. A.D. Linde, *Phys. Lett.* **70B** (1977) 306; **100B** (1981) 37; *Nucl. Phys.* **B216** (1983) 421.
23. L. McLerran, M. Shaposhnikov, N. Turok and M. Voloshin, *Phys. Lett.* **B256** (1991) 451.
24. M. Dine, P. Huet and R. Singleton Jr., *Nucl. Phys.* **B375** (1992) 625.
25. J.M. Moreno, M. Quirós and M. Seco, [hep-ph/9801272], *Nucl. Phys.* **B** to appear.
26. E.W. Kolb and M.S. Turner, *The Early Universe* (Addison-Wesley, 1990).
27. A.D. Sakharov, *Zh. Eksp. Teor. Fiz. Pis'ma* **5** (1967) 32; *JETP Lett.* **91B** (1967) 24.
28. Particle Data Group, Review of Particle Properties, *Phys. Rev.* **D54** (1996) 1.
29. E.W. Kolb and S. Wolfram, *Nucl. Phys.* **B172** (1980) 224; *Phys. Lett.* **B91** (1980) 217.
30. V.A. Kuzmin, V.A. Rubakov and M.E. Shaposhnikov, *Phys. Lett.* **B155** (1985) 36; *Phys. Lett.* **B191** (1987) 171.
31. A.G. Cohen, D.B. Kaplan and A.E. Nelson, *Annu. Rev. Nucl. Part. Sci.* **43** (1993) 27.
32. G. t' Hooft, *Phys. Rev. Lett.* **37** (1976) 8; *Phys. Rev.* **D14** (1976) 3432.

33. M. Kobayashi and M. Maskawa, *Prog. Theor. Phys.* **49** (1973) 652.
34. N.S. Manton, *Phys. Rev.* **D28** (1983) 2019; F.R. Klinkhamer and N.S. Manton, *Phys. Rev.* **D30** (1984) 2212.
35. J. Kunz, B. Kleihaus and Y. Brihaye, *Phys. Rev.* **D46** (1992) 3587.
36. Y. Brihaye and J. Kunz, *Phys. Rev.* **D48** (1993) 3884.
37. J.M. Moreno, D.H. Oaknin and M. Quirós, [hep-ph/9605387], *Nucl. Phys.* **B483** (1997) 267.
38. P. Arnold and L. McLerran, *Phys. Rev.* **D36** (1987) 581; S.Yu. Khlebnikov and M.E. Shaposhnikov, *Nucl. Phys.* **B308** (1988) 885.
39. J. Ambjorn, M. Laursen and M. Shaposhnikov, *Phys. Lett.* **B197** (1989)49; J. Ambjorn, T. Askaard, H. Porter and M. Shaposhnikov, *Phys. Lett.* **B244** (1990) 479; *Nucl. Phys.* **B353** (1991) 346; J. Ambjorn and K. Farakos, *Phys. Lett.* **B294** (1992) 248; J. Ambjorn and A. Krasnitz, *Phys. Lett.* **B362** (1995) 97; P. Arnold, D. Son and L.G. Yaffe,*Phys. Rev.* **D55** (1997) 6264; G.D. Moore, [hep-ph/9805264].
40. L. Carson, Xu Li, L. McLerran and R.-T. Wang, *Phys. Rev.* **D42** (1990) 2127.
41. M.E. Shaposhnikov, *Nucl. Phys.* **B287** (1987) 757; *Nucl. Phys.* **B299** (1988) 797; A.I. Bochkarev and M.E. Shaposhnikov, *Mod. Phys. Lett* **A2** (1987) 417.
42. A.I. Bochkarev, S.V. Kuzmin and M.E. Shaposhnikov, *Phys. Rev.* **D43** (1991) 369.
43. P. Fendley, *Phys. Lett.* **B196** (1987) 175.
44. J.R. Espinosa, M. Quirós and F. Zwirner, *Phys. Lett.* **B291** (1992) 115.
45. D.J. Gross, R.D. Pisarski and L.G. Yaffe, Rev. Mod. Phys. **53** (1981) 43.
46. R.R. Parwani, Phys. Rev. **D45** (1992) 4695.
47. P. Arnold and O. Espinosa, *Phys. Rev.* **D47** (1993) 3546.
48. M.E. Carrington, Phys. Rev. **D45** (1992) 2933.
49. K. Jansen, *Nucl. Phys. (Proc. Supl.)* **B47** (1996) 196 [hep-lat/9509018].
50. A.G. Cohen and A.E. Nelson, *Phys. Lett.* **B297 (1992)** 111.
51. J.R. Espinosa, M. Quirós and F. Zwirner, *Phys. Lett.* bf B307 (1993) 106.
52. A. Brignole, J.R. Espinosa, M. Quirós and F. Zwirner, *Phys. Lett.* bf B324 (1994) 181.
53. G.F. Giudice, *Phys. Rev.* **D45** (1992) 3177; S. Myint, *Phys. Lett.* **B287** (1992) 325.
54. M. Carena, M. Quiros and C.E.M. Wagner, *Phys. Lett.* **B380** (1996) 81
55. M. Carena and C.E.M. Wagner, *Nucl. Phys.* **B452** (1995) 45
56. J.R. Espinosa, *Nucl. Phys.* **B475** (1996) 273; B. de Carlos and J.R. Espinosa, [hep-ph/9703212], *Nucl. Phys.* **B503** (1997) 24.
57. D. Bodeker, P. John, M. Laine and M.G. Schmidt, *Nucl. Phys.* **B497** (1997) 387
58. M. Carena, M. Quirós and C.E.M. Wagner, [hep-ph/9710401] *Nucl. Phys.* **B**, to appear.
59. P. Janot, talk given at the International Europhysics Conference on High Energy Physics, Jerusalem, August 1997.
60. M. Laine and K. Rummukainen, hep-ph/9804255; hep-lat/9804019

The Highest Energy Cosmic Rays, Photons and Neutrinos

Enrique Zas

*Departamento de Física de Partículas,
Universidade de Santiago de Compostela, E-15706 Santiago, Spain.*

Abstract. In these lectures I introduce and discuss aspects of currently active fields of interest related to the production, transport and detection of high energy particles from extraterrestrial sources. I have payed most attention to the highest energies and I have divided the material according to the types of particles which will be searched for with different experimental facilities in planning: hadrons, gamma rays and neutrinos. Particular attention is given to shower development, stochastic acceleration and detection techniques.

INTRODUCTION

The commonly used term "Cosmic Ray" was introduced earlier in this century when it was found that a substantial part of the recently discovered natural radioactivity was actually due to penetrating radiation coming from outer space. Cosmic rays became the source the highest energy particles that could be observed and the study of their interactions marked the onset of developments that have dominated fundamental physics research during the whole century. Later on in the fifties cosmic rays were replaced by particles from man made accelerators allowing a more systematic study of the High Energy interactions. "Astroparticle Physics" is the modern term to describe a wide branch of physics dedicated to the study of all particle radiation directly or indirectly due to extraterrestrial phenomena. Their interactions and their detection provide an important link to particle physics while their production and transport are also quite generally related to astrophysics and cosmology.

The majority of the ionizing particles detected in the early days of cosmic rays are now known to be due to nuclei that bombard the top of the Earth's atmosphere from all directions at a rate of about one thousand particles per m^2 per second and per stereoradian [1–5]. Because they are ionized they wander in the magnetic fields associated to the plasma in the interstellar and intergalactic media, loosing directional information of their production site. Their concrete composition is one of the important clues towards their understanding. The nuclei are predominantly protons but there are also heavier elements in proportions suggestive of solar system abundances. At ordinary observation levels on the Earth's surface the atmosphere acts as a shield and only allows the detection of the secondary particles produced in their interactions with atmospheric nuclei. These secondary particles are mostly electrons, photons, muons and neutrinos.

We now maintain the term cosmic rays to refer to the primary nuclei that reach the atmosphere but they are by no means the only particles that come to us from outside, not even the most numerous. Basically there are also incident fluxes of all stable particles, charged particles such as electrons and positrons that have also lost directional information and have fluxes a small fraction of the cosmic ray flux, and neutral particles such as photons and neutrinos that point to their origin. From the early days of mankind photons have been the object of Astronomy which now spans over 22 orders of magnitude in wavelength, from the kilometer scale in radio astronomy to below 10^{-19} m for high energy gamma rays (photons above 10 TeV). On the other hand the detection of neutrinos from the Sun [6] and from supernova SN1987A [7] has recently paved the way to neutrino astronomy, a formidable challenge requiring the instrumentation of enormous detector volumes.

These particles come in energy spectra which typically span tens of decades of energy. Their differential energy spectra is highly non thermal, typically characterized by a power law $E^{-\gamma}$. Cosmic rays have provided the highest energy particles to be observed in nature, always higher than those obtained by artificial acceleration, and it is likely that it will continue to be so for still quite a few generations to come. Now they are known to reach over 10^{20} eV. This is certainly one of the most interesting aspects of the study of such particles. High energy photons are included in astroparticle physics, partly because of the borrowed detection techniques and partly because of their intimate relation to cosmic rays and neutrinos as will be discussed below.

In these lectures I will discuss hadrons, photons and neutrinos. The discussion is by no means attempting to be complete, the subject is vast and rapidly changing and it is impossible to do justice to all the subfields. Most recent developments are due to the success of new detection techniques; for detecting high energy photons and hadrons these techniques inevitably rely on detecting air showers. Shower development, in air or in denser media such as water, is also likely to be relevant for the detecion of the highest energy neutrinos, and I will discuss it in conection to detection techniques in quite some detail. There are projects in development to extend detection of hadrons, photons and neutrinos to higher energies where new discoveries must take place in the near future. For this reason I will stress the prospects for detection of the highest end of the particle spectra, trying to address relations between the fluxes of the different particle species. There will inevitably be many omissions for which I apologize in advance. Many of the topics addressed here can be found in more detail in excellent books on cosmic rays and Astroparticle Physics such as Refs. [2-4], or on High Energy Astrophysics such as Ref. [5], and many specific lectures and reviews on more specific topics, for instance on cosmic ray acceleration [8], gamma ray astronomy [9] and high energy neutrinos [10-12].

I have benefited from reading these and many of the treatments given here are largely inspired in their approaches to the discussed topics.

COSMIC RAYS: NUCLEI

Most of the incident cosmic rays on the Earth are protons of energy in the GeV range. The rate going through a 10 cm² surface is about one per second. It has been measured in ballon flights and with particle detectors in satellites. The most remarkable characteristic of these cosmic rays is actually their featureless energy spectrum, basically described by a power law over more than ten orders of magnitude in energy:

$$\phi_p(E_n) = \frac{d^4 N_p}{dE_n \, dA \, dt \, d\Omega} = B \left[\frac{E_n}{1 \text{ GeV}}\right]^{-\gamma} \quad (1)$$

Here the *spectral index* γ is close to 2.7 for energies up to the 10^{15} eV region changing to about 3 above this energy. The flux of nuclei is given per unit area A, time t, solid angle Ω and kinetic energy per nucleon of the incident particle E_n. Fig. 1 illustrates the integral of this flux for energies above a given threshold. The spectral index change at about 10^6 GeV, which is not always easy to appreciate in a logarithmic plot, is usually referred as *the knee*. The normalization constant for protons $B = \phi_p(1 \text{ GeV})$ has the value 1 cm^{-2}s^{-1}sr^{-1}GeV^{-1} for protons in the region between 10 GeV and 1000 TeV with $\gamma = 2.7$. The spectra of other nuclei essentially follow the proton spectrum with some different relative normalization. For Helium the α particle flux can be approximated by $B_{He} = \phi_{He}(1 \text{ GeV}) = 0.036$ cm^{-2}s^{-1}sr^{-1}(GeV/nucleon)$^{-1}$. Similar curves exist for heavier nuclei which are less abundant and some of them do show some deviations from the 2.7 proton spectral index.

The relative abundances of these cosmic rays are overall quite similar to solar system abundances, displaying abundance peaks of Carbon, Oxygen, Nitrogen and Iron with respect to other elements of similar atomic number and an even versus odd atomic number nuclei enhancement (*even-odd effect*). However the cosmic ray abundances vary less drastically. The even-odd effect is mitigated and the cosmic ray abundances of the heavier elements are closer to that of hydrogen than in he solar system. For some elements, such as Boron, Beryllium, Lithium and those just below the Iron peak, their cosmic ray abundances with respect to protons are strikingly larger (over 7 orders of magnitude for Beryllium) than in the solar system where they are particularly sparce. In their propagation, the more abundant Carbon, Nitrogen, Oxygen and Iron group elements produce lower atomic number elements in low energy interactions with the interstellar medium referred as spallation processes. The amount of matter traversed on average by cosmic rays to produce the observed low Z element abundances is induced to be about 5-10 g cm^{-2} which implies a lower bound for the time of residence in the galaxy of about 3-6 million years, assuming a density of 1 proton per cubic centimeter of interstellar matter. During this time the cosmic rays wander in the magnetic field of the Galaxy in a diffusive like process since a light ray only needs 100,000 years to cross the Galaxy along a diameter.

Satellite detection becomes impossible at energies above the few hundred TeV range because of the steep spectrum. Above this energy the atmosphere can fortunately be used as a calorimeter by placing different types of detectors on the Earth's surface. This is possible because the high energy particles produce many secondaries in the atmosphere which travel almost parallel and close to the speed of light constituting an Extensive Air Shower. The observation of these showers by a variety of methods has allowed the study of cosmic rays up to energies exceeding 10^{11} GeV. The spectrum at highest energies displays some interesting subtle features, a change of spectral index at about 10^{18} eV and above 10^{20} there are hints of yet another mild slope change. Unfortunately the statistics is poor at these energies.

A steep power law is not only characteristic of cosmic rays. Byproducts of these cosmic rays such as photons and neutrinos have also steep power law spectra as well as gamma rays from a number of other sources. The overall spectral shape is very characteristic of this field, setting most strict limitations for detection simply because as the energy rises, the detection area required to measure a significant flux in a reasonable amount of time increases dramatically. For this reason the understanding of showers becomes crucial for the study of the highest energy cosmic rays. It turns out that high energy photons (and electrons) initiate air showers in

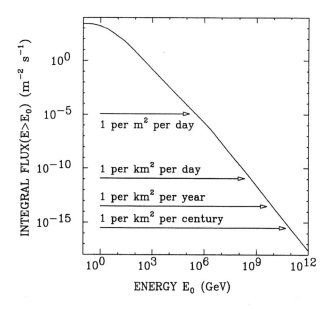

FIGURE 1. Integral flux of cosmic rays above an energy threshold

a very similar fashion to cosmic rays and neutrinos on the other hand also initiate showers, both in the air and in dense media. The study of shower development is thus very relevant for both high energy gamma ray and neutrino Astronomy.

Extensive Air Showers

High energy particles interact in the atmosphere distributing their energy among secondary particles which in turn also interact to produce after several generations a large number of secondaries. In the center of mass frame of each interaction the scattering angles of the emitted particles are characterized by a transverse momentum distribution of few hundred MeV. For high energy primaries, these secondary particles get boosted in the lab frame (atmosphere) to very small angles because of the high Lorentz factors involved. The shower particles travel through the atmosphere in a relatively thin disk moving at the speed of light. The directionality of the initial particle is pretty well maintained by the bulk of the secondaries so that the disk increases its radius slowly when compared to the distance of atmosphere it covers. Particles typically reach to few hundred meters away from the shower axis and only in the most energetic events a significant number of them reach out to distances over a kilometer. The natural effective area of a detector put on the Earth surface becomes the air shower transverse active surface (typically greater than 10^{4-5} m^2) instead of the actual detector area.

One of the best ways to understand the main characteristics of shower development is a toy model devised by Heitler [13] which gives reasonable results in spite of the gross oversimplifications which implies. Each particle undergoes a branching process (interaction) into two other particles, each carrying half of the progenitor's energy. The process is carried on until particles reach the so called critical energy E_c, which corresponds to the energy at which the particle is expected to loose all its energy in an interaction length through ionization losses. It reflects the fact that ionization losses become dominant at lower energies and they do not give rise to extra particles in the shower. Clearly the number of branchings b for a particle of energy E_0 is given by:

$$E_0 \, 2^{-b} = E_c \qquad\qquad b = \frac{ln\left[\frac{E_0}{E_c}\right]}{ln 2} \qquad (2)$$

When the particles reach E_c the number of shower particles (shower size) reaches a maximum N_{max} that is simply given by:

$$N_{max} = 2^b = \frac{E_0}{E_c} \qquad (3)$$

and which happens b interaction lengths away from the point where the initial particle was in the beginning.

The model displays the two most important features of air showers. Firstly that the depth into the material where shower maximum takes place only depends on

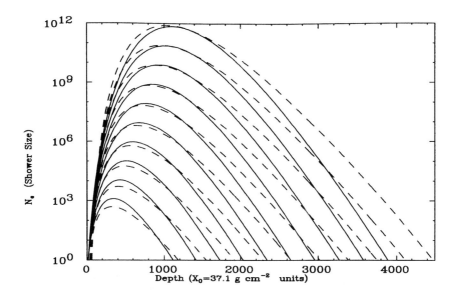

FIGURE 2. Electromagnetic shower depth distributions using Greisen's parameterization (full) and hadronic showers according to Gaisser's (dashed). From top to botom they correspond to primary particles of energy $10^{12}, 10^{11}, 10^{10}, 10^9, 10^8, 10^7, 10^6, 10^5, 10^4, 10^3$ GeV.

the primary energy in a logarithmic way and, secondly, that the number of particles at shower maximum is proportional to the primary energy. Both these properties hold closely in hadronic showers (initiated by cosmic rays) and in electromagnetic showers (initiated by photons or electrons) as shown in Fig. 2. It is the essence of what happens in real electromagnetic showers. For a high energy photon (electron), pair production (bremsstrahlung) in the fields of the nuclei is by far the hard process with highest cross section. Both these processes result in a distribution of the primary particle energy in between the two other particles. While for pair production the energy is distributed pretty evenly between the electron and the positron in bremsstrahlung this is not the case, the cross section becoming singular in the low photon energy limit. As both pair production and bremsstrahlung interactions are related processes their cross sections are both characterized by the same *radiation length*, $X_0 = 37.1$ g cm^{-2} in air (X_0 is the depth of absorber over which the electron attenuates to $1/e$ of its initial energy). The interaction length does not change dramatically with energy except for very high energies, where several effects suppress both cross sections [14]. Collective effects of nuclei are most effective in suppressing the processes in dense media, this is the Landau-Pomeranchuck-Migdal -LPM- effect [15].

Interactions with nuclei are also responsible for the main features of the transverse structure of the electromagnetic showers what is usually referred to as the lateral particle distribution. In this case elastic (Rutherford) scattering is responsible for the largest deflections. Because of the well known $\sin^{-4}\theta$ behavior for the differential cross section, the combination of multiple scattering at low scattering angles dominates the single large angle scattering processes. The theory of multiple scattering has been studied long ago under a number of different approximations yielding pretty equivalent results [16–19]. Molière's theory uses the screened atomic potentials in his formalism which is the most commonly used. The mean deflection angle for an electron of energy E traversing a depth t measured in radiation lengths is given approximately by a gaussian of width [16]:

$$\theta_{\rm MS} \simeq \frac{E_{\rm MS}}{E} \sqrt{t}, \tag{4}$$

$$\text{where } E_{\rm MS} = m_e \sqrt{\frac{4\pi}{\alpha}} \simeq 21 \text{ MeV}. \tag{5}$$

The scale of the transverse spread of the shower is given in terms of the Molière radius ($R_M = E_{\rm MS}/E_c$ in radiation length units), that is $R_M \sim 0.25 X_0 = 9.1$ gcm^{-2} using $E_c = 80$ MeV for air. About 50% of the particles in an electromagnetic shower are within the Molière radius, which is of order 100 m at sea level.

It may be even more surprising to note that hadronic showers do also show very similar features in spite of very different nature of the particles and interactions involved. The vast majority of particles produced in hadronic type interactions are pions and depending on the energy of the interaction they can be produced in large numbers (multiplicities). Hadronic showers can be in fact understood as a superposition of electromagnetic showers produced at different depths by the almost immediate decays of the neutral pions produced in high energy proton-proton and pion-proton interactions, together with a significant flux of muons and neutrinos which mainly come from the charged pion decay. The hadronic core is more penetrating and feeds electromagnetic subshowers at higher depths. The lateral distribution is somewhat harder than that of electromagnetic showers (there are more particles further away from the shower axis) because pions in the core have more transverse momentum and lower Lorentz factors than electrons for the same energy. The differences in the electron and photon lateral distributions are however not as large as one may naively expect from this fact because in the end it is the electromagnetic subshowers that are responsible for it.

The transport theory of this particles corresponds to cascade theory which have been studied long ago and solved for different conditions and making several approximations [1,20]. The solution of these complicated equations is presently replaced with simulation techniques which are more precise and flexible to compare to experiment. However the full simulation of the highest energy showers is presently unapproachable with Montecarlo because the number of particles in the shower can reach a trillion (10^{12}) and statistical techniques have to be used to obtain results in a reasonable time. This makes results more uncertain and in particular the study of fluctuations of these showers more problematic. For these reasons the analytical approach is also being reconsidered at present [21].

Detection Techniques

Several techniques have been developed to detect extensive air showers from the ground. Air Shower Arrays are particle detectors which are distributed over the ground surface to sample the particle density at different points of the shower front. There are convenient approximate parametrizations of the average number of electrons and positrons (shower size) as a function of depth in a medium, what is usually referred as the longitudinal development of both hadronic and electromagnetic showers. They are shown in Fig.2. The figure illustrates the relations between shower size, primary energy and depth from the first interaction point for air showers neglecting the LPM effect. Using the curves in Fig 2 it is relatively easy to see that for a shower to have its maximum at sea level, corresponding to 1000 g cm^{-2} of atmosphere for a vertical incident particle, it must be due to primaries of close to 10^{11} GeV while for an altitude of 2500 meters, corresponding to 700 g cm^{-2} it is reduced to about 10^7 GeV. The result is quite similar for proton and photon primaries. The observation of shower particles at a given altitude cannot be effectively done below some critical energy because the particle densities become too small. Typically air shower arrays cannot effectively perform ground particle density sampling for energies below the 10 TeV range.

The shower direction can be reconstructed from the timing because the shower front propagates as a relatively thin disk (few nanoseconds of width) moving at the speed of light. The energy is deduced from comparisons with detailed simulations and the type of particle is deduced in a similar fashion. The small differences between showers initiated by different primaries and the fluctuations, inherent to shower development, make it quite difficult to extract the primary composition. The distribution of muons plays an important role in this respect. In electromagnetic showers muons are mostly due to pair production which is a highly suppressed process. In hadronic showers the number of produced muons is dependent on the fraction of primary energy carried by each of the nucleon so that the relative electron to muon content of a shower is a tool for extracting information on its primary composition.

The detectors may be of different types such as scintillators or Cherenkov detectors. Cherenkov detectors measure the light emitted as the particles go through a medium such as water. Large arrays have been constructed using simple water tanks with photodetectors [22]. Recently there have been successful experiments which measure the Cherenkov light directly produced in the atmosphere [23]. The refraction index of the atmosphere is slightly greater than one and hence the shower particles travel through it at a speed greater than the speed of light in the medium, provided they have an energy above the Cherenkov energy threshold, which is typically of 80 MeV.

Many air shower arrays have contributed to the knowledge on cosmic rays. The largest one is AGASA in Akeno, Japan spanning an area of about 100 km^2. One event of energy above 10^{20} eV has been detected by this array in 1991 [24]. A large international effort is being made to construct the largest of the detectors of this type covering an overall area of 6000 km². This is the Pierre Auger Project [25] named after the scientist who first realized that cosmic radiation on the ground was spatially correlated and therefore due to higher energy particles, discovering air showers.

For lower energies when the number of ionizing particles at ground level becomes too small for to be effectively detected by an Air Shower Array it is possible to detect the Cherenkov light emitted by the shower particles before they get absorbed in the atmosphere. Because of the low refraction index of the atmosphere the light is emitted in a narrow Cherenkov cone which broadens slowly as the shower develops in lower and denser regions of the atmosphere. Part of this light, which is typically emitted in a 1° cone around the incident particle direction, can be collected several kilometers below, in Cherenkov telescopes on the Earth surface. These are like ordinary telescopes instrumented with a photomultiplier tube in the focal plane, sensitive to very small numbers of photons. They can detect the Cherenkov light from distant air showers in the few hundred GeV range and above. Recently these detectors have developed to imaging Cherenkov telescopes with several phototubes in the focal plane so that the angular distribution of the incoming Cherenkov light is sampled. With the imaging technique it has been possible to extract many shower parameters and particularly the shower direction to a much better degree of accuracy. The technique has made it possible to detect TeV gamma rays from point sources, allowing the separation of an excess number of showers from a given direction from the background of randomly directed cosmic ray showers. Several TeV sources have been positively identified in this decade opening one of the most interesting new windows into the Universe: TeV Gamma ray astronomy [26,27]. The prototype of this detector is the Whipple telescope in Arizona, a 10 m Telescope that has discovered several TeV sources.

At ultra high energies there is the possibility of detecting the fluorescence light of Nitrogen produced as the ionizing secondary particles of the Air showers go through the atmosphere. The efficiency of light emission is extremely low since only about 4 photons are isotropically emitted per track meter of ionizing particle. However when the shower energy is in the 10^9 GeV range and above the number of particles is typically also 10^9 and it becomes possible to detect the light from showers passing several kilometers away from the detector. The prototype of this detector is Fly's Eye in Utah and consists of a number of photomultiplier tubes each viewing a different solid angle fraction of the sky and covering a very wide field of view. The shower size development can be deduced from the light signal in each of the phototubes and from the timing information. These detector types have an important advantage in that they really do a calorimetric measurement of the shower as it develops through the atmosphere. In that way they reduce some uncertainties in the reconstruction of the characteristics of the incoming particle inherent to Air Shower Arrays that only sample the shower at a particular depth. They must however operate in dark conditions like Cherenkov telescopes. Fly's eye has observed the highest energy event, an intriguing cosmic ray of 3 10^{20} eV [28,29]. Fig 3 displays the cosmic ray flux measured at the highest energies.

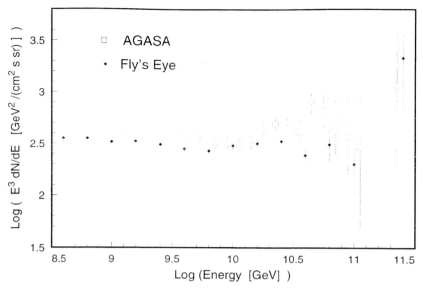

FIGURE 3. Highest energy cosmic ray measurements from Fly's eye and AGASA. No relative normalization factor has been allowed for.

Lastly it is also possible to detect radio pulses from the Air showers with antennas. Several mechanisms take part in the radio signal. The charged particles in the shower develop dipole moments as they propagate, the positive and negative charges separate in the geomagnetic field and there is an excess number of electrons in the shower because of the absence of positrons in the atmosphere. All these effects contribute to generating radio pulses as the shower propagates in the atmosphere. Unfortunately this potentially cheap and simple method has shown to be very difficult to systematize because it is largely affected by atmospheric conditions and weather conditions [30]. Nevertheless prospects for detecting the highest energies with such techniques are under reconsideration [31].

The increasing interest in the observation of the highest energy events with more statistics has resulted in an international effort to construct the largest detector of this type combining the two techniques that have lead to the highest energy event discoveries: extensive air shower array and the fluorescence detectors. It is Pierre Auger Project aiming to construct two 3000 km^2 arrays, one in each hemisphere, consisting of an array of Cherenkov water tanks in an triangular grid of 1.5 km separation combined with fluorescence light "eyes". The project was very well discussed in these lectures by Humberto Salazar and the principle is sketched in Fig 4.

Interactions

While cosmic rays are accelerated and while they propagate they are subject to many interactions. Those with the magnetic fields are fundamental since they are responsible for the diffusion like processes mentioned above. We recall the basics of charged particles in a constant magnetic field. The rate of change of particle momentum is given by the Lorentz Force and particles of charge Z with a velocity βc at a *pitch angle* θ (relative to the magnetic field **B**) describe a helix type motion of radius R (gyroradius) and drift velocity along the B field $v\cos\theta$:

$$\frac{d\mathbf{p}}{dt} = \gamma\beta m\frac{d\mathbf{v}}{dt} \qquad (6)$$

$$\frac{p}{c}\frac{v_\perp^2}{R} = eZBv_\perp \qquad (7)$$

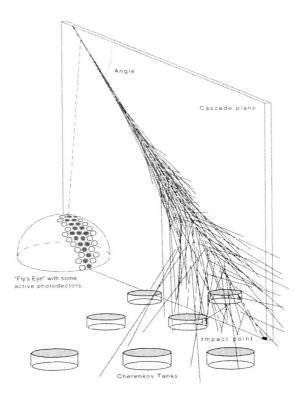

FIGURE 4. Schematic representation of the Pierre Auger Hybryd concept. Showers can be detected in the water tanks at ground level as particles go through it, and alternatively they can be detected through the Fluorescence light emitted by the atmospheric Nitrogen. Each optical module in the Fly's Eye detector sees a small fraction of the 2π sr of the upper hemisphere.

$$R = \frac{p\beta \sin\theta}{eZB} < \frac{E}{eZBc} \qquad (8)$$

When the magnetic field, as seen by the moving particle, changes slowly with respect to the period of circular motion in the plane perpendicular to $\vec{B}$ (gyroperiod), it can be shown that the magnitude R^2B remains constant: it is an *adiabatic invariant*. The particles follow the field line configuration adapting their gyroradius and $v_\perp$ to conserve this invariant quantity. The magnetic fields have some irregular components which can cause random scatterings. If these irregularities in the magnetic field are small perturbations which change over a time scale much smaller than the gyroperiod, the particle is hardly affected by them and behaves as if there was a constant or slowly varying B field (given by the average). A fairly efficient scattering can happen however when the change in the B field due to the irregularities lasts a time interval which is comparable to the particle period [5]. The random scattering has a mean free path which is related to the energy of the particle and the spectrum of the irregularities in the field, making charged particles behave in a diffusive manner and effectively trapping them.

At high enough energies the gyroradius may become larger than the distances between the sources and the observer or the scale of the magnetic field region itself and the charged particles do not scatter. In that case it is the regular component of the field over a large distance what causes the deviation of the particle direction. If the angle of deflection α is small it can be obtained as the ratio of the distance travelled to the gyroradius:

$$\Delta\alpha \simeq \frac{d}{R} \simeq \frac{deZBc}{E} = \frac{d}{30kpc}\frac{B_{nG}}{E_{18}} \qquad if \quad \frac{d}{R} \ll 1 \qquad (9)$$

In the last equality, which is for a proton, the magnetic field strength B_{nG} must be in units of nG while the energy E_{18} in units of 10^{18} eV (EeV). In the nominal field of the galaxy $B = 3$ μG a proton of 10^{20} eV does not deviate more than 0.1 radians (6°) over a distance of 10 kpc. We recall that the galaxy is only a few hundred pc wide so that deviations should be a lot smaller than 0.1 radians. On the other hand for an intergalactic field of 1 nG the distance that can be covered with such an angular deflection is 30 Mpc. This implies that such high energetic particles point in the direction of their sources to some accuracy if they come from relatively nearby sources as we argue below. If this is so high energy cosmic rays allow the possibility of doing astronomy with charged particles. Moreover it has been suggested that the analysis of the energies and directions of these particles if they come from point sources can allow determinations of the magnetic field in between the sources and us.

Another very important process for high energy cosmic rays is the interactions of protons with photons of the cosmic microwave background radiation. Because these photons are everywhere and there are about 400 per cm^3 they prevent the highest energy cosmic rays from propagating over large distances in cosmological scales. The photoproduction cross section becomes important above the threshold for the production of the $\Delta^+(1232)$ where it has a narrow resonance. The photon energy threshold for the reaction (ϵ_γ) is inversely proportional to the cosmic ray proton energy (E_p):

$$2\epsilon_\gamma E_p(1 - cos\theta) = s - (m_p c^2)^2 > (m_\Delta c^2)^2 - (m_p c^2)^2 \tag{10}$$

For protons around $\sim 5\ 10^{19}$ eV the threshold is just at the peak of the microwave spectrum and the photon density is high enough to prevent the protons from propagating. Higher energy protons also get attenuated. The spectrum should reflect this interaction as a flux suppression above these energies what is usually referred as the Greisen-Zatsepin-Kuz'min (GZK) cutoff [32]. The mean free path at these energies is about 5 Mpc and typically the proton loses a tenth of its energy, so that the distance for all energy loss is of order 50 Mpc. The reaction proceeds with the Δ decay into a pion and a nucleon. The Δ^+ corresponds to the total $|J, m_J>$ isospin state $|3/2, 1/2>$ which can be decomposed as a superposition of two particle states in the $|J_1, m_{J_1}; J_2, m_{J_2}>$ basis in the following manner:

$$\left|\frac{3}{2},\frac{1}{2}\right\rangle = \frac{1}{\sqrt{3}}\left[\left|\frac{1}{2},-\frac{1}{2};1,1\right\rangle + \sqrt{2}\left|\frac{1}{2},\frac{1}{2};1,0\right\rangle\right] = \frac{1}{\sqrt{3}}\left[|n,\pi^+> + \sqrt{2}|p,\pi^0>\right] \tag{11}$$

The π^0 produces almost instantaneously two photons while the π^+ decays into a muon and a muon neutrino and the muon subsequently decays into two neutrinos and a positron:

$$p\gamma \to \Delta^+(1232) \to n\ \pi^+$$
$$\pi^+ \to \nu_\mu + \mu^+$$
$$\mu^+ \to \overline{\nu}_\mu + \nu_e + e^+ \tag{12}$$

where each of the four stable leptons in the π^+ decay chain carries approximately about a fourth of the pion energy [10]. The reaction is a source of high energy gamma rays that also interact with the microwave background to produce a photon cascade until the photon energy falls below the threshold for producing electron positron pairs (~ 200 TeV). Several important consequences can be extracted from the decay chain. Firstly that the ratio of energy into photons to that of energy into neutrinos is about 3. One can also conclude that the ratio of muon neutrinos to electron neutrinos is about 2 and finally that there is a deficit of electron antineutrinos. Although they do not appear as part of the Δ^+ resonance decay chain, they can be produced in π^- decays, from reactions at energies above the Δ resonance.

The origin of the highest cosmic rays observed, reaching energies in excess of 10^{20} eV is quite difficult to understand because of this interaction. They should be produced in the cosmological neighborhood of the solar system, within a radius of about 50 Mpc, yet few sources are known within it that can have the potential to accelerate particles to such high energies. Quite accidentally heavier nuclei such as Iron also suffer a cutoff at similar energies. Although the energy per nucleon is a factor 56 below the energy of the total nucleus and 10^{20} eV Iron nuclei are below threshold for photoproduction by the microwave background, these photons induce photodesintegration reactions which in effect produce a very similar cutoff.

The implication of the cutoff interactions is that events above the cutoff must display spectral structure because of the sensitivity of the cross section to the primary particle energy and to the distance travelled. For very nearby sources (few Mpc) the distribution after interactions with the microwave photons represents some depletion of the source flux for energies above 10^{20}. At large distances it shows a sharp cutoff [33]. The GZK cutoff rises in energy as the microwave photons redshift in the expansion of the Universe so that the detected energy spectrum is also sensitive to the source distribution. Moreover there is spectral structure associated to composition due to the different upper limits expected in source acceleration and the differential attenuation. In summary the spectral study of the region around and above the GZK cutoff is of great interest and can reveal a lot of information on the sources distributions and their time (redshift) evolution. The Pierre Auger Project has been designed to detect particles in this energy region [25].

Acceleration

One of the most successful theories for acceleration of cosmic rays was proposed by Enrico Fermi in 1949 as an stochastic process. The attractive rests on its simplicity and its ability to reproduce power law spectra. The basic idea is that there are simple and common "events" after which charged particles gain on average a small amount of energy. It is through the repetition of many of these events that particles are accelerated to high energies. One crucial property of the acceleration mechanism can be obtained from the assumption that these events occur through ordinary interactions of particles with electromagnetic fields. Clearly for charged particles to be accelerated to high energies they must be contained for sufficient time in the region (of size L) where these events take place:

$$R = \frac{p\cos\theta}{ZeB} < L \quad E \simeq pc < ZeBcL \tag{13}$$

This means that the particles must have gyroradius smaller than the acceleration region or that for a given size accelerator with a given magnetic field there is a maximum energy that can be reached through stochastic acceleration. This incidentally is the same condition that applies in circular particle accelerators making them so expensive. The condition is illustrated in a plot due to Hillas shown in Fig. 5 which shows the sizes of possible accelerators versus the magnetic field strength. The different regions correspond to known astrophysical environments. The lines represent the condition of Eq (13) for particles of 10^{20} eV. Candidate accelerators must lie above the curve to satisfy the requirement. The three lines correspond to iron nuclei, protons and protons with a 10^{-2} efficiency. It is clear from this plot that the acceleration to such energies is marginally satisfied by few sources such as Active Galactic Nuclei (AGN), Neutron Stars (NS), Radio Galaxy Lobes (RGL), clusters

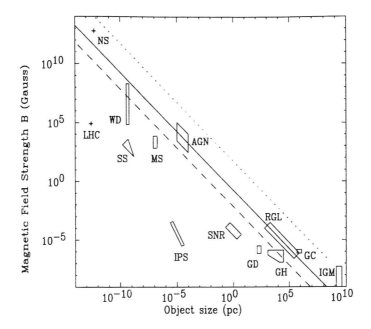

FIGURE 5. Hillas Plot of the typical size of a possible accelerator L versus its magnetic field strength B. From upper left to lower right the astrophysical objects correspond to Neutron Stars, White Dwarfs, Sun Spots, Magnetic Stars, Active Galactic Nuclei, Inter Planetary Space, Supernovae Remnants, Radiogalaxy Lobes, Galactic Disk, Galactic Halo, Clusters of Galaxies and the Inter Galactic Medium. Also shown is the point corresponding to the largest accelerator in planning LHC. The straight lines represent the limits given by Eq (13) for protons (full), Iron nuclei (dashed) and for protons assuming a 10% efficiency (dots).

of Galaxies (GC) and the whole intergalactic medium (IGM). This last candidate is unlikely in view of the necessity that the cosmic rays are produced within 50 Mpc. It also shows that very close to maximum efficiencies would be required. Acceleration to energies above the GZK cutoff presents a theoretical challenge.

Regardless of what the process is, it is easy to show that a power law spectrum is obtained with a constant fractional energy increase (on average):

$$\epsilon = \left\langle \frac{\Delta E}{E} \right\rangle \qquad (14)$$

and an energy independent probability that the particle escapes the accelerating region (p_{esc}) during the time interval between two successive events.

Assuming particles of energy E_0 (injection energy) are readily available for acceleration, the number of these events needed to accelerate particles to an energy E is given by:

$$E = E_0(1+\epsilon)^k \qquad k = \frac{ln(E/E_0)}{ln(1+\epsilon)} \qquad (15)$$

The probability of having energy above E corresponds to k or of more than k events and is given by:

$$\sum_{m=k}^{\infty} (1-p_{esc})^k = \frac{(1-p_{esc})^k}{p_{esc}} \qquad (16)$$

The number of particles reaching energies equal or greater that E is then:

$$N(>E) \propto \frac{(1-p_{esc})^{\frac{ln(E/E_0)}{ln(1+\epsilon)}}}{p_{esc}} = \frac{1}{p_{esc}} \left[\frac{E}{E_0}\right]^{\frac{ln(1-p_{esc})}{ln(1+\epsilon)}} \propto E^{-\gamma} \qquad (17)$$

where the spectral index γ is given by:

$$\gamma = -\frac{ln(1-p_{esc})}{ln(1+\epsilon)} \simeq \frac{p_{esc}}{\epsilon} \simeq \frac{1}{\epsilon} \frac{T_{cycle}}{T_{esc}} \qquad (18)$$

Here the last two expressions are correct for small values of ϵ and p_{esc}. p_{esc} has been estimated as the ratio of the time between events T_{cycle} and the residence time of cosmic rays within the acceleration region T_{esc}.

Related to the maximum energy there is a maximum acceleration rate. If particles obtain a net relative energy gain ϵ per event and the time interval between events is at least the inverse of the gyrofrequency of the particles in the magnetic field, it follows that the maximum acceleration rate must be:

$$\frac{dE}{dt} \simeq \frac{\Delta E}{\Delta t} < \frac{\epsilon E}{2\pi R/c} = \frac{\epsilon}{2\pi} ZeBc^2 \qquad (19)$$

Note that the energy dependence in both the gyroradius and the energy gain drops when taking the ratio.

The various forms of stochastic acceleration follow the above simple arguments and they differ basically in the nature of the events themselves. It is easy to show that these events can simply take place as particles move in and out of well defined plasma regions that have different bulk velocity than the rest. The accelerator can be characterized by two regions R and R' each having different plasma variables, in particular different velocities. We can visualize a plane separating them to make it simple. Such situations can arise by magnetic clouds that are moving through space with random velocities as was proposed originally by Fermi, or by contact discontinuities in plasma flows such as those produced by shock waves.

Crucial to the argument is that the random scattering of the cosmic rays in each of the two regions are produced in the *rest frame* of the corresponding plasma, that is in a system which moves along with the overall plasma velocity. We will refer to S and S' as the rest systems of R and R' respectively. After these random scattering the particle is "isotropized" in this frame while its energy is maintained because the interactions are magnetic. Every time the particle changes from one of the regions to the other a Lorentz transformation is required between S and S'. The transformations are characterized by γ and β corresponding to the relative velocity (βc) of the plasma in one region with respect to the other. Considering a particle that enters from S into S' with energy E_1 and momentum $\mathbf{p_1}$ at an angle θ_1 to the normal of the plane, it is straightforward to see that in S' its energy is:

$$E'_1 = \gamma(E_1 - \beta p_1 cos\theta_1) \simeq \gamma E_1(1 - \beta cos\theta_1) \qquad (20)$$

In R' the particle maintains its energy E'_1 but changes its direction randomly until it crosses again the front (there is a finite probability that this happens) with an angle θ'_2 to the normal. A new transformation is needed back to the system S:

$$E_2 = \gamma(E'_1 + \beta p'_1 cos\theta'_2) \simeq \gamma E'_1(1 + \beta cos\theta'_2) \qquad (21)$$

The process of crossing to R', becoming isotropic in S' and crossing back to R corresponds to an event and putting both transformations together the energy after this sequence is given by:

$$E_2 \simeq \gamma^2 E_1(1 - \beta cos\theta_1)(1 + \beta cos\theta'_2) \qquad (22)$$

Different geometries affect now the averaging process. In Fermi's proposal magnetic clouds are moving in random directions with velocity βc which gives the Lorentz transformation factors. The average over the angle θ'_2 is simply zero because particles are expected to escape the cloud in any direction with equal probability. We note that this is not the case for the average of θ_1 because the probability that a particle enters a cloud is proportional to the difference of the velocities of the cloud and the particle, "head on" collisions being more probable than "catching up" collisions with the result $<cos\theta_1> = \beta/3$. It is clearly seen that this process leads to an average energy gain:

$$\epsilon = \left\langle \frac{\Delta E}{E} \right\rangle = \frac{4}{3}\beta^2 \qquad (23)$$

The resulting acceleration is a second order effect in β and it is called second order acceleration. The scattering of the magnetic clouds originally proposed and easier to visualize is modernly replaced by scattering in the irregularities of the magnetic fields propagating as Alfvèn waves.

First order acceleration occurs in strong shock arrangements such as that produced in a supernova explosion. The particles in the interstellar medium are always swept by the blast wave preceded by a shock front. After diffusing in the expanding plasma they may emerge again out of the shock plane into the interstellar medium and that constitutes the event cycle. When energy gain is averaged over incoming and outgoing angles, it can be shown that the energy gain per cycle is of first order in the blast wave velocity:

$$\epsilon = \left\langle \frac{\Delta E}{E} \right\rangle = \frac{4}{3}\beta \qquad (24)$$

with βc being the velocity of the plasma behind the blast wave. In this case it is interesting to note that the probability of escape is related to this quantity. The point is that for a monoatomic gas the non relativistic blast wave dynamics imply that there is a shock front ahead moving at speed $U = 4/3\ \beta c$. In the rest frame of the shock itself the particles of the interstellar medium enter with velocity U and come out with velocity $1/4\ U$ behind the shock. If the particle density behind the shock front is n, the particle flow away from the shock is $1/4\ Un$ while the particle flow across the shock into the interstellar medium is $1/4\ nc$ from kinetic theory of gases. The ratio of the two rates is the probability of escape from the region and is also given by:

$$p_{esc} = \frac{U}{c} = \frac{4}{3}\beta \qquad (25)$$

p_{esc} and ϵ are equal. In the non relativistic limit they are both small and we can apply Eq (18) to obtain that strong shock acceleration naturally gives power law spectra with a universal spectral index $\gamma = 1$.

The simple treatment presented here does not reflect the richness of Fermi acceleration with many variations depending on the orientation of the regular B field component with respect to the shock front, on the strength of the shock, its velocity and many other important variables that give a range of spectra depending on the conditions [34].

Stability

The acceleration process is stable as long as the acceleration rate exceeds losses by any other process. At high energies synchrotron losses are important. The energy loss by synchrotron radiation of a particle of mass number A, charge Z, mass $M = Am_p$ and energy $E = \gamma M c^2$ is given by:

$$-\frac{dE}{dt} = \frac{4}{3}\sigma_T \left[\frac{Z^2 m_e}{M}\right]^2 \frac{B^2}{2\mu_0}\gamma^2 c \qquad (26)$$

We can equate the acceleration rate of Eq (19) to the losses to obtain in the case of synchrotron losses:

$$\eta Z e B c^2 = \kappa B^2 E^2 Z^4 \qquad (27)$$

Here η is an efficiency factor for the acceleration process and its value is limited to a maximum $\eta = \epsilon/2\pi$ from Eq. (19). κ is a constant depending on M and Z of the accelerated particle. The condition sets a maximum reachable energy in a region of a given B field:

$$E_{max} = C\sqrt{\eta}\sqrt{\frac{1T}{B}}\left[\frac{A^2}{Z^{\frac{3}{2}}}\right] \tag{28}$$

where $C = 6\ 10^2$ GeV for electrons and $C = 2\ 10^9$ GeV for nuclei and the last fraction in brackets is only to be considered for nuclei. Electrons suffer most losses and their energy limit is well below that of protons.

If we equate the acceleration rate to the photoproduction losses described above and use the fact that they do not depend on the magnetic field we obtain a very different limitation:

$$\eta Z e B c^2 = constant \tag{29}$$

The standard view of acceleration presents supernova explosions as the engines for accelerating cosmic rays to energies below few hundred TeV or so. The cosmic rays below this energy would be produced locally and the total energy density in cosmic rays is comparable to 1% of the energy release by supernova explosions assuming a rate of explosion within the Galaxy of about one every 30 years and 10 million year average residence time for the cosmic rays in the Galaxy. The energy balance seems about right and moreover the acceleration has recently been observed in such sites [35]. The problem still remains for the higher energy cosmic rays because a smooth spectrum is observed up to 10^{18} eV. Combining the finite lifetime of a typical supernova blast wave and the maximum acceleration rate discussed above it can be shown that it is difficult to accelerate particles beyond 100 TeV with a single supernova shock. Maybe there is reacceleration at other supernova shocks, a sort of galactic "pinball" effect [36]. Cosmic rays of energies above the GZK cutoff are no longer trapped within the Galaxy and travel in close to straight lines through it. It is believed that such cosmic rays must be extragalactic because of the anisotropy they display [37] and because of the measured change of slope above 10^{19} eV. The possibility that part of the lower energy cosmic rays is extragalactic too has been recently addressed [38].

The origin of cosmic rays above the GZK cutoff is however quite a mystery because of the few objects satisfying the dimensional limitations of Fermi acceleration, even fewer are close enough to reach us without being absorbed by the microwave background. It is for this reason that a wide range of possibilities has been discussed in the literature.

It is remarkable that many of the suggestions are based on shock acceleration with different shock scenarios. Many of them involve shocks at the scale of galaxies, such as in the galactic halo, in clusters of galaxies, collisions of galaxies or active galactic nuclei. The low B fields are compensated by large acceleration regions to achieve the highest energies. Gamma Ray Bursts (GRB), or rather the phenomena that cause them which are still under heavy debate, have also been proposed as regions where stochastic mechanisms can accelerate cosmic rays to the highest energies [39]. Alternatives to shock acceleration are provided by pulsars and there is the more speculative possibility that they are produced by decays of other types of objects that have been postulated but never observed. The category includes objects like primordial black holes (PBH) [40], Weakly Interactive Massive Particles (WIMPS) and decays of Topological Defects (TD) [41]. These latter models have been given a lot of attention in the last years because they escape the problem of acceleration and can arise naturally in grand unification theories.

All of these proposed mechanisms for cosmic ray acceleration have implications for the production not only of cosmic rays but also of gamma rays and neutrinos at least. They are the products of pion decays which are inevitably produced by the collisions of the cosmic rays with the microwave background, with matter at the acceleration site or in the case of PBH's and TD's pions are produced by quark fragmentation. The observation of these particles will reveal information on the sources and many efforts are being made with the aim of improving high energy γ ray astronomy and to develop neutrino astronomy. The neutrinos travel to us practically unaffected except for possible resonant interactions with the cosmic relic background of neutrinos. Since massless relic neutrinos have energies in the mili-eV range the interacting neutrinos must have extremely high energies to produce a Z boson in the s channel [42]. If neutrinos have small but non-negligible masses the energy threshold for resonant interactions is lowered. In fact this mechanism has been proposed as an explanation of the highest energy cosmic rays. High energy neutrinos behave as messangers and interact with massive relic neutrinos in our galactic halo to give rise to the observed cosmic rays [43].

Photons on the other hand interact with the infrared and the cosmic microwave backgrounds to produce electron positron pairs. The energy threshold for the reaction depends on the target photon energy. For the most numerous microwave background photons it is about 200 TeV. Photons produced together with the highest energy cosmic rays cascade down to lower energies to produce a diffuse γ ray flux. The γ ray observations and neutrino bounds have already proved useful in constraining some of these models [44,45].

GAMMA RAYS

One of the most exciting developments in astroparticle physics in the last ten years took place in high energy gamma ray astronomy. In April 1991 the Compton Gamma Ray Observatory (CGRO) was launched with four gamma ray detectors, OSSE, BATSE, COMPTEL and EGRET each optimized for different energies and signal characteristics. The EGRET instrument was aimed for the higher energy photons in the range 20 MeV–30 GeV, by detecting electron positron pair conversions. Since then it has detected high energy gamma-ray emission from over 100 sources [46] most of them new sources. Of these sources 16 have been tentatively, and 42 solidly identified with radio counterparts. They are all classified as "Blazars", a subclass of Active Galactic Nuclei, mostly Flat Spectrum Radio Quasars and BL-Lac objects [47].

The BATSE (Burst And Transient) instrument was designed as a wide angle gamma ray detector to look for short gamma ray flashes. This detector has been steadily detecting Gamma Ray Bursts (GRB) approximately at a rate of one burst per day. GRBs were long ago discovered by military satellites surveying nuclear explosions, but since the deployment of CGRO the amount of data accumulated has shown an initially puzzling isotropic distribution suggestive of a cosmological distribution. The cosmological hypothesis has been very recently confirmed after redshift observations of "afterglows" from these bursts implying that the energy budget for these objects is very large, comparable to that of a supernova explosion. The optical, radio and X-ray observations of these afterglows has been possible thanks to the excellent GRB positioning information provided by a new satellite BeppoSax operating since 1997. The nature of GRBs is still a matter of intense debate and has attracted a considerable amount of attention in the last years. Although GRB's typically emit in the keV to MeV range, photons have been detected up to the GeV range and it is possible that particles are accelerated to much greater energies.

Sometime before the CGRO launch, the Whipple telescope, the first one to use the imaging Cherenkov technique, discovered TeV photons from the Crab nebula [48], a supernova remnant. The birth of TeV gamma ray astronomy from ground observations was possible because of the good angular resolution of the imaging Cherenkov telescopes as discussed above. The establishment of the technique together with the discoveries made by EGRET gamma rays from Blazars paved the way for the future discoveries. Four other TeV sources have been identified by Whipple and other similar detectors since then. Three of them are also Blazars and in some cases photons have been observed up to energies of 10 TeV. These exciting developments in gamma ray astronomy have put both AGNs and GRBs at the forefront. We will concentrate here on AGN which are the sources of the highest gamma rays observed.

Active Galactic Nuclei

AGN are most remarkable objects being the brightest sources in the Universe and coming from distances approaching the limits of the observable Universe. They are characterized in general by a continuum band of electromagnetic radiation with a very broad spectrum which has been observed in over 13 orders of magnitude from radio frequencies corresponding to μeV to the highest observed gamma rays reaching over 10 TeV, thanks to CGRO and Whipple. For these objects the spectral characteristics can be rather different giving rise to various classifications and a large number of subspecies. The observed emission in each of the regions of the electromagnetic spectrum is dominated by different processes which are by no means well understood, particularly at the highest energies. Many AGN have broad emission lines indicating emission from high speed gas.

Quite commonly they have highly relativistic and confined jets of particles, which are seen in radio observations. These jets show radio structures (knots) which apparently move at velocities higher than light, what was long ago interpreted as the emission from a system approaching to us at relativistic speeds [49]. Sometimes large "lobes" are also observed at the end points of these jets. According to the amount of radio emission observed they are classified in Radio Loud and Radio Quiet.

Another quite common characteristic is the rapid time variations of the observed fluxes which are typically of order days and becomes larger at higher energies. Time variability implies that not only AGN engines must be powerful, but extremely compact. Their high energy luminosities are observed to have flares in which the observed flux rises by over an order of magnitude over time periods of order a day [50]. Only sites in the vicinity of black holes, a billion times more massive than our sun, can possibly satisfy the energetic demands of the observation of such distant objects. It is believed that energy is provided by the accretion of matter onto the black hole and that the jets are beams of particles accelerated near the black hole. Models for these objects have particles further accelerated by shocks produced in the accretion flow, in the jets or in the lobes (larger structures at the end of the jets) which provide the highest energy particles. High energy particles are then dumped on the radiation in the galaxy which consists of mostly thermal photons with densities of order $10^{14}/cm^3$. The multi-wavelength spectrum, from radio waves to TeV gamma rays, is produced in the interactions of the accelerated particles with the magnetic fields and ambient light in the galaxy.

A unified scheme of Radio Loud AGN has been recently proposed based on a large anisotropy of the emission [52]. The opposite jets suggest an axial symmetry possibly induced by an accretion disk [53]. Depending on the relative orientation to the observer, the emission is different. Blazars correspond to Radio Loud AGN viewed from a position illuminated by the cone of their relativistic jet. Several of the blazar sources observed by EGRET have shown strong variability, by a factor of 2 or so over a time scale of several days [50]. Time variability is more spectacular at higher energies. On May 7, 1996 the Whipple telescope observed an increase of the TeV-emission from the blazar Markarian 421 by a factor 2 in 1 hour reaching, eventually, a value 50 times larger than the steady flux. At this point the telescope registered 6 times more photons from the Markarian blazar than from the Crab supernova remnant [51], which is closer to us by a factor 10^5.

In fact it has been also checked that the observations of an extragalactic diffuse gamma ray flux is consistent with this interpretations and would be due in part to the accumulated gamma rays emitted by all the AGN population [54]. AGNs have also been suggested as possible acceleration sites for the highest energy cosmic rays. If this was the case only nearby AGN would contribute to the local cosmic ray spectrum and any peculiar distribution of the nearby AGN should be reflected in the cosmic ray distribution above the GZK cutoff. Indeed several authors have looked for an enhancement in both the cosmic and gamma ray signals from the "supergalactic plane", where the nearest AGN are concentrated [55].

Blazar Models

While earlier models proposed acceleration near the AGN cores, observational evidence and the unified model for AGN suggest that the acceleration of the highest energy particles is taking part in the jets themselves. Because of their relativistic bulk motion the emission gets boosted to high energies in the forward direction and for that reason only Blazars are seen in high energy gamma rays. It is likely that absorption effects explain why Markarian 421, the closest blazar on the EGRET list at a redshift of $z = 0.031$ corresponding to a distance of ~ 150 Mpc, produces the most prominent TeV signal. Although the closest, it is one of the weakest; the reason that it is detected whereas other, more distant, but more powerful, AGN are not, must be that the TeV gamma rays suffer absorption in intergalactic space by interaction with background infra-red light [56]. This most likely provides the explanation why much more powerful quasars with significant high energy components such as 3C279 at a redshift of 0.54 have not been identified as TeV sources.

In the Blazar models if the accelerated particles are electrons the multi-wavelength spectrum consists of three components: synchrotron radiation produced by the electron beam on the B-field in the jet, synchrotron photons Compton scattered to high energy by the electron beam and, finally, UV photons Compton scattered by the electron beam to produce the highest energy photons in the spectrum [57]. The seed photon field can be either external, e.g. radiated off the accretion disk, or result from the synchrotron radiation of the electrons in the jet, so-called synchrotron-self-Compton models. This is schematized in Fig. 6. The energetic gamma rays will subsequently lose energy by electron pair production in photon-photon interactions with the radiation field of the jet or the galactic disk. An electromagnetic cascade is thus initiated which, via pair production on the magnetic field and photon-photon interactions, contributes to the emerging gamma-ray spectrum at lower energies. The lower energy photons, observed by conventional astronomical techniques, are, as a result of the cascade process, several generations removed from the primary high energy beams.

The picture has a variety of problems. In order to reproduce the observed high energy luminosity, the accelerating bunches have to be positioned very close to the black hole. The photon target density is otherwise insufficient for inverse Compton scattering to produce the observed flux. This is a balancing act, because the same dense target will efficiently absorb the high energy photons by $\gamma\gamma$ collisions. The balance is difficult to arrange, especially in light of observations showing that the high energy photon flux extends beyond TeV energy [26]. The natural cutoff occurs in the 10–100 GeV region [57]. Finally, in order to prevent the electrons from losing too much energy before producing the high energy photons, the magnetic field in the jet has to be artificially adjusted to less than 10% of what is expected from equipartition with the radiation density.

In the alternative models protons as well as electrons are accelerated to overcome these difficulties. Because of reduced energy loss, protons can produce the high energy radiation further away from the black hole. The more favorable production-absorption balance far from the black hole makes it relatively easy to extend the high energy photon spectrum above 10 TeV energy, even with bulk Lorentz factors that are significantly smaller than in the inverse Compton models. Two recent incarnations of the proton blazar illustrate that these models can also describe the multi-wavelength spectrum of the AGN [58,59]. Because the seed density of photons is still much higher than that of target protons, the high energy cascade is initiated by the photoproduction of neutral pions by accelerated protons on ambient light via the Δ resonance. The protons collide either with synchrotron photons produced by electrons [60], or with the photons radiated off the accretion disk [61], as shown in Fig. 6.

The dimensional limitations of the models for hadronic acceleration in the blazar jets can be easily recognized making three simple assumptions [62]:

- Protons are accelerated in the jets with an E^{-2} spectrum as expected in shock acceleration.

- The maximum energy for the protons is 10^{20} eV

- The target photon density behaves as a negative power law $E^{-\alpha}$ (for AGN in the broad infrared to X-ray band α is typically around 1).

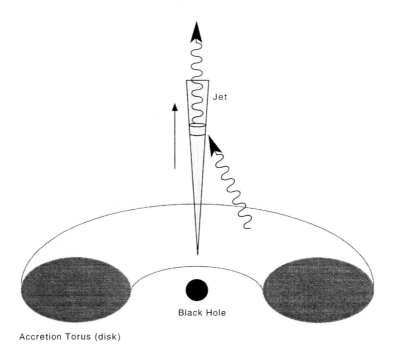

FIGURE 6. Schematic view of an AGN model in which photons are accelerated by high energy particles in the jets.

If higher energy photons are observed from these objects the proton blazar models will become more plausible. In the end the answer may come from neutrino observations. If photons come from neutral pion decays, inevitably the charged pions will have to produce neutrinos which are discussed below.

NEUTRINOS

Neutrinos are the third and last particle species I will discuss. Their detection constitutes a real challenge because of their low interaction cross section. For this reason they travel through large amounts of matter and can bring information from regions of the Universe that are otherwise shielded. Neutrinos are also a fundamental part of the Standard Model whose success has been discussed by Guido Altarelli in these lectures and their detection can help us complete the model. Neutrinos have been detected from the sun [63], from supernova 1987A [64] and from the interactions of cosmic rays with the atmosphere, atmospheric neutrinos [65]. Although few neutrinos were detected from SN 1987A a lot of information was obtained from them, ranging from the confirmation of the star collapse hypothesis to the establishment of the best bounds on neutrino mass [66]. The solar and atmospheric neutrino observations have both proven to be in disagreement with the theoretical expectations. This has motivated the concern for flavor oscillations and neutrino mass limits, which has also been discussed by Jorge Morfin in these lectures.

All of the neutrinos detected so far have low energies. We will concentrate here on the high energy expectations which are also of extreme interest both from the astrophysical and particle physics points of view. There are efforts on their way to build high energy detectors in the forthcoming years [10]. Two of these projects, AMANDA and Baikal, are already in operation. The "km^3" initiative, to instrument 1 km^3 of water or ice with photodetectors, is the natural extension of the lower scale prototypes in view of the expected neutrino fluxes [67]. Other projects NESTOR and ANTARES are well under development. The "Pierre Auger Project" to establish the flux of the highest energy cosmic rays, discussed above provides further motivation in a double way. As previously mentioned all proposed mechanisms for producing the highest energy cosmic rays must also produce high energy neutrinos. The detector on the other hand can be used to detect neutrinos of the highest energies by looking for horizontal showers [68] as discussed below, since at high zenith angles cosmic ray showers get absorbed in the atmosphere. The prospects for detecting both high energy cosmic rays and neutrinos in the near future has also stirred the theoretical activity in the field.

Candidate Sources

In the majority of mechanisms most neutrinos arise from the decay of charged pions (or kaons), produced in different type of high energy particle interactions.

The pions can be produced in proton-proton or photon-proton interactions or alternatively from direct fragmentation of quarks, in the same way they are produced routinely in electron positron colliders. For relativistic pions in flight it can be assumed that, on average, each of the four leptons produced in the reaction and subsequent muon decay carries one fourth of the parent energy [10].

The existence of high energy cosmic rays leaves little doubt about the actual production of neutrinos in their interactions with well understood targets. Atmospheric neutrinos fall in this category and are known to within about 10% certainty at energies below 1 PeV [10]. These neutrinos constitute the background for observation of other neutrinos sources. Their flux is zenith angle dependent because of the competition between interaction and decay of the parent pions. The vertical and horizontal atmospheric neutrino fluxes are shown in Fig 7. Note that here the flux in the "y-axis" is multiplied by E^2, not only to enhance the features but also because detection is typically more efficient at higher energies, so that peaks in these kind of plots roughly corresponds to peaks of expected events. At high enough energies the Lorentz expanded lifetime of the pion leads to more pion interactions decreasing the relative number of neutrinos to their parent pions. Because of this at some unknown energy above 100 TeV "prompt" neutrinos from the decays of charmed particles (that have considerably shorter lifetimes) dominate the atmospheric flux. A typical prompt neutrino prediction [69] is also illustrated in Fig. 7. The prompt component is very uncertain due to the poorly known charm production cross section. This problem is related to the observations of prompt atmospheric muons which are also produced in the charmed particle decays and it is of great interest as an alternative way to experimentally constrain open charm production in the Standard Model [69,70].

Cosmic rays must also interact with nucleons in the galaxy, such as dust, molecular clouds or compact objects like the sun. In their interactions with the galactic disk matter they produce a secondary neutrino flux which is reliably predicted to be detectable with future detectors in planning [11]. The calculated fluxes [72] are also shown in Fig. 7 where it is apparent that these neutrinos dominate the conventional atmospheric flux at the same energy at which prompt neutrinos are expected to dominate the atmospheric neutrino flux. This difficults their possible identification but it is nevertheless hoped that the prompt neutrinos can be also indirectly determined by measuring the atmospheric prompt muons which are also produced in the decays of charmed particles [69].

The interactions of cosmic rays with the cosmic microwave background is also an unavoidable source of neutrinos assuming the higher energy cosmic rays are of extragalactic origin and hence universal. This is supported by the cosmic ray anisotropy measured at high energies. I will refer to these as GZK neutrinos to stress their relation to the cosmic ray energy cutoff. Several groups have calculated these fluxes [10,73,33] and their results are within a couple of orders of magnitudes depending on the assumptions made. For this calculation the cosmic ray spectrum has to be estimated at the production site. This implies extrapolating to energies above the maximum currently observed ($\sim 3 \cdot 10^{20}$ eV) and making assumptions about the evolution of cosmic ray luminosity with cosmological time. The production mechanism has to be integrated over time (or redshift) up some earlier epoch (z_{ult}) which is expected to be provided by the Galaxy formation era (z_{ult}=2-4). These flux predictions are all fairly flat because the proton-photon interaction cross section has a threshold behavior at the resonant Δ production. Most neutrinos are produced with energies about a factor 20 (see next section) below the Greisen-Zatsepin-Kuz'min cutoff energy $\sim 10^{20}$ eV. Depending on (z_{ult}) the interactions of the highest energy neutrinos with the cosmic neutrino background can play a more important role altering their shapes in the highest energy region [42]. Fig. 8 includes some of these calculated fluxes indicating the levels of uncertainty.

Regardless of the uncertainties in the GZK neutrinos, all these mechanisms are certain, at least in the sense that if by some means they were found not to be there, the hypothetical implications of such non-discovery would have a larger impact in physics and/or cosmology than their actual observation. Other candidate sources are astrophysical objects that we know exist, where it is plausible that protons (or nuclei) are accelerated and that some of them collide to produce pions. Some of them can be galactic such as accretion in binary systems, supernova remnants,

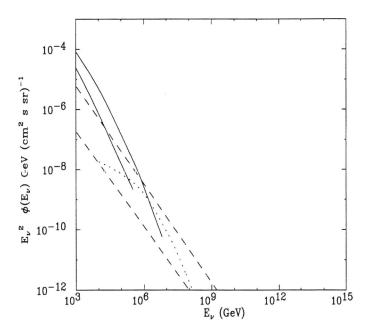

FIGURE 7. (from top to bottom where listed) Conventional Atmospheric Neutrino flux predictions from pion and kaon decays for 90^0 and 0^0 zenith angles (solid) and a prompt flux calculation from charmed particles (dotted) [71] compared to neutrino flux predictions from CR interactions with the galactic matter for 0^0 and 90^0 galactic latitude (dashes) [72].

but those reaching to highest energies are likely to be extragalactic. The most representative of this class are Active Galactic Nuclei (AGN). The γ-rays from the π^0 decays and their subsequent cascade would give rise to the high energy photons which have been detected (see previous section). The neutrinos come from the charged pion decays and constitute a signature of proton acceleration. Flux predictions from AGN cores and jets are show in Fig. 8 for comparison. The mechanisms responsible for the mysterious Gamma Ray Bursts (GRB), still under heavy debate, are also possible candidates to produce high energy neutrino fluxes in spite of the fact that little is know with certainty from them.

More exotic candidates are Primordial Black Holes (PBH), decays of topological defects (TD) or WIMP annihilation. These sources are however not yet known to exist. TD models arise naturally in grand unified theories of particle interactions with spontaneous symmetry breakdown, when some field vacuum goes through a phase transition to a new degenerate vacuum as the Universe cools down in its expansion. Different regions of space go to different vacua and the net distribution may evolve later into a vacuum field with non-trivial topology, surrounding a point (monopole), a line (string) or a surface (domain wall). These cosmological objects accumulate energy and when they interact with themselves or with other objects of

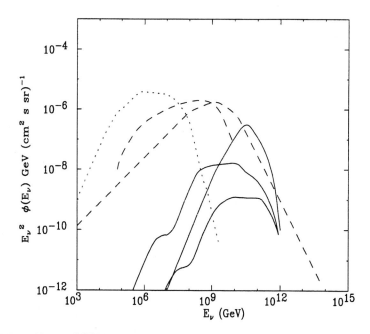

FIGURE 8. Fluxes of GZK neutrinos for $z_{ult} = 2.2$ [73], and $z_{ult} = 4$ and $z_{ult} = 2$ for ref. [33] (solid from top to bottom) compared to those from AGN jets in refs. [59] and [58] (dashed from left to right) and to earlier predictions from AGN cores [74] (dotted)

their own nature, they annihilate liberating large amounts of energy in the form of X particles, the Gauge bosons of the underlying grand unified theory. Topological defect (TD) scenarios have been given a lot of attention recently because they avoid the difficulties inherent to the acceleration of cosmic rays to the highest energies.

The models are however very uncertain because they are significantly affected by a variety of parameters besides an overall unknown normalization. Several authors have normalized injection rates to cosmic or gamma ray measurements and bounds [41,75,44] obtaining neutrino flux predictions. The injection rates of X particles per comoving volume are usually parameterized as t^{p-4} with $p = 0$ for superconducting cosmic strings, $p = 1$ for monopoles and cosmic strings and $p = 2$ for constant injection. The flux predictions extend to very high energies dictated by the mass of the assumed X particles, expected to be of order 10^{14} – 10^{16} GeV. The spectral index of the neutrinos is governed by the combination of the parameter p and the fragmentation model used and quite generally they are pretty flat. Fig. 9 illustrates some predictions for different p, different M_X and using the injection rates compatible with observations [44]. Some of the highest predictions have already been ruled out by experiments [45,44]. Unfortunately even the injection rate normalization to cosmic or gamma rays is also affected by

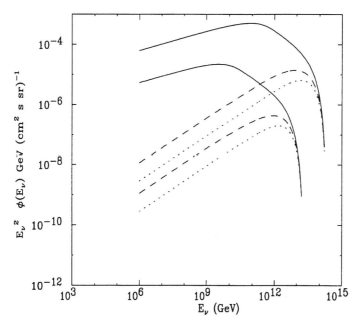

FIGURE 9. Neutrino flux predictions from models with decays of topological defect with evolution parameters $p = 0$ (full), 1 (dashed) and 2 (dots). The upper (lower) set corresponds to $M_X = 10^{15}$ GeV ($M_X = 10^{14}$ GeV).

uncertainties, mostly due to the interactions with the poorly known extragalactic B fields [44].

Although it is presently impossible to predict the most interesting energy range for detecting high energy neutrinos because of the number of open possibilities, it is interesting to remark the improved chances for EeV neutrinos. The exotic but attractive possibility of TD models, the association of the highest energy γ-rays to Blazars and models with acceleration in the jets and the unavoidable neutrino source, the GZK cutoff, all extend up and beyond the EeV region. This has important implications for detection which are discussed below.

Interactions

The peculiar behavior of the neutrino interactions has important implications for the detection of these particles. In the Standard Model neutrinos only have weak interactions which are suppressed by the massive gauge bosons propagator exchange (W^{+-} and Z^0). Their cross sections with fermions are very low, rising linearly with the center of mass energy until the four momentum transfer becomes comparable to the boson mass. Their interactions with hadrons are given by the standard Deep Inelastic Scattering cross sections in terms of convolutions with the parton distributions of the nucleons. We will refer to these as DIS hadronic interactions. At high energies propagator effects become relevant and the differential cross section for the charged current interaction (mediated by W^{+-} exchange) is the standard convolution of the parton distribution functions [77,76]:

$$\frac{d\sigma^{cc}}{dy} = \frac{G_F^2 m_p E_\nu}{\pi} \int_0^1 dx \left[\frac{M_W^2}{M_W^2 + 2m_p E_\nu xy}\right]^2 \quad (30)$$

$$\left[\left(1 - y - \frac{m_p xy}{2E_\nu}\right) F_2^{cc}(x, Q^2) + y^2 x F_1^{cc}(x, Q^2) + y\left(1 - \frac{y}{2}\right) x F_3^{cc}(x, Q^2)\right]$$

where y is the fraction of energy transferred to the nucleon in the lab frame, Q^2 is minus the square of the 4-momentum transfer and x is related to them through $Q^2 = 2m_p E xy$. The structure functions at zeroth order for charged current interactions in the case of isoscalar targets are simply:

$$F_2^{\nu\ cc}(x, Q^2) = x(u + d + 2s + 2b + \bar{u} + \bar{d} + 2\bar{c} + 2\bar{t}) \quad (31)$$

$$xF_3^{\nu\ cc}(x, Q^2) = x(u + d + 2s + 2b - \bar{u} - \bar{d} - 2\bar{c} - 2\bar{t}) \quad (32)$$

$$F_2^{\bar{\nu}\ cc}(x, Q^2) = x(u + d + 2c + 2t + \bar{u} + \bar{d} + 2\bar{s} + 2\bar{b}) \quad (33)$$

$$xF_3^{\bar{\nu}\ cc}(x, Q^2) = x(u + d + 2c + 2t - \bar{u} - \bar{d} - 2\bar{s} - 2\bar{b}) \quad (34)$$

$$F_2^{cc}(x, Q^2) = 2xF_1^{cc}(x, Q^2) \quad (35)$$

where x is to be interpreted as the momentum fraction carried by the partons. The dependence of the parton distribution functions on x and Q^2 has been omitted for clarity. For neutral current interactions the corresponding expressions neglecting target mass contributions are:

$$\frac{d\sigma^{nc}}{dy} = \frac{G_F^2 m_p E_\nu}{\pi} \int_0^1 dx\, x \left[\frac{M_Z^2}{M_Z^2 + 2m_p E_\nu xy}\right]^2 \quad (36)$$

$$\left[[A_1(z)^2 + B_1(z)^2(1-y)^2]\left(\frac{1}{2}(u+d+\bar{u}+\bar{d}) + s + b + \bar{s} + \bar{b}\right)\right.$$

$$\left. + [A_2(z)^2 + B_2(z)^2(1-y)^2]\left(\frac{1}{2}(u+d+\bar{u}+\bar{d}) + c + t + \bar{c} + \bar{t}\right)\right]$$

Here $A_{1,2}(z)$ and $B_{1,2}(z)$ are four first order polynomials in $z = sin^2\theta_w$, and θ_w is the electroweak mixing angle:

$$A_1^\nu(z) = B_1^{\bar{\nu}}(z) = -\frac{1}{2} + \frac{1}{3}sin^2\theta_w \quad (37)$$

$$B_1^\nu(z) = A_1^{\bar{\nu}}(z) = \frac{1}{3}sin^2\theta_w \quad (38)$$

$$A_2^\nu(z) = B_2^{\bar{\nu}}(z) = \frac{1}{2} - \frac{2}{3}sin^2\theta_w \quad (39)$$

$$B_2^\nu(z) = A_2^{\bar{\nu}}(z) = -\frac{2}{3}sin^2\theta_w \quad (40)$$

At very high energies the propagator becomes effecting in suppressing the cross section when Q^2 becomes larger than the square of the boson mass. As a result of the convolution in x the main contribution to the cross section is due to the values of $Q^2 = 2m_p Exy \lesssim M_{W(Z)}^2$, that is:

$$x \lesssim \frac{M_{W(Z)}^2}{2m_p Ey} \quad (41)$$

As the neutrino energy rises, the momentum fraction of the partons becomes lower and the cross section rises because of the well established increase in the parton densities at low x [83,84]. Calculations of the cross sections are unavoidably performed using extrapolations of the measured structure functions to kinematical regions in x and Q^2 which are inaccessible to accelerators. The uncertainties in the cross section become larger at high energies because the relative contributions of the unmeasured kinematical regions grow. Fig 10 shows cross section calculations for different extrapolations of the structure functions (according to different parton function parameterizations) illustrating the uncertainties. Except for energies above the ~PeV range neutrinos can travel through the Earth practically unattenuated.

Detection

At low energies (in the MeV-GeV range) although the cross section is low the fluxes from the sun, SN1987A and atmospheric neutrinos are quite high and it has

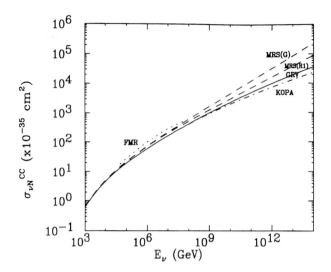

FIGURE 10. Total neutrino charged current cross sections using different structure functions; FMR,MRS(G),MRS(R1),GRV and KOPA from Refs [78-82].

been possible to detect few interactions in large scale underground detectors. At higher energies the flux expectations become a great dealer smaller as explained above, and the detectors have to become extremely large to have any chance to detect neutrinos, in spite of the growth of the cross section.

Cherenkov Detectors Underground

The developping projects to detect high energy neutrinos try to use large natural environments to detect throughgoing muons produced in charged current neutrino interactions. This is sometimes referred to as the "conventional" technique. Large volumes of water or ice are presently being instrumented with phototubes to detect the Cherenkov light from muons [10]. If the direction of these muons can be determined, it is possible to select muons that travel upwards in those detectors and that have to have been induced by neutrino interactions. Otherwise there is a large flux of downgoing muons produced by cosmic ray interactions in the atmosphere. High energy muons have ranges of several kilometers extending the interaction volume well beyond the region where the photomultipliers are instrumented. Two experiments are already taking data, AMANDA using Antarctic ice and BAIKAL using water of lake Baikal and there are two other projects ANTARES and NESTOR which will use ocean water.

Besides detecting muons the neutrino induced showers can also be detected. Conventional detectors in water or ice may, in addition, detect Cherenkov light from the showers with the advantage of being also sensitive to the electron neutrino interactions. This may become most important at high energies (above $\sim$PeV) when the showers will be very large and the Earth becomes opaque to neutrinos. They still require good directional and shower reconstruction capabilities to separate neutrino showers from those induced by atmospheric muons.

Coherent pulses

There have been suggestions for alternative techniques to detect neutrino induced showers and all of them become most interesting at the highest energies for different reasons. One possibility to detect neutrino showers consists in using hydrophones to detect the coherent sound pulse emitted by very large showers. This can be done in water and it may be possible to use existing sonar arrays. The technique may become competitive for the highest energies [85] if it can be developed because it is a coherent effect. The wavelength of the emission is much larger than the separation between particles so that the signal from each shower particle adds coherently and the total scales with the square of the primary energy. The detection of radio pulses from the showers is also coherent in the same respect.

A second possible alternative is the radio technique, proposed almost 40 years ago and consisting on detecting the radio waves emitted by the showers [86]. The Cherenkov light emitted is not restricted to the optical band but extends in frequency from the radio waves as long as the medium is transparent to the radiation. Moreover the emission from each particle becomes coherent as long as the wavelength of the radiation is larger that the dimension of the shower, and the electric field becomes proportional to the excess charge. Showers develop an excess negative charge because of the absence of positrons in matter. The annihilation of the positrons in the shower, and the acceleration of matter electrons into it contribute to about a 20% excess negative charge. As a result the energy in radio emission becomes proportional to the square of the shower energy, making the method most advantageous for the highest energies.

Ice has been studied as a possible medium for this experiment where radio waves have attenuation lengths exceeding 1 kilometer; both sand and salt have also been considered. For the detector the phototubes are simply replaced by cheaper and more sparsely spaced radio antennas that would sample the Cherenkov radio cone [88]. Many experimental difficulties are however expected [90] and presently the method is being tested together with AMANDA in Antarctica (RICE) [91].

Typically the electric field spectrum of the radio pulse displays a complex diffraction like pattern which depends on the frequency considered, as shown in Fig 11. The maximum of the emission is emitted along the Cherenkov cone with an angular width that depends linearly on frequency, about a degree at 1 GHz to ten degrees at 100 MHz [87]. At very high energies the electromagnetic showers behave very differently in dense media such as ice or the Moon surface, because of the Landau-

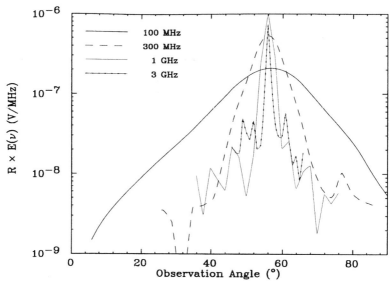

FIGURE 11. Fourier components of the Electric field strength of a radio pulse generated by a 100 TeV shower as a function of observation angle to the shower axis. Note the width of the Cherenkov pulse and its dependence on frequency.

Pomerančuck- Migdal (LPM) effect. This is a drastic reduction in the cross section for pair production and bremsstrahlung, the main interactions that govern shower development. As a result the showers can become very long (hundreds of radiation lengths) and the radio emission becomes narrower [89]. This effect is incidentally also important for all other techniques relying on detecting showers in dense media.

An interesting possibility has been suggested to detect radio waves from showers in the Moon surface [92]. Although the attenuation length is much shorter it is possible that radio signals from very high energy showers produced close to the surface can be detected from the Earth using radiotelescopes. The sensitive volume exceeds that achievable in the conceivable extensions of the other alternatives. This may prove to be also a technique for detecting ultra high energy cosmic ray showers. The cosmic rays in fact can constitute a problem for detecting the neutrino showers. Because of the refraction index of the Moon surface, cosmic rays must hit the surface of the Moon tangentially and only those hitting the rim as seen by us can be detected from the Earth. Depending on the angular spread of the pulse this condition can be relaxed and the rim around the Moon becomes larger for the lower observational frequencies. The inner region of the Moon, excluding this cosmic ray rim, can be used for searching the neutrino showers. Calculation of expected rates are very sensitive to the telescope parameters and the noise which has to be determined experimentally. As a guide some reference calculations are tabulated in Table 1 for different frequencies both for cosmic rays and for neutrinos

TABLE 1. Yearly neutrino event rates expected in a radiotelescope for different operational frequencies. The shower energy threshold is estimated for the case of electromagnetic showers assuming a noise level of 250 K, a 10% bandwidth and a 200 m dish. Number in brackets indicates the neutrinos expected in the central region of the Moon, where no cosmic ray signals are expected.

ν_{op} (MHz)	E_{th} (TeV)	Cosmic Rays	ν (AGN)Ref. [58]	ν (GZK) [73]
300	$2\,10^8$	2,000	0 (0)	0 (0)
500	$4\,10^7$	14,000	1 (0)	8 (0)
10^3	$9\,10^6$	80,000	24 (1)	19 (1)
$3\,10^3$	$3\,10^6$	60,000	17 (5)	3 (1)
$5\,10^3$	$3\,10^6$	20,000	6 (2.5)	1 (0.5)
10^4	$4\,10^6$	3,000	1 (0)	0 (0)

from a number of predictions [93].

Horizontal Showers.

Air shower arrays, Cherenkov telescopes and fluorescence detectors, described in the first section can also be used to detect showers in the atmosphere induced by all flavor neutrinos. It has been known for long that Horizontal Air Showers (HAS) measurements are sensitive to deeply penetrating particles such as muons and neutrinos [94]. At large zenith angles the electromagnetic component of cosmic ray showers gets absorbed in the slant atmospheric depth that can reach up to 1000 radiation lengths and only muons and neutrinos from the shower reach ground level. At large zenith angle the cosmic ray showers practically only have muons and a small electromagnetic component associated to bremsstrahlung of the highest energy muons which makes them significantly different from ordinary air showers both of electromagnetic and hadronic type. The cosmic ray shower rate decreases rapidly with zenith angle. Shower size determinations can be used as an energy threshold.

Existing air shower array data have already been used to constrain models for diffuse neutrino fluxes [95,96] but it is generally accepted that at intermediate shower sizes (energies $\sim$1 PeV) these measurements will be more effective for establishing the prompt component of the atmospheric muon flux [69]. Muons interact through bremsstrahlung, electron positron pair production (virtual photon bremsstrahlung) and hadronic interactions which are approximately equally relevant as energy loss mechanisms. Bremsstrahlung is the hardest interaction and it becomes the dominant mechanism of horizontal showers up to the highest measured shower sizes of about 10^5 particles. Fig. 12 illustrates the horizontal shower rates expected for some of the discussed neutrino fluxes compared to the old data from the university of Tokyo [97] and to some simple 90% confidence upper bounds that could be established on the basis of non observation of horizontal events in a year [45]. They are estimated on theoretical grounds for three different array sizes 0.1 km² (A), 100 km² (B) and 10,000 km² (C).

The atmospheric muon flux has a very steep spectral index between 3 and 4 and at very high energies its contribution to the horizontal air shower rate becomes undetectable. For this reason it is expected that future projects for detecting the highest energy showers such as the Pierre Auger Observatory, the Telescope array and HiRes will play an important role also as neutrino telescopes. Fig. 13 displays the expected neutrino acceptance of the Pierre Auger Observatory for neutrino induced showers above $\theta_{th} = 75^\circ$ [68,98]. This detector will be particularly well suited for neutrino detection because it consists of water cherenkov tanks which are similarly effective for detecting particles from vertical and horizontal showers. In table2 we give some results of the amount of expected neutrino event rates that can be expected in the Pierre Auger project for a number of neutrino flux predictions from different possible types of neutrino sources [68].

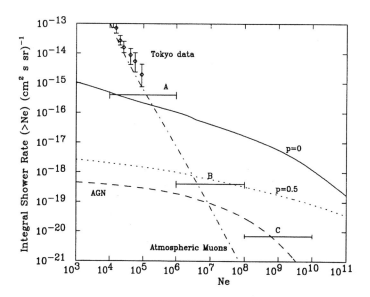

FIGURE 12. Integral horizontal shower rate as a function of shower size (Ne) for different models. TD neutrinos with $p = 0$ (solid line), $p = 0.5$ (dotted), AGN neutrinos (dashed) and atmospheric muons (dot-dashed). Diamonds represent the HAS data from ref 24. Horizontal lines marked as A, B and C represent the expected limit of non observation in a year made with ideal detectors of areas 10^5 m^2 (A), 10^8 m^2 (B), and 10^{10} m^2 (C). See text.

TABLE 2. Range of yearly neutrino event rates expected in one of the 3000 km² arrays in the Pierre Auger Project for several fluxes discussed above. The spread in the value is indicative uncertainties in the array acceptance and in the neutrino cross section.

	AGN-cores [99]	AGN-jets [58]	GZK $z_{ult}=4$ [33]	TD $p=1.5$ [41]
Yearly event rate	0.2-1.5	2-7	0.1-0.4	2-10

Implications of Existing Experimental Results

There are already some experimental results in the form of upper bounds provided by three types of experiments. One is from underground muon detectors, of which Fréjus provides the most stringent limit [100], the other two are from Extensive Air Shower detectors, particle detector arrays such as AKENO and EAS-TOP and a fluorescence light detector, Fly's Eye. Their results are not straightforward to convert to bounds on differential neutrino spectra because the conversion involves an assumption on the shape of the neutrino spectrum. Moreover there are important uncertainties in the high energy neutrino cross sections, besides the usual experimental uncertainties associated with each of the experiments. Fig. 14 compares

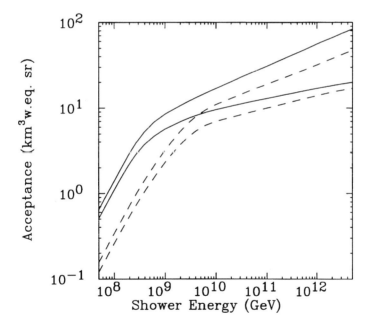

FIGURE 13. Expected acceptance of the Pierre Auger observatory as a function of shower energy for different horizontal showers.

atmospheric fluxes, some TD fluxes and fluxes from AGN jets to these bounds. The sets of parallel bounds for each experiment indicates the uncertainty in the number from different assumptions about the neutrino spectral index assumed in the conversion. Some of the fluxes are clearly in conflict with experiment. Some of the earlier models for diffuse neutrino fluxes from AGN and from the annihilations of Superconducting Cosmic Strings are ruled out by underground detectors and by the muon poor horizontal shower bound from AKENO [45].

I have stressed the importance that in my opinion the highest energy neutrinos have for the future of this field, particularly in the light of recent theoretical developments. One of the principal challenges is the detection of the low level neutrino fluxes from the GZK cutoff. Their observation would provide additional information on the origin of the high energy cosmic rays closing the loop between cosmic ray, γ-ray and neutrino astronomy. Hopefully in the near future we will have some neutrino events. Underground muon detectors with a 1 km^2 surface area will have very enhanced acceptances for muon neutrinos because of the long range of the muon produced in charged current interactions, and can detect neutrinos for energies starting from roughly 100 GeV or so. The Pierre Auger Project could have an acceptance comparable to 1 km^3 for contained events and electron neutrinos.

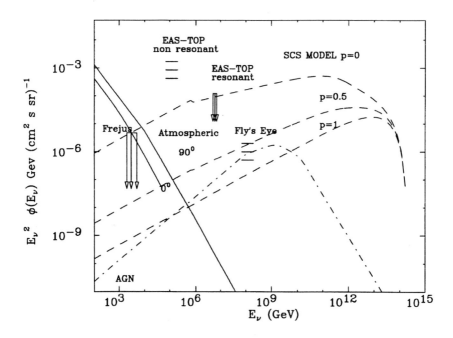

FIGURE 14. Neutrino flux predictions compared to existing bounds from experiment. Parallel lines indicate the uncertainty associated to the spectral index assumed for the conversion to a flux bound.

Lastly the radio technique may open new possibilities of exploring even larger energies and lower fluxes if difficulties are superated. The complementarity between each of the detector types would no doubt constrain any hypothetically detected flux and allow the extraction of much more precise information.

SUMMARY

We have discussed cosmic rays, gamma rays and neutrinos and important interdependences of these fields. The spectacular development of Astroparticle Physics in the last years is due to the development of new detection techniques. The great new improvements in the experimental facilities have lead to new discoveries and good perspectives for future ones. For these reason we have given a considerable amount of time to the study of the principles behind these techniques discussing in some detail several aspects related to detection possibilities. The future of the field looks promising. We know cosmic rays exist at the highest energies, we also know that there are very high energy $\gamma-$ rays from at least two types of objects and we know there have to be neutrinos of the highest energies at least from the interactions of cosmic rays with the microwave background. The experimental situation is also quite promising. For neutrinos we know of techniques which have the potential to be developed to measure high energy fluxes. For cosmic rays and photons we now have detector designs to measure the spectra of these high energy particles in great more detail. The possibilities of detecting other particles besides photons that keep directional information could well force the extension of the realm of Astronomy to neutrinos and the highest energy cosmic rays in the very near future, blurring the border between Astronomy and Astroparticle Physics. A lot of effort is however required to turn all these projects into successful advances in our knowledge of the Universe and the interactions that govern it.

ACKNOWLEDGEMENTS

I am indebted to Jaime Alvarez Muñiz, Gonzalo Parente and Ricardo Vázquez for many discussions and comments after reading the manuscript and for help with a number of figures and I thank CICYT (AEN96-1773) and Xunta de Galicia (XUGA-20604A96) for supporting my work in the field.

REFERENCES

1. B. Rossi; *High Energy Particles*, Prentice-Hall, Inc., Engelwood Cliffs, N.J. (1952).
2. T.K. Gaisser, *Cosmic Rays and Particle Physics*, Cambridge University Press.
3. P. Sokolsky; *Introduction to Ultrahigh Energy Cosmic Ray Physics*, Redwood City, Cal. Addison-Wesley (1989).
4. Berezinsky, V.S. et al., *Proc. Astrophysics of Cosmic Rays*, Elsevier, (1991).
5. M.S. Longair; *High Energy Astrophysics* Vol. 2, Cambridge Univ. Press (1994).
6. See for example Bahcall, J.; in *Proc. of the 18th Texas Symposium on Relativistic Astrophysics*, eds. A. Olinto, J. Frieman and D.Schramm; World Scientific, Singapore, (1997).
7. For a review see for instance, Schramm, D.N., *Proc. Les Rencontres de Physique de la Vallée d'Aoste* (La Thuile, Italy 1988); La Thuile Rencontres (1988), p. 183. Fermilab preprint Conf-88/45-A.
8. R.J. Protheroe in "Towards the Millennium in Astrophysics: Problems and Prospects", Erice 1996, eds. M.M. Shapiro and J.P. Wefel (World Scientific, Singapore); e-print Archive: astro-ph/9612212.
9. T.C. Weekes, *Phys. Rep.* **160** (1988) 1-1211.
10. T.K. Gaisser, F. Halzen and T. Stanev, Phys. Rep. 238 (1995) 173.
11. R.J. Protheroe in "Towards the Millennium in Astrophysics: Problems and Prospects", Erice 1996, eds. M.M. Shapiro and J.P. Wefel (World Scientific, Singapore); e-print Archive: astro-ph/9612213.
12. E. Zas, *Proc. of XXXIInd Rencontres de Moriond*, "High-Energy Phenomena in Astrophysics", Les Arcs, Jan 1997, eds. Giraud-Heraud, Y. and Tran Thanh Van, J. (Editions Frontieres, Paris), pp 37-42.
13. Heitler, W., *The Quantum Theory of Radiation*, Clarendon Press, Oxford (1954).
14. For a recent review see S.R. Klein, e-Print Archive: hep-ph/9802442.
15. L. Landau and I. Pomerančuk, *Dokl. Akad. Nauk SSSR* **92** 535 (1953); **92** 735 (1953). A.B. Migdal, *Phys. Rev.* **103** 1811 (1956); *Zh. Eksp. Teor. Fiz.* **32** 633 (1957) [*Sov. Phys. JETP* **5** 527 (1957)].
16. G. Molière, *Z. Naturforsch* **3a** 78 (1948).
17. H. Snyder and W.T. Scott; *Phys. Rev.* **85** 245 (1952).
18. W.T. Scott; *Phys. Rev.* **76** 220 (1949).
19. H.A. Bethe; *Phys. Rev.* **89** 1256 (1953)
20. J. Nishimura; *Theory of Cascade Showers*, (ed. Flugge, S.) Springer, Berlin (1967) *Handbuch der Physik* Bd. **XLVI**/2.
21. R.A. Vázquez, *Astropart. Phys.* **6** (1997) 411-422.
22. A.A. Watson, *Nuc. Phys. (Proc. Supp.)* **22B** (1991) 116.
23. A. Karle et al., *Astropart. Phys.* **3** (1995) 321.
24. M. Nagano et al., *J. Phys. G: Nucl. Part. Phys.* **18**(1992) 423.
25. For a recent review on the experimental status of high energy cosmic rays see *Design Report of the Pierre Auger Collaboration*, Fermilab Report, February 1997, and references therein.
26. M. Punch et al., *Nature* **358**, 477–478 (1992); J. Quinn et al., *Ap. J.* **456**, L83 (1996); F. Krennrich, et al., *Ap. J.* **481**, (1997) 758-763.

27. S.M. Bradbury et al., *Astron.Astrophys.* **320** (1997) L5-L8; D. Petry et al., *Astron.Astrophys.* **311** (1996) L13-L16.
28. D.J. Bird et al., *Phys. Rev. Lett.* **71** (1993) 3401.
29. F. Halzen, R.A. Vázquez, T. Stanev and CH. Vankov, *Astropart. Phys.* **3** (1995) 151-156.
30. For a good review on air shower and the radio technique see H.R. Allan, *Progress in Elementary Particles and Cosmic Ray Physics* **10** (1971) pp. 171-302 (North Holland Publ. Co.) and references therein.
31. M.T. Dova et al., *Proc. XXVth International Cosmic Ray Conference*, Durban, South Africa, Ed. M.S. Potgieter et al. (1997), Vol. 7, pp. 373-376.
32. K. Greisen; *Phys. Rev. Lett.*, **16** (1966), 748. G.T. Zatsepin and V.A. Kuz'min, *JETP Lett.*, **4** (1966), 78.
33. S. Yoshida and M. Teshima, Prog. of Theo. Phys. 89 (1993) 833.
34. L.O'C. Drury, 1990. *Proc. of Astrophysical aspects of the most energetic cosmic rays* Kofu (1990), Ed. M. Nagano and F. Takahara, World Scientific (1991) pp. 252-260; Matthew G. Baring, *Proc. of 32nd Rencontres de Moriond* "High-Energy Phenomena in Astrophysics", Les Arcs, Jan 1997, eds. Giraud-Heraud, Y. and Tran Thanh Van, J. (Editions Frontieres, Paris), p. 97. e-Print Archive: astro-ph/9711177.
35. T. Tanimori et al. Submitted to *Astrophys. J. Lett.* e-Print Archive: astro-ph/9801275.
36. W.I. Axford, in *Proc. of Astrophysical aspects of the most energetic cosmic rays* Kofu (1990), Ed. M. Nagano and F. Takahara, World Scientific (1991) p. 406. Kofu (1990) pp. 252-260
37. A.A. Wolfendale and J. Wdowczyk in *Cosmic Rays, Supernovae and the Interstellar Medium*, eds. M.M. Shapiro, R. Silberberg and J.P. Wefel, Kluwer Acedemic Press, Dordrecht (1991) p. 313.
38. R. Plaga To appear in *5th Workshop on TeV Gamma Ray Astrophysics: Towards a Major Atmospheric Cerenkov Detector*, Kruger National Park, South Africa, 8-11 Aug 1997. e-Print Archives: astro-ph/9712321, astro-ph/9711094.
39. E. Waxman and J. Bahcall, *Phys. Rev. Lett.* **78** (1997) 2292-2295.
40. J.H. MacGibbon and B.J. Carr, Astrophys. J. **371** (1991), 447; F. Halzen, B. Keszthelyi and E. Zas, *Phys. Rev* **D53** (1995) 3239-3247.
41. P. Battacharjee, C.T. Hill, D.N. Schramm, Phys. Rev. Lett. **69** (1992) 567.
42. T.J. Weiler, *Ap. J.* **285** (1984) 495.
43. T.J. Weiler e-Print Archive: hep-ph/9710431; D. Fargion, B. Mele and A. Salis, e-Print Archive: astro-ph/9710029.
44. R.J. Protheroe and T. Stanev, *Phys. Rev. Lett.*, **77**, 3708 (1996), and *Erratum* **78**, 3420 (1997).
45. J.J. Blanco-Pillado, R.A. Vázquez and E. Zas, *Phys. Rev. Lett.*, **78**, 3614 (1997).
46. D.J. Thompson et al., *Ap. J. Suppl.* **101** (1995) 259; *Ap. J. Suppl.* (1996).
47. J.R. Mattox et al., *Ap. J.* **481** (1997).
48. M.J. Lang et al., *Nucl. Phys. Proc. Suppl.* **14A** (1990) 165-168.
49. M.J. Rees, Nature **211**, 468 (1966).
50. M. Jang and H.R. Miller, *Ap. J.* **452** (1995) 582.

51. D. J. Macomb et al., *Ap. J.* **438**, 59; **446**, 99 (1995).
52. C.M. Urry and P. Padovani, *Pub. Astr. Soc. Pacif.* **107**, 803–845 (1995).
53. for a recent review of jet formation in accretion disk see H.C. Spruit, to appear in *Physical Processes in Binary Stars* Ed. M.B. Davies and C.A. Tout, Kluwer Acedemic Press, Dordrecht (1996) (NATO ASI series), e-Print Archive: astro-ph/9602022.
54. J. Chiang et al., *Ap. J.* **452**, 156 (1996).
55. T. Stanev et al., *Phys. Rev. Letters* **75**, 3056 (1995). H. Meyer and S. Westerhoff (for the HEGRA collaboration), in *Proceedings of the IXth International Symposium on Very High Energy Interactions*, Karlsruhe (1996); W. Rhode, High Energy Neutrino Workshop, Aspen, Colorado (1996).
56. F. W. Stecker, O. C. De Jager and M. H. Salamon, *Ap. J.* **390**, L49 (1992).
57. M. Sikora, M. C. Begelman and M. J. Rees, *Ap. J. Lett.* **421**, 153 (1994) and references therein.
58. K. Mannheim, Astrop. Phys. 3 (1995) 295-302.
59. R.J. Protheroe, in Proc. IAU Colloq. 163, Accretion Phenomena and Related Outflows, ed. D. Wickramasinghe et al., in press (1996).
60. K. Mannheim, S. Westerhoff, H. Meyer and H. H. Fink, *Astron. Astrophys.* **315** (1996) 77-85.
61. R. J. Protheroe, Gamma Rays and Neutrinos from AGN Jets, Adelaide preprint ADP-AT-96-4 (1996).
62. F. Halzen and E. Zas, *Ap. J.* **488** (1997) 669-674.
63. J. Bahcall, to appear in *Proc. of the 18th Texas Symposium on Relativistic Astrophysics* eds. A. Olinto, J. Frieman and D.Schramm (World Scientific, Singapore, 1997). e-Print Archive: astro-ph/9702057.
64. See for instance D.N. Schramm, *Comments Nucl. Part. Phys.* **17** 5 (1987) pp 239-278 and references therein.
65. K.S. Hirata et al., *Phys. Lett.* **B280** (1992) 146-152.
66. L.F. Abbott, A. de Rújula and T.P. Walker *Nucl. Phys.* **B299** (1988) 734-756.
67. F. Halzen, The Case for a Kilometer-Scale Neutrino Detector: 1996, in *Proc. of the Sixth International Symposium on Neutrino Telescopes*, ed. M. Baldo-Ceolin, Venice (1996); F. Halzen, High Energy Neutrino Astronomy and its Telescopes: the Case for a Kilometer-Scale Detector, in *Proceedings of 1994 Snowmass Summer Study: Particle and Nuclear Astrophysics in the Next Millenium*, ed. by E.W. Kolb and R.D. Peccei (World Scientific, Singapore, 1996) p. 256.
68. G. Parente and E. Zas, in *Proceedings of the 7th Int.Symposium on Neutrino Telescopes*. p. 345, ed. by M. Baldo Ceolin, Venice (1996).
69. E. Zas, F. Halzen and R.A. Vázquez, *Astropart. Phys.* **1**, 297 (1993).
70. P. Gondolo, G. Ingelman and M. Thunman, in *Proc. of the 4th International Workshop on Theoretical and Phenomenological Aspects of Underground Physics (TAUP 95)*, Toledo, Spain, 17-21 Sep 1995. Published in Nucl.Phys.Proc.Suppl.48:472-474,1996 e-Print Archive: hep-ph/9602402.
71. L.V. Volkova et al., *Nuovo Cim.* **10C** (1987) 465.
72. G. Domokos et al., *J. Phys. G: Nucl. Part. Phys.* **19** (1993) 899.
73. F.W. Stecker, C. Done, M.H. Salamon and P. Sommers, *Phys. Rev. Lett.* **66** (1991)

2697.
74. F.W. Stecker, C. Done, M.H. Salamon and P. Sommers, *Phys. Rev. Lett.* **69**, 2738(E) (1992).
75. G. Sigl, S. Lee and D.N. Schramm, *Phys. Lett.* **B392** (1997) 129–134.
76. R.M. Barnet *et al.*, "Particle Data Book" *Phys. Rev.* **D54** (1996) 1.
77. R. Gandhi, C. Quigg, M.H. Reno and I. Sarcevic, *Astropart. Phys.* **5** (1996) 81-110
78. G.M. Frichter, D.M. McKay and J.P. Ralston, *Phys. Rev. Lett.*, **74** (1995) 1508.
79. A.D. Martin, R.G. Roberts and W.J. Stirling, *Phys. Lett.* **B354** *(1995) 155.*
80. A.D. Martin, R.G. Roberts and W.J. Stirling, *Phys. Lett.* **B387** *(1996) 419.*
81. M. Glück, E. Reya and A. Vogt, *Z Phys. C Particles and Fields* **67** *(1995) 433.*
82. G. Parente and A.V. Kotikov, in *Proc. of low x Workshop* Miraflores de la Sierra, Spain, June 1997).
83. H1 Collaboration, *Nucl. Phys.* **B470** (1996) 3; ZEUS Collaboration *Zeit. Phys.* **C69** (1996) 607 and **C72** (1996) 399.
84. G. Parente and E. Zas, *Proc. XXVth International Cosmic Ray Conference*, Durban, South Africa, Ed. M.S. Potgieter *et al.* (1997), Vol. 7, pp. 109–112.
85. L.G. Dedenko *et al.*, *Proc. XXVth International Cosmic Ray Conference*, Durban, South Africa, Ed. M.S. Potgieter *et al.* (1997), Vol. 7, pp. 89–92.
86. G.A. Askar'yan, *Soviet Physics* JETP **14,2** 441–443 (1962); **48** 988–990 (1965).
87. F. Halzen, E. Zas, E. and T. Stanev, *Phys. Lett.* **B257** 432 (1991).
88. M.A. Markov and I.M. Zheleznykh, *Nucl. Inst. Methods* **A 248** 242–251 (1986).
89. J. Alvarez-Muñiz and E. Zas, *Phys. Lett.* **B411** (1997) 218-224.
90. J.V. Jelley, *Astropart. Phys.* **5** (1996)255-261.
91. C. Allen *et al.*, *Proc. XXVth International Cosmic Ray Conference*, Durban, South Africa, Ed. M.S. Potgieter *et al.* (1997), Vol. 7, pp. 85–88.
92. R.D. Dagkesamansky and I.M. Zheleznykh, *Pis'ma Zh. Eksp. Teor. Fiz. (JETP Lett.* **50** (1989) 233; *Proc. of Astrophysical aspects of the most energetic cosmic rays* Kofu (1990), Ed. M. Nagano and F. Takahara, World Scientific (1991) p. 373.
93. J. Alvarez-Muñiz and E. Zas, *Proc. XXVth International Cosmic Ray Conference*, Durban, South Africa, Ed. M.S. Potgieter *et al.* (1997), Vol. 7, pp. 309–312..
94. V.S. Berezinsky and G.T. Zatsepin *Phys. Lett.* **B28** (1969) 423; P. Kiraly *et al.*, *J. Phys. A:Gen. Phys.* **4** (1971) 367. V.S. Berezinsky and A. Yu. Smirnov, *Astrophys. Space Science* **32** (1975) 461.
95. E. Zas *et al.*, *Astropart. Phys.* **1** (1993) 297.
96. R.M. Baltrusaitis *et al.*, *Phys. Rev* **D31** (1985) 2192.
97. M. Nagano *et al.*, *J. Phys. Soc. Japan.* **30** (1971) 33; S. Mikamo *et al.*, *Lett. al Nuov. Cim.* **34** (1982) 273.
98. J. Capelle, J.W. Cronin, G. Parente and E. Zas; *Astropart. Phys.* **8** (1998) 321.
99. F.W. Stecker and M.H. Salamon, "TeV Gamma-ray Astrophys" (1994) 341-355 (QCD953:I48:1994) e-Print Archive: astro-ph/9501064.
100. W. Rhode, K. Daum, *et al.* Proc. 24th ICRC, Rome 1995. vol. 1, p. 781. Also, W. Rhode, *et al. Astropart. Phys.* **4** (1996) 217.

The Width and Height of Interferometry

Alejandro Ayala

Instituto de Ciencias Nucleares, Universidad Nacional Autónoma de México A.P. 70-543, México D.F. 04510, México

Abstract. The Hanbury-Brown Twiss effect in high-energy, heavy-ion collisions is discussed as an interference phenomenon, in terms of the quantum mechanical wave functions representing the particles produced after a reaction. We look into how these wave functions are detected by means of high-energy particle detectors and establish the conditions for the occurrence of interference. We introduce a simple analytical model to illustrate the consequences of this formulation in the computation of the two-particle correlation function and show that the scale parameter in the correlation function contains information on the size and coherence lengths as well as on the spectrum produced by the source. We also discuss the effect of multiple scattering during the detection process in the measured particle correlation properties

INTRODUCTION

Two particle correlation analyses in high energy nucleus-nucleus collision experiments are a powerful tool to obtain information on the space-time structure of the collision volume. Particle correlation studies are in a unique class of high energy physics experiments that are sensitive to the wave mechanics of the particles at the detectors.

Intensity (or second order) or HBT interferometry was pioneered in the mid 1950's by R. Hanbury-Brown and R. Twiss [1] who applied it to the measurement of the angular sizes of radio and optical stellar sources. While their understanding of the effect was in terms of classical waves, they also recognized that it could be explained by the bosonic nature of photons [2]. In the realm of particle physics, the same principle was incorporated by G. Goldhaber, S. Goldhaber, W. Lee and A. Pais in the description of the angular correlations of two like-charge pions produced in $p\bar{p}$ annihilation [3].

In recent years, the advent of more powerful accelerators and the necessity of an accurate characterization of the hadronic system formed after a high-energy collision have stimulated more thorough particle correlation analyses [4]– [7]. The picture that emerges is that, in addition to quantum statistics, the origin of correlations includes also dynamical as well as kinematical effects, coming from initial or final state interactions and conservation laws.

A fundamental property of the HBT effect is the fact that interference is a consequence of wave mechanics performed by particle detectors; the effect does not depend on the history of the interfering particles but only on the characteristics of their wave functions at the detector.

In this paper, we adopt this last point of view to look into the nature of the two like-particle correlation function. We study how we can represent these wave functions and how they are detected by means of typical high energy particle detectors. We find that a necessary condition for the occurrence of two-particle interference is that the wave functions, of each of the two particles, overlap at the two elements carrying the detection, within a time interval given by the inverse of the atomic frequency spread for ionization.

We argue that any model for the description of two-particle properties of a multiparticle system has to be also consistent with the description of the single particle properties. We review how in the absence of long range Coulomb interactions and under the source chaoticity assumption, the object that determines the single and two–particle properties is the correlation of two source currents. We introduce a representation of the source correlation to illustrate how its properties are revealed by interferometry.

Since extracting the information in the two-particle correlation function from its general momentum dependence is obscured by the fact that the correlation is not necessarily only a simple transform of the source distribution, we present a simple one-dimensional model and identify the elements that determine the scale parameter measured by interferometry.

This work is organized as follows: in Section 2 we first review how the detection process carried out by particle detectors can be represented in terms of the elementary interactions between a fast moving particle and the sensitive elements of the detector. We start by expressing the probability of detecting a charged particle by means of a device whose sensitive part consist of a single atom. This allows us to address questions regarding the conditions under which it is possible to obtain the HBT interference effect. In Section 3 we study the characteristics of source correlations and their relation to the two particle probability. We introduce a representation for the ensemble average of the product of two source currents and use it to look more closely at the details of the source correlations. In Section 4, we introduce a simple one-dimensional model to identify the information regarding the source characteristics. In Section 5, we briefly discuss the influence of multiple scattering during the detection process in the particle's measured correlation properties. We finally summarize and conclude in Section 6.

PARTICLE DETECTION AND INTERFEROMETRY

HBT is an interference phenomenon whose ultimate origin is the wave nature of the interfering particles. The wave properties of individual particles can be described by the single particle wave functions. When we are interested on the interference properties of a multiparticle system produced after a reaction, it is qualitatively clear that the interfering wave functions correspond to particles belonging to the same *event*. Since these wave functions must carry the information on the space-time size and correlation lengths of the region producing the particles, interferometry will then reveal such information. This considerations are at the core of the standard interpretations of interferometry results.

The above scenario however, does not exhaust the possible occurrences of interference. One could imagine for instance, a situation in which the origin of the interfering particles is unknown and all we know is that they are detected within some time interval. We can ask how to quantify the extent of this time interval and what information, if any, will be revealed by interferometry. By answering these questions, one can give a more precise meaning to the notion that in collision experiments, interference happens only within particles in the same event.

In order to address these questions, it is necessary to pay a closer look to the way particles are detected in a typical high-energy experiment. For simplicity, we concentrate on the detection of charged particles.

The one atom detector

Charged particles are usually detected experimentally through a series of position measurements performed by multiwire proportional chambers (MWPC's). After particle identification, a momentum is inferred by reconstructing a track from the set of position measurements. MWPC's are filled with a mixture of gases that easily ionize when a fast charged particle hits the chamber. Upon a hit, a cascade of ions drifts towards the wires which are organized in different planes with a potential difference between them, thus allowing one to determine accurately the location of the hit.

The detection process is described by the electromagnetic interaction of a wave packet, representing a charged incoming particle and an atomic electron. The interaction can be described perturbatively to lowest order. We are interested in accounting for processes that lead from an initial state, consisting of an incoming particle and a bound atomic electron in its ground state $|ig\rangle = |i\rangle|g\rangle$, to a final state consisting of a final particle and an electron promoted to an atomic excited state $|fa\rangle = |f\rangle|a\rangle$. The probability of detecting a single particle making a transition between given initial and final states, by means of a single atom detector, can be expressed as

$$\mathcal{P}^{(1)}_{fi} = \int d^3x d^3x' dt dt' s(t - t')[\varphi_i^*(x')\varphi_f(x')][\varphi_f^*(x)\varphi_i(x)] \tag{1}$$

where φ_i, φ_f are the single particle wave functions corresponding to the initial and final particle states, respectively. The spatial integrations d^3x and d^3x' are performed over the location of the given excited detector atom. The function $s(t-t')$ is called the sensitivity function [9] and determines the time extent over which the atom carries out the particle detection.

The detection is not necessarily a local process in time, it becomes more so as the detector becomes sensitive to a wider range of frequencies, $\Delta\omega$, since Δt, the characteristic spread in $t-t'$, is of order $1/\Delta\omega$. The frequency ω corresponds to the energy lost by the particle in its interaction with the atom. The average energy loss per unit distance traversed is a well known quantity that depends on the particular fill gas and particle's velocity range [10]. The fluctuations around this average for a single interaction correspond precisely to the one-atom detector bandwidth. The average energy loss per ion pair created at normal temperature and presure in argon is about 26.4 eV [11]. However, in occasional collisions, an energetic electron is knocked off with an energy on the order of a few KeV (delta electrons) [12]. Nevertheless, these are rare events contributing only to the tail energy loss distribution. The majority of ionization events are distributed around the average energy loss with a width corresponding to the atom's spread in binding energies, usually on the order of 10 eV.

The detection process is thus not exactly instantaneous since $\Delta t \sim 1/10$ eV$^{-1} \sim 10^{-16}$ sec, an eternity compared to typical oscillation periods of high energy particles! What is then the effect of this detection time uncertainty on the interpretation of HBT measurements? As we discuss in the next subsection, the answer is found in its interplay with one more time scale important for the interference phenomenon, the delay time between the arrival to the detector of two wave packets.

The interference phenomenon

Armed with the expression for the transition probability, Eq. (1), we turn to the detection of two indistinguishable Bose particles by means of different one-atom detectors. The amplitude for the *direct* process is the product of the amplitudes for $I \to a$ and $II \to b$. Analogously, the amplitude for the *crossed* process is the product of the amplitudes for $I \to b$ and $II \to a$. Since the particles are indistinguishable, these amplitudes interfere and the total amplitude is the sum of the above. The probability $\mathcal{P}^{(2)}$ of detecting the pair, the absolute square of the total amplitude, can be written as the sum of a term representing the probability of detecting the two particles as if they were distinguishable and a term involving the interference of the two different processes. For simplicity, we concentrate on the time structure of this probability,

$$\mathcal{P}^{(2)} \sim \int dt dt' s_a(t-t')[\varphi_I^*(t')\varphi_a(t')][\varphi_a^*(t)\varphi_I(t)]$$

$$\times \int dt dt' s_b(t-t')[\varphi_{II}^*(t')\varphi_b(t')][\varphi_b^*(t)\varphi_{II}(t)]$$
$$+ \int dt dt' s_a(t-t')[\varphi_I^*(t')\varphi_a(t')][\varphi_a^*(t)\varphi_{II}(t)]$$
$$\times \int dt dt' s_b(t-t')[\varphi_I^*(t')\varphi_b(t')][\varphi_b^*(t)\varphi_{II}(t)]. \quad (2)$$

As we see from the second term in eq. (2), the interference is proportional to the *overlap* of the both incident particle wave functions at *both* detectors. Note that there is no requirement that the interfering particles originate from the same space-time region. Let us concentrate on the last term in Eq. (2) and focus on one of its factors. Also, take the time dependence of the final state wave function to be that of a particle with a definite energy, E_a. The term under scrutiny then looks like

$$\mathcal{P}_{int}^{(2)} \sim \int dt dt' s_a(t-t') e^{iE_a(t-t')}[\varphi_I^*(t')\varphi_{II}(t)]. \quad (3)$$

Now, to represent the wave function of a particle approaching the detector, we can, in general, consider a wave packet of the form

$$\varphi_I(t) \sim \int dE f_I(E) e^{-iEt}. \quad (4)$$

The specific characteristics of the function f_I are not important for the present considerations. If for instance, the particles originate from the same collision volume, these characteristics will reflect the properties of the production region. But, as we have pointed out, all we need is that their wave functions coincide at the detector. Let us now proceed to quantify this last statement.

In addition to the time dependence given by eq. (4), let us also consider the influence of an explicit time delay t_d between the arrival of the wave packets. By means of eq. (4), we can rewrite eq. (3) as

$$\mathcal{P}_{int}^{(2)} \sim \int d\omega \tilde{s}(\omega) \int dt dt' \int dE dE' e^{i(E+E')t_d/2}$$
$$e^{i(\omega+E_a-E)t} e^{-i(\omega+E_a-E')t'} f_I^*(E') f_{II}(E) \quad (5)$$

where we have written

$$s(t-t') = \int d\omega \tilde{s}(\omega) e^{i(t-t')\omega}. \quad (6)$$

Performing the time and energy integrations in eq. (5), we get

$$\mathcal{P}_{int}^{(2)} \sim e^{iE_a t_d} \int d\omega \tilde{s}(\omega) e^{i\omega t_d} f_I^*(E_a+\omega) f_{II}(E_a+\omega). \quad (7)$$

Recall that the frequency spread of the sensitivity function corresponds to a bandwidth of ~ 10 eV whereas the typical energy of a particle can be on the order

$E_a \sim 1$ GeV. It is therefore safe to ignore the ω dependence on the functions f_I, f_{II}. By doing this, we see that eq. (5) can be written as

$$\mathcal{P}_{int}^{(2)} \sim e^{iE_a t_d} s(t_d) f_I^*(E_a) f_{II}(E_a). \tag{8}$$

The result is then that, even for particles in similar states ($I \sim II$), interference is significant only when the delay in the arrival time t_d is within the time interval given by the inverse of the detector bandwidth, $\Delta t \sim 1/\Delta w$ ($\sim 10^{-16}$) sec.

In a high energy collision, since the time delay between the arrival of particles of same energy in the same event is on the order of the production time interval ($\sim 10^{-21}$) sec, the function $s(t-t')$ can be approximated by $s(0)$, that is, a constant. On the other hand, the function s cuts off the possible interference between arrivals separated by more than Δt. This is the physical content of the intuitive picture discussed at the beginning of this section. In passing, we note that even for particles in the same event, a delay in the arrival time introduced by random interactions, such as multiple scattering in air or in the gas mixture of MWPC's, could in principle be large enough to produce a loss of interference signal.

PARTICLE INTERFEROMETRY AND SOURCE CORRELATIONS

The power of two-particle interferometry rests on the relationship that the conditional probability for detecting two indistinguishable particles bears with the properties of the region where the particles originate. As we have argued, interference between two indistinguishable particles happens when their single particle wave functions overlap at the detector (within the time corresponding to the inverse of the atomic bandwidth), regardless of the history of these wave functions. However, when these wave functions come from the system formed after a high energy reaction, they will carry information about the emission region such as its space-time extent and possible correlation lengths, as well as on the spectrum produced. In this case, interferometry is one mean to reveal such information.

The two-particle properties are just one aspect of the multiparticle field produced after a typical high energy reaction. It is important to keep in mind that the two-particle properties are not independent, for example, of the single particle properties of the field and that any description of the former has to be consistent with a description of the latter. One way of accomplish a consistent description is to use the language of the n-particle density matrix.

Density matrix and source correlations

For definitiveness, let us consider the interfering particles to be pions. The outgoing pion field, $\phi(x)$, after decoupling and in the absence of long range Coulomb interactions, can be thought of as originating from a source $J(x)$, whose details depend on the particular dynamics of the collision. The pion field obeys the Klein-Gordon equation

$$\left(\Box^2 + m_\pi^2\right)\phi(x) = -J(x), \tag{9}$$

and is equivalently given in terms of the source by

$$\phi(x) = \int d^4x' D_r(x - x') J(x'), \tag{10}$$

where $D_r(x)$ is the retarded free propagator.

The single particle properties of the field ϕ are given in terms of the single particle density matrix, or Green's function, which is the ensemble average of the product of $\phi^\dagger$ and ϕ

$$\langle \phi^\dagger(x_2)\phi(x_1)\rangle = \int d^4x_2' d^4x_1' D_r^\dagger(x_2 - x_2') D_r(x_1 - x_1') \langle J^\dagger(x_2') J(x_1')\rangle. \tag{11}$$

For freely propagating particles, the spatial Fourier transform of the single particle density matrix describes the number of pions $2E_k d^3N/d^3k$ per invariant unit volume of momentum space, $d^3k/[(2\pi)^3 2E_k]$, where

$$\frac{d^3N}{d^3k} = 2E_k \int d^3x d^3x' e^{ik(x'-x)} \langle \phi^\dagger(x)\phi(x')\rangle. \tag{12}$$

Similarly, the two particle properties of the field are given in terms of the two-particle density matrix,

$$\langle \phi^\dagger(x_4)\phi^\dagger(x_3)\phi(x_2)\phi(x_1)\rangle = \int d^4x_4' \ldots d^4x_1' D_r^\dagger(x_4 - x_4') \ldots D_r(x_1 - x_1')$$
$$\langle J^\dagger(x_4') J^\dagger(x_3') J(x_2') J(x_1')\rangle, \tag{13}$$

describing the two particle properties in terms the ensemble average of the product of four source currents. The distribution of pion pairs is given in terms of the two-particle density matrix by

$$\frac{d^6N}{d^3k d^3k'} = (2E_k)(2E_{k'}) \int d^3x_1 \ldots d^3x_4 e^{ik(x_1-x_4)} e^{ik(x_2-x_3)}$$
$$\langle \phi^\dagger(x_4)\phi^\dagger(x_3)\phi(x_2)\phi(x_1)\rangle \tag{14}$$

Let us now assume that particle production is *completely chaotic*, so that the sources of the outgoing pions factor as

$$\langle J^\dagger(x_4) J^\dagger(x_3) J(x_2) J(x_1)\rangle = \langle J^\dagger(x_4) J(x_1)\rangle \langle J^\dagger(x_3) J(x_2)\rangle$$
$$+ \langle J^\dagger(x_4) J(x_2)\rangle \langle J^\dagger(x_3) J(x_1)\rangle. \tag{15}$$

Equation (15) corresponds to a scenario in which possible initial phase information is lost through subsequent scattering in the interaction region. The relative positive sign symmetrizing the two terms in Eq. (15) reflects the bosonic nature of the system. In contrast, the completely *coherent* scenario, represented by just one of the terms in the right hand side of Eq. (15), corresponds to a situation in which a particular phase is preserved and applies for instance, to the case of a laser or a Bose-Einstein condensed source. Eqs. (11) and two particle characteristics for sources satisfying the factorization property, is the ensemble average, or correlation of the product of two currents $\langle J^\dagger(x_2)J(x_1)\rangle$.

On the other hand, as we have seen in the previous section, it proves convenient to describe the detection of particles in the field in terms of the elemental interactions of single particle wave functions with the detector. These wave functions correspond to the single particle approximation of the many body system formed after the reaction. One can also use these functions to represent the n-particle density matrix and thus, to describe the n-particle properties of the field. This representation provides a comprehensive picture of the characteristics of the multiparticle field. Moreover, it also makes it clear that the average in expressions such as eqs. (11) or (13) is taken over a mixed ensemble, which enphasizes the relationship between the multiparticle properties of the field with the spectrum produced after the reaction.

The single particle wave functions

We begin by representing the single particle density matrix in terms of single particle wave functions

$$\langle \phi^\dagger(x_2)\phi(x_1)\rangle = \sum_i F_i \varphi_i^*(x_2)\varphi_i(x_1) \qquad (16)$$

where $\{i\}$ stands for a set of states, F_i is the weight that corresponds to the state with label i and $\varphi_i(x)$ is identified with the single particle wave function in state i. Notice that the set of states is not entirely at our disposal since in order for eq. (16) to be a valid representation, the wave functions need to be eigenfunctions of the density operator. For the problem at hand, after the system of particles leave the production region and when ignoring long range interactions, we can chose the set of states to be the momentum basis and we call $\varphi_{\vec{p}}(x)$ the single particle wave packet. Eqs. (11) and (16) then show that the form of the wave packets is determined by the form of the ensemble average of the product of $J^\dagger$ and J. In order to make this relation explicit, we introduce a similar representation to eq. (16) for the current ensemble average

$$\langle J^\dagger(x_2)J(x_1)\rangle = \sum_{\vec{p},x_0} G(x_1,x_0)F_{\vec{p}}(x_0)G(x_2,x_0)e^{ip(x_1-x_0)}e^{-ip(x_2-x_0)} \qquad (17)$$

where we introduced the space-time point x_0 to label the location of the wave emitters within the source space-time volume and the ensemble average correspondingly includes also an average over these locations. $G(x,x_0)$ represents the region of wave formation around the center x_0. Notice that the model, eq. (17), describes the production of waves from the formation regions which are themselves distributed inside the overall space-time source volume. The distribution of the emitter centers is also a property of the source distribution which we have indicated by the dependence of $F_{\vec{p}}$ on x_0. We are thus describing a source with two intrinsic volumes of different nature: a space-time volume that can be called a *coherence* volume corresponding to the wave formation region and a space-time volume that corresponds to the overall source's size. Each formation region represents a distinct domain where the pions are produced by separate inelastic collisions. Their spatial extent correspond, roughly, to the size of a nucleon and their duration is given by the time during which two nucleons with large relative speed interact, roughly, the size of a nucleon divided by the speed of light. Thus, even for a heavy ion collision, these domains are only a few times smaller than the overall source's size and cannot in principle be neglected.

THE INFORMATION IN THE CORRELATION FUNCTION

Recall that experimentally, the two-pion correlation function, as a function of the relative momentum $\vec{q} = \vec{k}_1 - \vec{k}_2$, is defined by [14]

$$C_2(\vec{q}) = \frac{\left\{\langle n_{\vec{K}+\vec{q}/2} n_{\vec{K}-\vec{q}/2}\rangle\right\}_{\vec{K}}}{\left\{\langle n_{\vec{K}+\vec{q}/2}\rangle\langle n_{\vec{K}-\vec{q}/2}\rangle\right\}_{\vec{K}}} \qquad (18)$$

where the numerator represents the number of pions pairs with relative momentum $\vec{q}$ in a given event, averaged over all of the events and over the average pair momentum $\vec{K} = (\vec{k}_1 + \vec{k}_2)/2$, and the denominator is the number of pairs from two different events, also averaged over the events and over $\vec{K}$. Under the factorization assumption (15) for indistinguishable particles and in the absence of any other correlation effects, $C_2(\vec{q})$ should approach 2 as $\vec{q} \to 0$. In contrast, if the two particles are distinguishable, this function should be flat and equal to 1. This corresponds to the statement that it is twice more likely to find two identical Bose particles scattered into the same group of final states than it is for two distinguishable particles.

In terms of the probabilities discussed in the previous section, the correlation function can be represented as the ratio of the two-particle probability to the product of two one-particle probabilities.

$$C_2(\vec{k}_2, \vec{k}_1) = \frac{\mathcal{P}^{(2)}(\vec{k}_2, \vec{k}_1)}{\mathcal{P}^{(1)}(\vec{k}_2)\mathcal{P}^{(1)}(\vec{k}_1)}. \qquad (19)$$

Let us restrict the discussion to the case where we ignore possible initial state correlations. According to the results of the previous section, under the factorization assumption and in the absence of final state interactions, the above can be written as

$$C_2(\vec{k}_2, \vec{k}_1) = 1 + \frac{\left|I(\vec{k}_2, \vec{k}_1)\right|^2}{I(\vec{k}_2, \vec{k}_2)I(\vec{k}_1, \vec{k}_1)} \qquad (20)$$

where we define I as

$$I(\vec{k}_2, \vec{k}_1) \equiv \int d^4x_2 d^4x_1 e^{-i(k_2 x_2 - k_1 x_1)} \langle \phi^\dagger(x_2)\phi(x_1)\rangle. \qquad (21)$$

Recall that the spatial integrations over $\vec{x}_1$ and $\vec{x}_2$ are restricted to a small volume centered around the direction of motion. The solid angle is subtended by the atomic transverse area, at a macroscopic distance away from the interaction region. We can thus approximate the above by taking $\vec{k}_2$ and $\vec{k}_1$ parallel to $\vec{x}_2$ and $\vec{x}_1$, respectively. The integrations yield an overall multiplicative constant and we finally get

$$I(\vec{k}_2, \vec{k}_1) \sim \frac{1}{|\vec{x}_1||\vec{x}_2|} \int d^4x'_2 d^4x'_1 e^{-i(k_2 x'_2 - k_1 x'_1)} \langle J^\dagger(x'_2) J(x'_1)\rangle. \qquad (22)$$

Eq. (22) is correct within the far field approximation. However, its dependence on the relative momentum is not necessarily trivial since the ensemble average can in general depend on the spectrum produced by the source as well as on its coherence and overall volumes. In order to identify the different elements that determine the behavior of C_2 as a function of the relative momentum, let us look again at the model introduced in eq. (17) where for simplicity we consider only a one dimensional case. The ensemble average of the product of two currents is then

$$\langle J^\dagger(x_2)J(x_1)\rangle = \int dp\, dx_0\, f_p e^{-x_0^2/2R^2} e^{(x_1-x_0)^2/2R_f^2} e^{(x_2-x_0)^2/2R_f^2} e^{ip(x_2-x_1)} \qquad (23)$$

where we have taken

$$F_p(x_0) = f_p e^{-x_0^2/2R^2}. \qquad (24)$$

Performing the x_0 integral we get

$$\langle J^\dagger(x_2)J(x_1)\rangle \simeq e^{-(x_2-x_1)^2/R_f^2} e^{-(x_2+x_1)^2/4(2R^2+R_f^2)} \int dp\, f_p e^{2ip(x_2-x_1)} \qquad (25)$$

For definitiveness, take the source spectrum to be $f_p = e^{-\beta p^2}$. It is now easy to show that the correlation function for small $q = k_2 - k_1$ becomes

$$C_2(q) \simeq 2 - \left[R^2 + \frac{2R_f^4}{(\beta + 4R_f^2)}\left\{1 + \frac{\overline{(k_2+k_1)^2}\beta^2}{2(\beta+4R_f^2)}\right\}\right]q^2, \qquad (26)$$

where $\overline{(k_2 + k_1)^2}$, indicates the average value of $(k_2 + k_1)^2$. Eq. (26) exemplifies a general property of the correlation function, namely, that the scale parameters measured by HBT experiments are not independent of the spectrum produced by the source. It also shows that the extracted Gaussian parameters are in general made out of a combination of the overall source size and coherence lengths. For a heavy-ion reaction, the source volume is only a few times larger than the size of the characteristic formation region. It is then not justified to ignore this last by assuming that the Gaussian parameters reflect just the overall source dimensions.

MULTIPLE SCATTERING OF PARTICLES IN AIR

An other important issue that needs to be addressed in connection with the interpretation of the parameters associated with the correlation function is the influence of multiple scattering of the correlated particles during the detection process.

A typical high-energy experiment involves an spectrometer several meters in length. The trajectories of the detected particles pass through a wide layer of air and the particles experience a considerable amount of scattering with the air molecules. Since the linear dimension of the scatterers r_0 is on the order of Angstroms and the de Broglie wave length λ of the high-energy produced particles is on the order of a tenth of a fermi, $\lambda \ll r_0$ and we can consider the scattering process in the optical limit. Let $Q(\vec{s}_f, \vec{s}_i, z)$ be the probability that a pion ends up traveling in a direction $\vec{s}_f$ after having traveled a distance z, when it started moving in a direction $\vec{s}_i$. Let $f(\vec{s}_i)$ be the probability of producing a pion traveling in direction $\vec{s}_i$. Then, the probability of having a pion traveling in direction $\vec{s}_f$ after a distance z is given by the folding of the above probabilities,

$$Q(\vec{s}_f, z) = \int d^2 s_i f(\vec{s}_i) Q(\vec{s}_f, \vec{s}_i, z). \tag{27}$$

In the same manner, the joint probability of finding two pions traveling in directions $\vec{s}_1$ and $\vec{s}_2$ after traveling a distance z is

$$\langle Q(\vec{s}_1, z) Q(\vec{s}_2, z) \rangle = \int d^2 s_a d^2 s_b \langle f(\vec{s}_a) f(\vec{s}_b) \rangle Q(\vec{s}_1, \vec{s}_a, z) Q(\vec{s}_2, \vec{s}_b, z). \tag{28}$$

We see that in principle, the original correlation $\langle f(\vec{s}_a) f(\vec{s}_b) \rangle$ is modified. The probability in Eq. (27) can be expressed in terms of the probabilities of experiencing an arbitrary number of collisions

$$Q(\vec{s}_1, \vec{s}_a, z) = \sum_{n=0}^{\infty} P_n(z) q_n(\vec{s}_1, \vec{s}_a), \tag{29}$$

where $q_n(\vec{s}_1, \vec{s}_a)$ is the probability of finding a pion traveling in direction $\vec{s}_1$ when it started traveling in direction $\vec{s}_i$ after n collisions and P_n is a Poisson distribution. It can easily be checked that the general form of q_n is

$$q_n(\vec{s}_f, \vec{s}_i) = \left(\frac{a^2}{n\pi}\right) e^{-\frac{a^2}{n}(\vec{s}_f - \vec{s}_i)^2}, \tag{30}$$

where a^{-1} is the average scattering angle per collision. Let us take

$$\langle f(\vec{s}_a) f(\vec{s}_b) \rangle = 1 + e^{-\beta^2 (\vec{s}_a - \vec{s}_b)^2}, \tag{31}$$

with β^{-1} the original correlation angle. It is then easy to show that

$$\langle Q(\vec{s_1})Q(\vec{s_2})\rangle \simeq 1 + he^{-h\beta^2(\vec{s_1}-\vec{s_2})^2}, \qquad (32)$$

where $h = 1/[1 + 2N\frac{\beta^2}{a^2}]$ with N the average number of collisions. We then see that the original correlation angle widens, i.e. $\beta^{-1} \to h^{-1/2}\beta^{-1}$ whereas the *height* of the correlation function decreases, i.e. $1 \to h$. Taking an average total traveled distance $z = 15$m and for normal air density, $N \sim 10^5$. On the other hand, taking $a = r_0/\lambda \sim 10^5$, $\beta = d/\lambda \sim 10^2$ (where d is the linear size of the reaction volume, on the order of 10fm) then $h \sim 0.9$. which is not negligible at all. It is important to notice that the effect is independent of the particle's mass and thus it should be taken into account for extracting the correlation parameters both from kaon or pion correlation measurements.

SUMMARY AND CONCLUSIONS

In this paper, we have studied the way interference measurements are carried out by particle detectors. We have established that interference happens when two single particle wave functions coincide at the sensitive element of the detector, a single atom, within the time ΔT given by the inverse of the atomic bandwidth, on the order of $\Delta T \sim 10^{-16}$ sec. This quantifies the notion that interference happens for particles detected within the same event.

We have argued that for sources satisfying the factorization property and in the absence of final state correlations, the object that determines the behavior of the two particle density matrix is the source correlation or ensemble average of the product of two source currents $\langle J^\dagger(x_2)J(x_1)\rangle$. We illustrate the properties of the source distribution that are measured by interferometry introducing a model for the source correlation.

We have also argued that, since extracting the information on the correlation function from its general behavior as a function of the relative momentum is obscured by the fact that the interference term is not just the square of the source's distribution Fourier transform, it is then instructive to look at a one dimensional version of the model in eq. (17), to illustrate that this yields as the scale parameter, a combination of the overall source size and coherence lengths, as well as the information on the spectrum produced after the collision.

It is important to emphasize that by looking at the details of the particle detection process as applied to interferometry, we have also accomplished to point out to a possible mechanism that could be responsible for the loss of interference signal, namely multiple scattering in air during the detection processes. Thus, we face the situation in which even a completely chaotic source looks, after interferometry, as only partially chaotic when in fact, the loss of signal in not a property of the source. This issues deserve a careful analysis and are the subject of a following up work [13].

ACKNOWLEDGMENTS

Support for this work has been received in part by the US NSF under grant NSF PHY94-21309 and by CONACyT México.

REFERENCES

1. R. Hanbury Brown and R.Q. Twiss, *Phil. Mag.* **45**, (1954), 663, *Nature* **177** (1956) 27, *Nature* **177**, 27 (1956), *Nature* **178**, (1956), 1046.
2. R. Hanbury Brown and R.Q. Twiss, *Proc. Roy. Soc. A*, **242**, (1957), 300.
3. G. Goldhaber, S. Goldhaber, W. Lee, and A. Pais, *Phys. Rev.* **120**, (1960), 30.
4. G.I. Kopylov, *Phys. Lett.* **B 50**, (1974), 472.
5. M. Gyulassy, S.K. Kauffmann, and L.W. Wilson, *Phys. Rev.* **C 20**, (1979), 2267.
6. S. Pratt, *Phys. Rev. Lett.* **53**, (1984), 1219; *Phys. Rev.* **D 33**, (1986), 72; *Phys. Rev.* **D 33**, (1986), 1314; *Phys. Rev.* **C 42**, (1990), 2646.
7. G.F. Bertsch, G.E. Brown, V. Koch and B.-A. Li, *Nucl. Phys.* **A490**, (1988), 745; G.F. Bertsch, *Nucl. Phys.* **A498**, (1989), 173c; G.F. Bertsch, M. Gong and M. Tohymaya, *Phys. Rev.* **C 37**, (1988), 1896; G.F. Bertsch, P. Danielewicz and M. Herrmann, *Phys. Rev.* **C 49**, (1994), 442; M. Herrmann and G.F. Bertsch, *Phys. Rev.* **C 51**, (1995), 328.
8. G. Baym and P. Braun-Munzinger, *Nucl. Phys.* **A610**, (1996), 286c.
9. In the case of an optical field, the analysis is been made by R.J. Glauber, *Quantum Optics and Electronics* Les Houches 1964 summer school edited by C. DeWitt, A. Blandin and C. Cohen-Tannoudji, pp. 63.
10. See for example, Rev. Part. Properties, *Phys. Rev.* **D 50** (1994), 1251.
11. W.P. Jesse and J. Sadaukis, *Phys. Rev.* **107** (1957), 766.
12. P. Rice-Evans, *Spark, Streamer, Proportional and Drift Chambers* (Richelieu Press Ltd. 1974).
13. A. Ayala, G. Baym, B.V. Jacak, J.L. Popp and B. Vanderheyden, work in preparation
14. See, e.g., J. Barrette et al., E814 Collaboration, *Phys. Lett.* **B 333**, (1994), 33.

III. Theory

Brane Engineering

César Gómez* and Rafael Hernández[†]

*Instituto de Matemáticas y Física Fundamental, CSIC,
Serrano 123, 28006 Madrid, Spain
[†]Instituto de Física Teórica (UAM-CSIC), C-XVI, Universidad Autónoma de Madrid,
Cantoblanco 28049, Madrid, Spain

I INTRODUCTION TO BRANES

A Nambu-Goto Action.

By p-branes we will generically mean extended objects with p space and one time dimensions [1] [1]. The dynamics of these objects is defined through the condition of minimizing the worldvolume

$$S = -T_p \int d^{p+1}\xi \sqrt{-\det \partial_i X^\mu \partial_j X^\nu \eta_{\mu\nu}}. \tag{I.1}$$

For $p = 1$ this is the well known Nambu-Goto action for the bosonic string. The equations of motion that follow (I.1) can be equivalently obtained from the covariant action

$$S = -T_p \int d^{p+1}\xi [\sqrt{-\gamma}\gamma^{ij}\partial_i X^\mu \partial_j X^\nu \eta_{\mu\nu} - (p-1)\sqrt{-\gamma}], \tag{I.2}$$

with γ^{ij} the worldvolume metric, and $\gamma = |\det \gamma^{ij}|$. The string is special, among extended objects, because only for $p = 1$ the action (I.2) is invariant under Weyl scaling transformations, $\gamma^{ij} \to \Lambda \gamma^{ij}$. In the string case, the action (I.2) can be generalized by replacing the flat Minkowski metric $\eta_{\mu\nu}$ by a generic metric, $g_{\mu\nu}$. Besides, the string can be coupled to other backgrounds, preserving the renormalizability of the theory. These backgrounds correspond to the massless spectrum of the bosonic string, and are the massless spectrum $B_{\mu\nu}(X)$ and the dilaton field $\Phi(X)$,

$$S = -T_1 \int d^2\xi [\sqrt{-\gamma}\gamma^{ij}\partial_i X^\mu \partial_j X^\nu g_{\mu\nu} + \epsilon^{ij}\partial_i X^\mu \partial_j X^{nu} B_{\mu\nu}(X) + \frac{1}{4\pi}\frac{\sqrt{-\gamma}}{T_1}\Phi(X)R^{(2)}], \tag{I.3}$$

---

[1] We apologize for the extremely incomplete set of references included in these notes. More details can be found in a number of general reviews on branes and non perturbative string theory or reference [2], to appear.

with $R^{(2)}$ the scalar curvature of the worldsheet. In order to preserve Weyl invariance for the sigma model (I.3), we should require the beta functions β_g, β_Φ and β_B to be equal zero. Recalling now the definition of the topological Euler number,

$$\frac{1}{4\pi}\int \sqrt{-\gamma} R^{(2)} = 2 - 2g, \qquad (I.4)$$

we observe directly from (I.3) that the vacuum expectation value $<\Phi(X)>$ of the dilaton field defining the background fixes the magnitude of the string coupling constant, g_s. In fact, a Riemann surface Feynman diagram of genus g contains $2g - 2$ string vertices, and contributes to the partition function as $\exp[(2g - 2) < \Phi >]$, so that we can define the string coupling constant g_s as

$$g_s = e^{<\phi>}. \qquad (I.5)$$

The term $\epsilon^{ij}\partial_i X^\mu \partial_j X^\nu$ can be interpreted as a minimal coupling of the string to the $B^{\mu\nu}(X)$ field, with the gauge transformations of $B^{\mu\nu}$ defined by

$$\delta B^{\mu\nu} = \partial^\mu \Lambda^\nu - \partial^\nu \Lambda^\mu. \qquad (I.6)$$

This can be interpreted claming that the string is a source for the 2-form gauge field $B^{\mu\nu}$. In the generic case of a p-brane extended object, the equivalent to equation (I.3) is

$$S = -T_p \int d^{p+1}\xi [\sqrt{-\gamma}\gamma^{ij}\partial_i X^\mu \partial_j X^\nu g_{\mu\nu}(X) +$$

$$\epsilon_{i_1...i_{p+1}} A^{\mu_1...\mu_{p+1}}\partial_{i_1} X^{\mu_1} ... \partial_{i_{p+1}} X^{\mu_{p+1}} - (p-1)\sqrt{-\gamma}], \qquad (I.7)$$

with A a $p+1$ form, and the p extended object as a source of the field A.

A particularly interresting theory is eleven dimensional supergravity, which contains, in addition to gravitons and gravitinos, a 3-form field, $C_{\mu\nu\rho}$. In this case it is natural to postulate an extended object of dimension two (a membrane) as the source for the $C_{\mu\nu\rho}$ field. The corresponding membrane action would be

$$S = -T_2 \int d^3\xi[\sqrt{-\gamma}\gamma^{ij}\partial_i X^\mu \partial_j X^\nu g_{\mu\nu}(X) +$$

$$\epsilon_{ijk} C^{\mu\nu\rho}\partial_i X^\mu \partial_j X^\nu \partial_k X^\rho - \sqrt{-\gamma}]. \qquad (I.8)$$

However, this action is unfortunately not Weyl invariant, and hence we can not connect (I.8) with the eleven dimensional supergarvity action through a similar argument to the one employed in string theory, where the string action can be related to ten dimensional general relativity equations by impossing Weyl invariance in the form of vanishing beta functions. In the membrane case, the equivalent argument is known as kappa symmetry, and requires an essential ingredient in order to impose both worldvolume and spacetime supersymmetry. Using kappa symmetry, it is possible to derive, from the membrane action (I.8), the equations of motion of elevan dimensional supergravity. In spite of the fact that eleven dimensional supergravity is not renormalizable, there is one aspect of this theory that makes it, in a certain sense, more fundamental than string theory. In fact, as shown above, the string coupling constant is a free parameter depending on the vacuum expectation value of the dilaton; however, as we move to eleven dimensions this dilaton field becomes part of the eleven dimensional metric tensor, and the string action (I.3) can be obtained through *double dimensional reduction* from the membrane action.

By double dimensional reduction we mean a compactification of both spacetime and worldvolume. More precisely, of the membrane coordinates are decomposed as (X^μ, Y), with $\mu = 0, \ldots, 9$, the double reduction is defined in terms of

$$\partial_q Y = \partial_Y C = 0,$$
$$Y = \xi_3,$$
$$\partial_{1,2} Y = 0,$$
$$\partial_3 X = 0, \quad\quad\quad (I.9)$$

with ξ_3 the compactified worldvolume coordinate.

If we do not care about trouble related to quantization of membranes, we can continue working some classical facts of membrane dynamics in eleven dimensions. In particular, if we consider the membrane as a source for the 3-form field $C^{\mu\nu\rho}$, we can wonder about its Hodge dual object: in eleven dimensions, the Hodge dual of the strength tensor 4-form, $F^4 = dC$, is a seven form, whose source, to be interpreted as the magnetic dual to the membrane, is an extended object of dimension equal five. We will refer to this object as the M-theory fivebrane. These objects play an interesting role in the dynamics, as ordinary two dimensional membranes can end on them: for a membrane ending on the M-theory fivebrane, the boundary on the fivebrane worldvolume is a string. In the six dimensional worldvolume of the fivebrane, a string is (Hodge) self dual, as it is the source of a 2-form field. Moreover, on the worldvolume of the fivebrane there is a 2-form field, that can couple to the self dual string. Thus, configurations with fivebranes and membranes as the one represented in Figure 1 are allowed, where the membrane can be thought of as an open membrane. Figure 1 determines an special type of boundary conditions on the membrane fields $x^\mu(\xi_i)$. If we take the fivebrane with worldvolume coordinates x^0, x^1, x^2, x^3, x^4 and x^5 and located at the spacetime positions $x^6 = x^7 = x^8 = x^9 = x^{10} = 0$, then the membrane fields X^μ, with $\mu = 6, \ldots, 10$ are fixed to zero at the boundary of the membrane, so that we are impossing on the membrane fields Dirichlet boundary conditions for X^μ with $\mu = 6, \ldots, 10$.

B Membranes and D-Branes.

The vertex defined in Figure 1 can be dimensionally reduced twice, compactifying one spacetime coordinate and one worldvolume coordinate for the M-theory fivebrane and the membrane. Then, some kind of fourbrane should be expected to arise from the

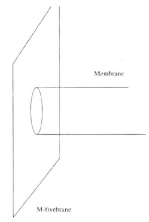

FIGURE 1. A membrane with its boundary on a fivebrane.

fivebrane, besides from a string coming from the membrane; this string will now end on the fourbrane obtained through double dimensional reduction of the fivebrane. If all these steps are consistent, what we will get is a four dimensional object that should now be considered as part of the ten dimensional string spectrum. Moreover, the string will be allowed to end on this four dimensional object. In order to understand the nature of this object we must recall some well known facts about superstring theory. We will focus the discussion on type IIA and type type IIB superstrings. The field content of these theories in the Ramond-Ramond sector is

$$\begin{aligned} \text{IIA} &\to A^1, \quad A^3, \\ \text{IIB} &\to A^2, \quad A^4. \end{aligned} \qquad (I.10)$$

As ten dimensional type IIA supergravity can be obtained through dimensional reduction eleven dimensional supergravity, we should search for the four dimensional extended object in type IIA string theory. A four dimensional object is a source for a 5-form field which, in ten dimensions, is the Hodge dual to the 3-form field, A^3. Thus, a good candidate for the double dimensional reduction of the M-theory fivebrane should be a source for the Ramond-Ramond field Hodge dual to A^3. Moreover, on this brane strings can end, which means, by the same argument as above, that on the worldvolume of the brane a 1-form field A^1 must be allowed to exist. As discussed above, the boundary conditions for the string in the $x^6, \ldots, x^{10}$ directions are of Dirichlet type (recall that by double dimensional reduction we have compactified a worldvolume coordinate of the fivebrane). The previous set of properties define what is meant as s Dirichlet-brane; in the example under consideration, a Dirichlet-fourbrane. Hence, the above discussion can be summarized through the diagram in Figure 2.

Moreover, when the sources of all Ramond-Ramond fields in (I.10) are interpreted as

FIGURE 2. Correspondence between eleven dimensional and ten dimensional objects.

Dirichlet, the picture in Figure 3 arises, where the arrows represent the Hodge duality relation in ten dimensions.

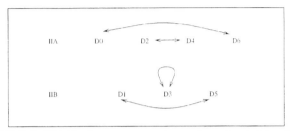

FIGURE 3. Dirichlet-brane spectrum in ten dimensional string theory.

As type IIA and type IIB string theories are, after compactification on S^1 down to nine dimensions, related by T-duality, we can naturally wonder about the action of T-duality on the D-brane spectrum. One of the main abilities of D-branes is defining regions of spacetime where Dirichlet boundary conditions can be impossed for strings. Hence, if a D-fourbrane is located at the $x^5 = \cdots = x^9 = 0$ values, the string worldvolume coordinates at the boundary satisfy the Dirichlet conditions $x^5 = \cdots = x^9 = 0$, while ordinary Neumann boundary conditions, $\partial_{\vec{n}} x^i = 0$ are satisfied for $i = 1, 2, 3, 4$. We can then consider one of the Dirichlet x^j coordinates to live on a circle, S^1. Under T-duality we transform, along this direction, Neumann into Dirichlet boundary conditions, and reciprocally. Thus, under T-duality along one of the $x^5, \ldots, x^9$ directions we transform the four dimensional brane into a five dimensional brane, and under T-duality along one of the worldvolume directions, $x^1, \ldots, x^4$, we transform a D-fourbarne in type IIA into a D-threebrane in type IIB string theory. The whole set of connections between D-branes under T-duality is given in Figure 4.

The set of tools for brane engineering is completed once we known the kind of branes that can be obtained in ten dimensions from the M-theory membrane and fivebrane when performing direct dimensional reduction. In this case, the M-theory membrane goes into the D-twobrane of type string theory, and the fivebrane becomes the ten dimensional solitonic fivebrane[2]

The final image arises when $Sl(2; \mathbf{Z})$ duality of type IIB string theory is included:

---

[2] The three different categories of branes are classified in terms of their tension: order one for fundamental objects, order $\frac{1}{g_s}$ for Dirichlet branes, and order $\frac{1}{g_s^2}$ for the solitonic fivebrane.

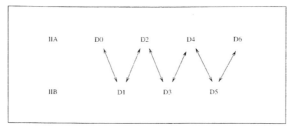

FIGURE 4. T-duality relates type IIA and type IIB D-branes.

$Sl(2;\mathbf{Z})$ S duality exchanges the two form in the R-R sector with the two form field $B_{\mu\nu}$ in the NS-NS sector, apperaing in the string action (I.3); thus, it exchanges the D-string with the fundamental string, and the D-fivebrane with the solitonic fivebrane. The whole discussion is summarized in Figure 5.

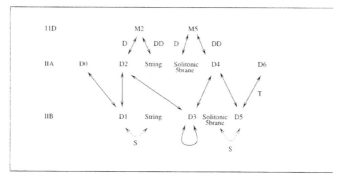

FIGURE 5. Set of correspondences for eleven dimensional and ten dimensional extended objects.

C Brane Vertices.

In this section we will define brane vertices through dimensional reduction and T and S dualities, starting from the generic vertex between a Dirichlet brane and a fundamental string, shown in Figure 6.

This is an allowed vertex in type IIA or type IIB string depending on the even or odd value of p. We will consider the $p=3$ value of type IIB strings. Then, by S-duality we can pass from a vertex between the D-threebrane (which is self-dual) of type IIB string theory and the string, to a vertex where a D-string ends on the D-threebrane. Now, T-duality transformations in the directions orthogonal to the worldvolumes of the D-3brane and the D-1brane lead to a type IIA vertex between a D-4brane and a D-2brane and, finally, a vertex in type IIB string theory where a D-3brane ends on D-5brane. as shown in Figure 7.

FIGURE 6. A D-brane defined as a boundary for a fundamental string.

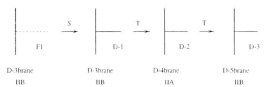

FIGURE 7. T and S-duality determine the allowed ten dimensional vertices.

S-duality on the last vertex of Figure 7 provides a new vertex in type IIB string theory, between the solitonic fivebrane and a D-3brane.

More vertices can yet be obtained upon dimensional reduction. We will now start with the eleven dimensional vertex between the M-theory fivebrane and the membrane. By dimensional reduction, we get the type IIA vertex between the solitonic fivebrane and the D-2brane. Now, we can again perform two T-duality transformations. The result, shown in Figure 8, is a vertex between the solitonic fivebrane and a D-4brane in type IIA.

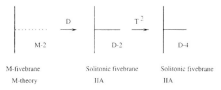

FIGURE 8. A vertex from eleven dimensions.

In the next section we will define brane configurations in flat spacetime using the previous set of brane vertices [3]- [20].

II THREE DIMENSIONAL FIELD THEORIES WITH $N = 4$ SUPERSYMMETRY.

Let us now consider some brane configurations build up using the vertices $(5,3)$ and $(5^{NS}, 3)$ in type IIB theory [3]. In particular, we will consider solitonic fivebranes, with worldvolume coordinates x^0, x^1, x^2, x^3, x^4 and x^5, located at some definite values of x^6, x^7, x^8 and x^9. It is convenient to organize the coordinates of the fivebrane as

$(x^6, \vec{\omega})$ where $\vec{\omega} = (x^7, x^8, x^9)$. By construction of the vertex, the D-3brane will share two worldvolume coordinates, in addition to time, with the fivebrane. Thus, we can consider D-3branes with worldvolume coordinates x^0, x^1, x^2 and x^6. If we put a D-3brane in between two solitonic fivebranes, at x_2^6 and x_1^6 positions in the x^6 coordinate, then the worldvolume of the D-3brane will be finite in the x^6 direction (see Figure 9).

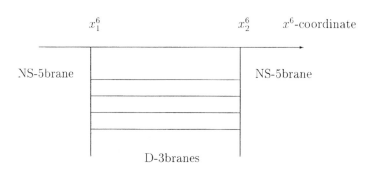

FIGURE 9. Solitonic fivebranes with n Dirichlet threebranes stretching along them.

Therefore, the macroscopic physics, i. e., for scales larger than $|x_2^6 - x_1^6|$, can be effectively described by a $2+1$ dimensional theory. In order to unravel what kind of $2+1$ dimensional theory, we are obtaining through this brane configuration, we must first work out the type of constraint impossed by the fivebrane boundary conditions. In fact, the worldvolume low energy lagrangian for a D-3brane is a $U(1)$ gauge theory. Once we put the D-3brane in between two solitonic fivebranes we impposse Neumann boundary conditions, in the x^6 direction, for the fields living on the D-3brane worldvolume. This means in particular that for scalar fields we impposse

$$\partial_6 \phi = 0 \qquad (II.1)$$

and, for gauge fields,

$$F_{\mu 6} = 0, \quad \mu = 0, 1, 2. \qquad (II.2)$$

Thus, the three dimensional $U(1)$ gauge field, A_μ, with $\mu = 0, 1, 2$, is unconstrained which already means that we can interpret the effective three dimensional theory as a $U(1)$ gauge theory for one D-3brane, and therefore as a $U(n)$ gauge theory for n D-3branes. Next, we need to discover the amount of supersymmetry left unbroken by the brane configuration. If we consider Dirichlet threebranes, with worldvolume coordinates x^0, x^1, x^2 and x^6, then we are forcing the solitonic fivebranes to be at positions $(x^6, \vec{\omega}_1)$ and $(x^6, \vec{\omega}_2)$, with $\vec{\omega}_1 = \vec{\omega}_2$. In this particular case, the allowed motion for the D-3brane is reduced to the space $\mathbf{R}^3$, with coordinates x^3, x^4 and x^5. These are the coordinates on the fivebrane worldvolume where the D-3brane ends. Thus, we have defined on the D-3brane three scalar fields. By condition (II.1), the values of these scalar fields can be constrained to be constant on the x^6 direction. What this in practice means is that the two ends of the of the D-3brane have the same x^3, x^4 and x^5 coordinates. Now, if we combine these three scalar fields with the $U(1)$ gauge field A_μ, we get an $N = 4$ vector multiplet in three dimensions. Therefore, we can conclude that our effective three dimensional theory for n parallel D-3branes suspended between two solitonic fivebranes (Figure 1) is a gauge theory with $U(n)$ gauge group, and $N = 4$ supersymmetry. Denoting by $\vec{v}$ the vector (x^3, x^4, x^5), the Coulomb branch of this theory is parametrized by the v_i positions of the n D-3branes (with i labelling each brane). In addition, we have, as discussed in chapter II, the dual photons for each $U(1)$ factor. In this way, we get the hyperkähler structure of the Coulomb branch of the moduli. Hence, a direct way to get supersymmetry preserved by the brane configuration is as follows. The supersymmetry charges are defined as

$$\epsilon_L Q_L + \epsilon_R Q_R, \tag{II.3}$$

where Q_L and Q_R are the supercharges generated by the left and right-moving worldsheet degrees of freedom, and ϵ_L and ϵ_R are ten dimensional spinors. Each solitonic $p$brane, with worldvolume extending along $x^0, x^1, \ldots, x^p$, imposes the conditions

$$\epsilon_L = \Gamma_0 \ldots \Gamma_p \epsilon_L, \quad \epsilon_R = -\Gamma_0 \ldots \Gamma_p \epsilon_R, \tag{II.4}$$

in terms of the ten dimensional Dirac gamma matrices, Γ_i; on the other hand, the D-$p$branes, with worldvolumes extending along $x^0, x^1, \ldots, x^p$, imply the constraint

$$\epsilon_L = \Gamma_0 \Gamma_1 \ldots \Gamma_p \epsilon_R. \tag{II.5}$$

Thus, we see that NS solitonic fivebrane, with worldvolume located at x^0, x^1, x^2, x^3, x^4 and x^5, and equal values of $\vec{\omega}$, and Dirichlet threebranes with worldvolume along x^0, x^1, x^2 and x^6, preserve eight supersymmetries on the D-3brane worldvolume or, equivalently, $N = 4$ supersymmetry on the effective three dimensional theory.

The brane array just described allows a simple computation of the gauge coupling constant of the effective three dimensional theory: by standard Kaluza-Klein reduction on the finite x^6 direction, after integrating over the (compactified) x^6 direction to reduce the lagrangian to an effective three dimensional lagrangian, the gauge coupling constant is given by

$$\frac{1}{g_3^2} = \frac{|x_6^2 - x_6^1|}{g_4^2}, \tag{II.6}$$

in terms of the four dimensional gauge coupling constant. Naturally, (II.6) is a classical expression that is not taking into account the effect on the fivebrane position at x^6 of the D-3brane ending on its worldvolume. In fact, we can consider the dependence of x^6 on the coordinate $\vec{v}$, normal to the position of the D-3brane. The dynamics of the fivebranes should then be recovered when the Nambu-Goto action of the solitonic fivebrane is minimized. Far from the influence of the points where the fivebranes are located (at large values of x^3, x^4 and x^5), the equation of motion is simply three dimensional Laplace's equation,

$$\nabla^2 x^6(x^3, x^4, x^5) = 0, \tag{II.7}$$

with solution

$$x^6(r) = \frac{k}{r} + \alpha, \tag{II.8}$$

where k and α are constants depending on the threebrane tensions, and r is the spherical radius at the point (x^3, x^4, x^5). From (II.8), it is clear that there is a well defined limit as $r \to \infty$; hence, the difference $x_2^6 - x_1^6$ is a well defined constant, $\alpha_2 - \alpha_1$, in the $r \to \infty$ limit.

Part of the beauty of brane technology is that it allows to obtain very strong results by simply performing geometrical brane manipulations. We will now present one example, concerning our previous model. If we consider the brane configuration from the point of view of the fivebrane, the n suspended threebranes will look like n magnetic monopoles. This is really suggesting since, as described in chapter II, we know that the Coulomb branch moduli space of $N = 4$ supersymmetric $SU(n)$ gauge theories is isomorphic to the moduli space of BPS monopole configurations, with magnetic charge equal n. This analogy can be put more precisely: the vertex $(5^{NS}, 3)$ can, as described above, be transformed into a $(3, 1)$ vertex. In this case, from the point of view of the threebrane, we have a four dimensional gauge theory with $SU(2)$ gauge group broken down to $U(1)$, and n magnetic monopoles. Notice that by passing from the configuration build up ussing $(5^{NS}, 3)$ vertices, to that build up with the $(3, 1)$ vertex, the Coulomb moduli remains the same.

Next, we will work out the same configuration, but now with the vertex $(5, 3)$ made out of two Dirichlet branes. The main difference with the previous example comes from the boundary conditions (II.1) and (II.2), which should now be replaced by Dirichlet boundary conditions. We will choose as worldvolume coordinates for the D-5branes x^0, x^1, x^2, x^7, x^8 and x^9, so that they will be located at some definite values of x^3, x^4, x^5 and x^6. As before, let us denote this positions by $(\vec{m}, x^6)$, where now $\vec{m} = (x^3, x^4, x^5)$. An equivalent configuration to the one studied above will be now a set of two D-5branes, at some points of the x^6 coordinate, that we will again call x_1^6 and x_2^6, subject to $\vec{m}_1 = \vec{m}_2$, with D-3branes stretching between them along the x^6 coordinate, with worldvolume extending again along the coordinates x^0, x^1, x^2 and x^6 (Figure 10). Our task now will be the description of the effective three dimensional theory on these threebranes. The end points of the D-3branes on the fivebrane worlvolumes will now be parametrized by values of x^7, x^8 and x^9. This means that we have three scalar fields in the effective three dimensional theory. The scalar fields corresponding to the coordinates x^3, x^4, x^5 and x^6 of the threebranes are forzen to the constant values where the fivebranes are located. Next, we should consider what happens to the $U(1)$ gauge field on the D-3brane worldvolume. Impossing Dirichlet boundary conditions for this field is equivalent to

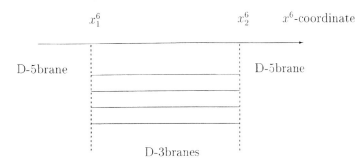

FIGURE 10. Dirichlet threebranes extending between a pair of Dirichlet fivebranes (in dashed lines).

$$F_{\mu\nu} = 0, \quad \mu, \nu = 0, 1, 2, \tag{II.9}$$

i. e., there is no electromagnetic tensor in the effective three dimensional field theory. Before going on, it would be convenient summarizing the rules we have used to impose the different boundary conditions. Consider a D-$p$brane, and let M be its worldvolume manifold, and $B = \partial M$ the boundary of M. Neumann and Dirichlet boundary conditions for the gauge field on the D-$p$brane worldvolume are defined respectively by

$$\begin{aligned} N &\longrightarrow F_{\mu\rho} = 0, \\ D &\longrightarrow F_{\mu\nu} = 0, \end{aligned} \tag{II.10}$$

where μ and ν are directions of tangency to B, and ρ are the normal coordinates to B. If B is part of the worldvolume of a solitonic brane, we will imposse Neumann conditions, and if it is part of the worldvolume of a Dirichlet brane, we will imposse Dirichlet conditions. Returning to (II.9), we see that on the three dimensional effective theory, the only non vanishing component of the four dimensional strenght tensor is $F_{\mu 6} \equiv \partial_\mu b$. Therefore, all together we have four scalar fields in three dimensions or, equivalently, a multiplet with $N = 4$ supersymmetry. Thus, the theory defined by the n suspended D-3branes in between a pair of D-5branes, is a theory of n $N = 4$ massless hypermultiplets.

There exits a different way to interpret the theory, namely as a magnetic dual gauge theory. In fact, if we perform a duality transformation in the four dimensional $U(1)$ gauge theory, and use magnetic variables $*F$, instead of the electric field F, what we get in three dimensions, after imposing D-boundary conditions, is a dual photon, or a magnetic $U(1)$ gauge theory.

The configuration chosen for the worldvolume of the Dirichlet and solitonic fivebranes yet allows a different configuration with D-3branes suspended between a D-5brane and a NS-5brane. This is in fact consistent with the supersymmetry requirements (II.4) and (II.5). Namely, for the Dirichlet fivebrane we have

$$\epsilon_L = \Gamma_0\Gamma_1\Gamma_2\Gamma_7\Gamma_8\Gamma_9\epsilon_R. \tag{II.11}$$

The solitonic fivebrane imposses

$$\epsilon_L = \Gamma_0\ldots\Gamma_5\epsilon_L, \quad \epsilon_R = -\Gamma_0\ldots\Gamma_5\epsilon_R, \tag{II.12}$$

while the suspended threebranes imply

$$\epsilon_L = \Gamma_0\Gamma_1\Gamma_2\Gamma_6\epsilon_R, \tag{II.13}$$

which are easily seen to be consistent. The problem now is that the suspended D-3brane is frozen. In fact, the position (x^3, x^4, x^5) of the end point of the NS-5brane is equal to the position $\vec{m}$ of the D-5brane, and the position (x^7, x^8, x^9) of the end point on the D-5brane is forced to be equal to the position $\vec{\omega}$ of the NS-5brane. The fact that the D-3brane is frozen means that the theory defined on it has no moduli, i. e., posseses a mass gap.

Using the vertices between branes we have described so far we can build quite complicated brane configurations. When Dirichlet threebranes are placed to the right and left of a fivebrane, open strings can connect the threebranes at different sides of the fivebrane. They will represent hypermultiplets transforming as $(k_1, \bar{k}_2)$, with k_1 and k_2 the number of threebranes to the left and right, respectively, of the fivebrane. In case the fivebrane is solitonic, the hypermultiplets are charged with respect to an electric group, while in case it is a D-5brane, they are magnetically charged. Another possibility is that with a pair of NS-5branes, with D-3branes extending between them, and also a D-5brane located between the two solitonic fivebranes. A massless hypermultiplet will now appear whenever the (x^3, x^4, x^5) position of the D-3brane coincides with the $\vec{m} = (x^3, x^4, x^5)$ position of the D-5brane.

So far we have used brane configurations for representing different gauge theories. In these brane configurations we have considered two different types of moduli. For the examples described above, these two types of moduli are as follows: the moduli of the effective three dimensional theory, corresponding to the different positions where the suspended D-3branes can be located, and the moduli corresponding to the different locations of the fivebranes, which are being used as boundaries. This second type of moduli specifies, from the point of view of the three dimensional theory, different coupling constants; hence, we can move the location of the fivebranes, and follow the changes taking place in the effective three dimensional theory. Let us then consider a case with two solitonic branes, and a Dirichlet fivebrane placed between them. Let us now move the NS-5brane on the left of the D-5brane to the right. In doing so, there is a moment when both fivebranes meet, sharing a common value of x^6. If the interpretation of the hypermultiplet we have presented above is correct, we must discover what happens to the hypermultiplet after this exchange of branes has been performed. In order to maintain the hypermultiplet, a new D-3brane should be created after the exchange, extending from the right solitonic fivebrane to the Dirichlet fivebrane. To prove this we will need D-brane dynamics at work. Let us start considering two interpenetrating closed loops, C and C', and suppose electrically charged particles are moving in C, while magnetically charged particles move in C'.

The linking number $L(C, C')$ can be defined using the standard Wilson and 't Hooft loops. Namely, we can measure the electric flux passing through C' or, equivalently, compute $B(C')$, or measure the magnetic flux passing through C, i. e., the Wilson line $A(C)$. In both cases, what we are doing is integrating over C' and C the dual to the field created by the particle moving in C and C', respectively. Let us now extend this simple result to the case of fivebranes. A fivebrane is a source of 7-form tensor field, and its dual is therefore a 3-form. We will call this 3-form H_{NS} for NS-5branes, and H_D for D-5branes. Now, let us consider the worldvolume of the two fivebranes,

$$\mathbf{R}^3 \times Y_{NS},$$
$$\mathbf{R}^3 \times Y_D. \tag{II.14}$$

We can now define the linking number as we did before, in the simpler case of a particle:

$$L(Y_{NS}, Y_D) = -\int_{Y_{NS}} H^D = \int_{Y_D} H^{NS}. \tag{II.15}$$

The 3-form H^{NS} is locally dB_{NS}. Since we have no sources for H^{NS}, we can use $H^{NS} = dB_{NS}$ globally; however, this requires B to be globally defined, or gauge invariant. In type IIB string theory, B is not gauge invariant; however, on a D-brane we can define the combination $B_{NS} - F_D$, which is invariant, with F_D the two form for the $U(1)$ gauge field on the D-brane. Now, when the D-5brane and the NS-5brane do not intersect, the linking number is obviously zero. When they intersect, this linking number changes, which means that (II.15) should, in that case, be non vanishing. Writing

$$\int_{Y_D} H^{NS} = \int_{Y_D} dB_{NS} - dF_D, \tag{II.16}$$

we observe that the only way to get linking numbers would be adding sources for F_D. These sources for F_D are point like on Y_D, and are therefore the D-3branes with worldvolume $\mathbf{R}^3 \times C$, with C ending on Y_D, which is precisely the required appearance of extra D-3branes.

III D-BRANE DESCRIPTION OF SEIBERG-WITTEN SOLUTION.

In the previous example we have considered type IIB string theory and three and fivebranes. Now, let us consider type IIA strings, where we have fourbranes that can be used to define, by analogy with the previous picture, $N = 2$ four dimensional gauge theories [21]. The idea will again be the use of solitonic fivebranes, with sets of fourbranes in between. The only difference now is that the fivebrane does not create a RR field in type IIA string theory and, therefore, the physics of the two parallel solitonic fivebranes does not have the interpretation of a gauge theory, as was the case for the type IIB configuration above described [6].

Let us consider configurations of infinite solitonic fivebranes, with worldvolume coordinates x^0, x^1, x^2, x^3, x^4 and x^5, located at $x^7 = x^8 = x^9 = 0$ and at some fixed value of the x^6 coordinate. In addition, let us introduce finite Dirichlet fourbranes, with worldvolume coordinates x^0, x^1, x^2, x^3 and x^6, which terminate on the solitonic fivebranes; thus, they are finite in the x^6 direction. On the fourbrane worldvolume, we can define a macroscopic four dimensional field theory, with $N = 2$ supersymmetry. This four dimensional theory will, as in the type IIB case considered in previous section, be defined by standard Kaluza-Klein dimensional reduction of the five dimensional theory defined on the D-4brane worldvolume. Then, the bare coupling constant of the four dimensional theory will be

$$\frac{1}{g_4^2} = \frac{|x_6^2 - x_6^1|}{g_5^2}, \tag{III.1}$$

in terms of the five dimensional coupling constant. Moreover, we can interpret as classical moduli parameters of the effective field theory on the dimensionally reduced worldvolume of the fourbrane the coordinates x^4 and x^5, which locate the points on the fivebrane worldvolume where the D-4branes terminate.

In addition to the Dirichlet fourbranes and solitonic fivebranes, we can yet include Dirichlet sixbranes, without any further break of supersymmetry on the theory in the worldvolume of the fourbranes. To prove this, we notice that each NS-5brane imposes the projections

$$\epsilon_L = \Gamma_0 \ldots \Gamma_5 \epsilon_L, \quad \epsilon_R = -\Gamma_0 \ldots \Gamma_5 \epsilon_R, \tag{III.2}$$

while the D-4branes, with worldvolume localized at x^0, x^1, x^2, x^3 and x^6, imply

$$\epsilon_L = \Gamma_0 \Gamma_1 \Gamma_2 \Gamma_3 \Gamma_6 \epsilon_R. \tag{III.3}$$

Conditions (III.2) and (III.3) can be recombined into

$$\epsilon_L = \Gamma_0 \Gamma_1 \Gamma_2 \Gamma_3 \Gamma_7 \Gamma_8 \Gamma_9 \epsilon_R, \tag{III.4}$$

which shows that certainly sixbranes can be added with no additional supersymmetry breaking.

The solitonic fivebranes break half of the supersymmetries, while the D-6brane breaks again half of the remaining symmetry, leaving eight real supercharges, which leads to four dimensional $N = 2$ supersymmetry.

As we will discuss later on, the sixbranes of type IIA string theory can be used to add hypermultiplets to the effective macroscopic four dimensional theory. In particular, the mass of these hypermultiplets will become zero whenever the D-4brane meets a D-6brane.

One of the main achievements of the brane representations of supersymmetric gauge theories is the ability to represent the different moduli spaces, namely the Coulomb and Higgs branches, in terms of the brane motions left free. For a configuration of k fourbranes connecting two solitonic fivebranes along the x^6 direction, as the one we have described above, the Coulomb branch of the moduli space of the four dimensional theory is parametrized by the different positions of the transversal fourbranes on the fivebranes. When N_f Dirichlet sixbranes are added to this configuration, what we are describing is the Coulomb branch of a four dimensional field theory with $SU(N_c)$ gauge group (in case N_c is the number of D-4branes we are considering), with N_f flavor hypermultiplets. In this brane representation, the Higgs branch of the theory is obtained when each fourbrane is broken into several pieces ending on different sixbranes: the locations of the D-4branes living between two D-6branes determine the Higgs branch. However, we will mostly concentrate on the study of the Coulomb branch for pure gauge theories.

As we know from the Seiberg-Witten solution of $N = 2$ supersymmetric gauge theories, the classical moduli of the theory is corrected by quantum effects. There are two types of effects that enter the game: a non vanishing beta function (determined at one loop) implies the existence, in the assymptotically free regime, of a singularity at the infinity point in moduli space, and strong coupling effects, which imply the existence of extra singularities, where some magnetically charged particles become massless. The problem we are facing now is how to derive such a complete characterization of the quantum moduli space of four dimensional $N = 2$ supersymmetric field theory directly from the dynamics governing the brane configuration. The approach to be used is completely different from a brane construction in type IIA string theory to a type IIB brane configuration. In fact, in the type IIB case, employed in the description of the preceding section of three dimensional $N = 4$ supersymmetric field theories, we can pass from weak to strong coupling through the standard $Sl(2,\mathbf{Z})$ duality of type IIB strings; hence, the essential ingredient we need is to know how brane configurations transform under this duality symmetry. In the case of type IIA string theory, the situation is more complicated, as the theory is not $Sl(2,\mathbf{Z})$ self dual. However, we know that the strong coupling limit of type IIA dynamics is described by the eleven dimensional M-theory; therefore, we should expect to recover the strong coupling dynamics of four dimensional $N = 2$ supersymmetric gauge theories using the M-theory description of strongly coupled type IIA strings.

Let us first start by considering weak coupling effects. The first thing to be noticed, concerning the above described configuration of N_c Dirichlet fourbranes extending along the x^6 direction between two solitonic fivebranes, where only a rigid motion of the transversal fourbranes is allowed, is that this simple image is missing the classical dynamics of the fivebranes. In fact, in this picture we are assuming that the x^6 coordinate on the fivebrane worldvolume is constant, which is in fact a very bad approximation. Of course, one physical requirement we should impose to a brane configuration, as we did in the case of the type IIB configurations of the previous section, is that of minimizing the total worldvolume action. More precisely, what we have interpreted as Coulomb or Higgs branches in term of free motions of some branes entering the configuration, should correspond to zero modes of the brane configuration, i. e., to changes in the configuration preserving the condition of minimum worldvolume action (in other words, changes in the brane configuration that do not constitute an energy expense). The coordinate x^6 can be assumed to only depend on the "normal" coordinates x^4 and x^5, which can be combined into the complex coordinate

$$v \equiv x^4 + ix^5, \tag{III.5}$$

representing the normal to the position of the transversal fourbranes. Far away from the position of the fourbranes, the equation for x^6 reduces now to a two dimensional laplacian,

$$\nabla^2 x^6(v) = 0, \tag{III.6}$$

with solution

$$x^6(v) = k \ln |v| + \alpha, \tag{III.7}$$

for some constants k and α, that will depend on the solitonic and Dirichlet brane tensions. As we can see from (III.7), the value of x^6 will diverge at infinity. This constitutes, as a difference with the type IIB case, a first problem for the interpretation of equation (III.1). In fact, in deriving (III.1) we have used a standard Kaluza-Klein argument, where the four dimensional coupling constant is defined by the volume of the internal space (in this ocasion, the x^6 interval between the two solitonic fivebranes). Since the Dirichlet four branes will deform the solitonic fivebrane, the natural way to define the internal space would be as the interval defined by the values of the coordinate x^6 at v equal to infinity, which is the region where the disturbing effect of the four brane is very likely vanishing, as was the case in the definition of the effective three dimensional coupling in the type IIB case. However, equations (III.6) and (III.7) already indicate us that this can not be the right picture, since these values of the x^6 coordinate are divergent. Let us then consider a configuration with N_c transversal fourbranes. From equations (III.1) and (III.7), we get, for large v,

$$\frac{1}{g_4^2} = -\frac{2kN_c \ln(v)}{g_5^2}, \tag{III.8}$$

where we have differentiated the direction in which the fourbranes pull the fivebrane. Equation (III.8) can have a very nice meaning if we interpret it as the one loop renormalization group equation for the effective coupling constant. In order to justify this interpretation, let us first analyze the physical meaning of the parameter k. From equation (III.7), we notice that if we move in v around a value where a fourbrane is located (that we are assuming is $v = 0$), we get the monodromy transformation

$$x^6 \to x^6 + 2\pi i k. \tag{III.9}$$

This equation can be easily understood in M-theory, where we add an extra eleventh dimension, x^{10}, that we use to define the complex coordinate

$$x^6 + ix^{10}. \tag{III.10}$$

Now, using the fact that the extra coordinate is compactified on a circle of radius R we can, from (III.9), identify k with R. From a field theory point of view, we have a similar interpretation of the monodromy of (III.8), but now in terms of a change in the theta parameter. Let us then consider the one loop renormalization group equation for $SU(N_c)$ $N=2$ supersymmetric gauge theories without hypermultiplets,

$$\frac{4\pi}{g_4^2(u)} = \frac{4\pi}{g_0^2} - \frac{2N_c}{4\pi} \ln\left(\frac{u}{\Lambda}\right), \qquad (\text{III.11})$$

with Λ the dynamically generated scale, and g_0 the bare coupling constant. The bare coupling constant can be absorved through a change in Λ; in fact, when going from Λ to a new scale Λ', we get

$$\frac{4\pi}{g_4^2(u)} = \frac{4\pi}{g_0^2} - \frac{2N_c}{4\pi} \ln\left(\frac{u}{\Lambda'}\right) - \frac{2N_c}{4\pi} \ln\left(\frac{\Lambda'}{\Lambda}\right). \qquad (\text{III.12})$$

Thus, once we fix a reference scale Λ_0, the dependence on the scale Λ of the bare coupling constant is given by

$$-\frac{2N_c}{4\pi} \ln\left(\frac{\Lambda}{\Lambda_0}\right). \qquad (\text{III.13})$$

It is important to distinguish the dependence on Λ of the bare coupling constant, and the dependence on u of the effective coupling. In the brane configuration approach, the coupling constant defined by (III.8) is the bare coupling constant of the theory, as determined by the definite brane configuration. Hence, it is (III.13) that we should compare with (III.8); naturally, some care is needed concerning units and scales. Once we interpret k as the radius of the internal S^1 of M-theory we can, in order to make contact with (III.13), identify g_5^2 with the radius of S^1, which in M-theory units is given by

$$R = g l_s, \qquad (\text{III.14})$$

with g the string coupling constant, and l_s the string length. Therefore, (III.1) should be modified to

$$\frac{1}{g_4^2} = \frac{x_2^6 - x_1^6}{g l_s} = -2N_c \ln(v/l_s), \qquad (\text{III.15})$$

which is dimensionless. We can fix the scale of the four dimensional theory, Λ, in such a way that $\frac{1}{g_4^2(u;\Lambda)} = -2N_c \ln\left(\frac{u}{\Lambda}\right)$. With this normalization the bare coupling constant at the string scale is given by (III.13) as

$$\frac{4\pi}{g_0^2(l_s^{-1})} = -\frac{2N_c}{4\pi} \ln\left(\frac{l_s^{-1}}{\Lambda}\right). \qquad (\text{III.16})$$

Comparing now (III.15) and (III.16), we notice that $\frac{1}{v}$ can be used as the scale of the theory.

Defining now an adimensional complex variable,

$$s \equiv (x^6 + ix^{10})/R, \qquad (\text{III.17})$$

and a complexified coupling constant,

$$\tau = \frac{4\pi i}{g^2} + \frac{\theta}{2\pi}, \qquad (\text{III.18})$$

we can generalize (III.15) to

$$-i\tau_\alpha(v) = s_2(v) - s_1(v), \qquad (\text{III.19})$$

for the simple configuration of branes defining a pure gauge theory. Now, we can clearly notice how the monodromy, as we move around $v = 0$, means a change $\theta \to \theta + 2\pi N_c$.

It is important stressing that nothing, a priori, enforces us to interpret the θ-parameter of the four dimensional field theory in terms of the extra dimension coming from M-theory. However, such an interpretation is very natural once the solitonic fivebrane of type IIA string theory is understood as the direct dimensional reduction of the M-theory fivebrane.

Let us now come back, for a moment, to the bad behaviour of $x^6(v)$ at large values of v. A possible way to solve this problem is modifying the configuration of a single pair of fivebranes, with N_c fourbranes extending between them, to consider a larger set of solitonic fivebranes. Labelling this fivebranes by α, with $\alpha = 0, \ldots, n$, the corresponding x^6_α coordinate will depend on v as follows:

$$x^6(v)_\alpha = R \sum_{i=1}^{q_L} \ln |v - a_i| - R \sum_{j=1}^{q_R} \ln |v - b_j|, \qquad (\text{III.20})$$

where q_L and q_R represent, respectively, the number of D-4branes to the left and right of the αth fivebrane. As is clear from (III.20), a good behaviour at large v will only be possible if the numbers of fourbranes to the right and left of a fivebrane are equal, $q_L = q_R$, which somehow amounts to compensating the perturbation created by the fourbranes at the sides of a fivebrane. The four dimensional field theory represented now by this brane array will have a gauge group $\prod_\alpha U(k_\alpha)$, where k_α is the number of transversal fourbranes between the $\alpha - 1$ and α^{th} solitonic fivebranes. Now, minimization of the worldvolume action will require not only taking into account the dependence of x^6 on v, but also the fourbrane positions on the NS-5brane, represented by a_i and b_j in (III.20), on the four dimensional worldvolume coordinates x^0, x^1, x^2 and x^3. Using (III.20), and the Nambu-Goto action for the solitonic fivebrane, we get, for the kinetic energy,

$$\int d^4x d^2v \sum_{\mu=0}^{3} \partial_\mu x^6(v, a_i(x^\mu), b_j(x^\mu)) \partial^\mu x^6(v, a_i(x^\mu), b_j(x^\mu)). \qquad (\text{III.21})$$

Convergence of the v integration implies

$$\partial_\mu(\sum_i a_i - \sum_j b_j) = 0 \qquad (\text{III.22})$$

or, equivalently,

$$\sum_i a_i - \sum_j b_j = \text{constant}. \qquad (\text{III.23})$$

This "constant of motion" is showing how the average of the relative position between left and right fourbranes must be hold constant. Since the Coulomb branch of the $\prod_\alpha U(k_\alpha)$ gauge theory will be associated with different configurations of the transversal fourbranes, constraint (III.23) will reduce the dimension of this space. As we know from our general discussion on D-branes, the $U(1)$ part of the $U(k_\alpha)$ gauge group can be associated to the motion of the center of mass. Constraint (III.23) implies that the center of mass is frozen in each sector. With no semi-infinite fourbranes to the right, we have that $\sum_i a_i = 0$; now, this constraint will force the center of mass of all sectors to vanish, which means that the field theory we are describing is $\prod_\alpha SU(k_\alpha)$, instead of $\prod_\alpha U(k_\alpha)$. The same result can be derived if we include semi-infinite fourbranes to the left and right of the first and last solitonic fivebranes: as they are infinitely massive, we can assume that they do not move in the x^4 and x^5 directions. An important difference will appear if we consider periodic configurations of fivebranes, upon compactification of the x^6 direction to a circle: in this case, constraint (III.23) is now only able to reduce the group to $\prod_\alpha SU(k_\alpha) \times U(1)$, leaving alive a $U(1)$ factor.

Hypermultiplets in this gauge theory are understood as strings connecting fourbranes on different sides of a fivebrane: therefore, whenever the positions of the fourbranes to the left and right of a solitonic brane become coincident, a massless hypermultiplet arises. As the hypermultiplets are charged under the gauge groups at both sides of a certain $\alpha + 1$ fivebrane, they will transform as $(k_\alpha, \bar{k}_{\alpha+1})$.

However, as the position of the fourbranes on both sides of a fivebrane varies as a function of x^0, x^1, x^2 and x^3, the existence of a well defined hypermultiplet can only be accomplished thanks to the fact that its variation rates on both sides are the same, as follows again from (III.22): $\partial_\mu(\sum_i a_{i,\alpha}) = \partial_\mu(\sum_j a_{j,\alpha+1})$. The definition of the bare massses comes then naturally from constraint (III.23):

$$m_\alpha = \frac{1}{k_\alpha}\sum_i a_{i,\alpha} - \frac{1}{k_{\alpha+1}}\sum_j a_{j,\alpha+1}. \qquad (\text{III.24})$$

With this interpretation, the constraint (III.23) becomes very natural from a physical point of view: it states that the masses of the hypermultiplets do not depend on the spacetime position.

The consistency of the previous definition of hypermultiplets can be checked using the previous construction of the one-loop beta function. In fact, from equation (III.19), we get, for large values of v,

$$-i\tau_\alpha(v) = (2k_\alpha - k_{\alpha-1} - k_{\alpha+1})\ln v. \qquad (\text{III.25})$$

The number k_α of branes in the αth is, as we know, the number of colours, N_c. Comparing with the beta function for $N=2$ supersymmetric $SU(N_c)$ gauge theory with N_f flavors, we conclude that

$$N_f = k_{\alpha-1} + k_{\alpha+1}. \qquad (\text{III.26})$$

so that the number of fourbranes (hypemultiplets) at both sides of a certain pair of fivebranes, $k_{\alpha+1} + k_{\alpha-1} \equiv N_f$, becomes the number of flavors.

Notice, from (III.24), that the mass of all the hypermultiplets associated with fourbranes at both sides of a solitonic fivebrane are the same. This implies a global flavor symmetry. This global flavor symmetry is the gauge symmetry of the adjacent sector. This explains the physical meaning of (III.24).

Let us now come back to equation (III.15). What we need in order to unravel the strong coupling dynamics of our effective four dimensional gauge theory is the u dependence of the effective coupling constant, dependence that will contain non perturbative effects due to instantons. It is from this dependence that we read the Seiberg-Witten geometry of the quantum moduli space. Next, we will see how M-theory can be effectively used to find the quantum moduli space.

A M-Theory and Strong Coupling.

From the M-theory point of view, the brane configuration we are considering can be interpreted in a different way. In particular, the D-4branes we are using to define the four dimensional macroscopic gauge theory can be considered as fivebranes wrapping the eleven dimensional S^1. Moreover, the trick we have used to make finite these fourbranes in the x^6 direction can be directly obtained if we consider fivebranes with worldvolume $\mathbf{R}^4 \times \Sigma$, where $\mathbf{R}^4$ is parametrized by the coordinates x^0, x^1, x^2 and x^4, and Σ is two dimensional, and holomorphically embedded in the four dimesional space of coordinates x^4, x^5, x^6 and x^{10}.

In order to get the Riemann surface Σ we will proceed as follows. Let us then define the single valued coordinate t,

$$t \equiv \exp{-s}, \qquad (\text{III.27})$$

and define the surface Σ we are looking for through

$$F(t,v) = 0. \qquad (\text{III.28})$$

From the classical equations of motion of the fivebrane we know the assymptotic behaviour for very large t,

$$t \sim v^k, \qquad (\text{III.29})$$

and for very small t,

$$t \sim v^{-k}. \tag{III.30}$$

Conditions (III.29) and (III.30) imply that $F(t,v)$ will have, for fixed values of t, k roots, while two different roots for fixed v. It must be stressed that the assymptotic behaviour (III.29) and (III.30) corresponds to the one loop beta function for a field theory with gauge group $SU(k)$, and without hypermultiplets. A function satisfying the previous conditions will be of the generic type

$$F(t,v) = A(v)t^2 + B(v)t + C(v), \tag{III.31}$$

with A, B and C polynomials in v of degree k. From (III.29) and (III.30), the function (III.31) becomes

$$F(t,v) = t^2 + B(v)t + \text{constant}, \tag{III.32}$$

with one undetermined constant. Equation (III.32) is a relation between dimensionless variables, and the constant in (III.32) can be killed through rescalings of v and t. In fact, a change in the constant is equivalent to a rescaling of v through

$$v \to v \left(\frac{\text{const}'}{\text{const}}\right)^{\frac{1}{2k}}. \tag{III.33}$$

If we define the dimensionless variable v, relative to the scale of the theory, Λ^{-1}, we observe that a change in Λ is equivalent to (III.33) when the constant is Λ^{2k}. Thus, we can identify the constant in (III.32) with the scale of the theory, Λ^{2k}

In order to kill this constant, we can rescale t to $t/\text{constant}$. The meaning of this rescaling can be easily understood in terms of of the one loop beta function, written as (III.29) and (III.30). In fact, these equations can be read as

$$s = -k \ln\left(\frac{v}{R}\right), \tag{III.34}$$

and therefore the rescaling of R goes like

$$s \to -k \ln\left(\frac{v}{R'}\frac{R'}{R}\right) \tag{III.35}$$

or, equivalently,

$$t \to t \left(\frac{R'}{R}\right)^k. \tag{III.36}$$

Thus, and based on the above discussion on the definition of the scale, we observe that the constant in (III.32) defines the scale of the theory. With this interpretation of the constant, we can get the Seiberg-Witten solution for $N = 2$ pure gauge theories, with gauge group $SU(k)$. If $B(v)$ is chosen to be

$$B(v) = v^k + u_2 v^{k-2} + u_3 v^{k-3} + \cdots + u_k, \tag{III.37}$$

we finally get the Riemann surface

$$t^2 + B(v)t + 1 = 0, \tag{III.38}$$

a Riemann surface of genus $k-1$, which is in fact the rank of the gauge group. Moreover, we can now try to visualize this Riemann surface as the worldvolume of the fivebrane describing our original brane configuration: each v-plane can be compactified to $\mathbf{P}^1$, and the transversal fourbranes cna be interpreted as gluing tubes, which clearly represents a surface with $k-1$ handles. This image corresponds to gluing two copies of $\mathbf{P}^1$, with k disjoint cuts on each copy or, equivalently, $2k$ branch points. Thus, as can be observed from (III.38), to each transversal D-4brane there correspond two branch points and one cut on $\mathbf{P}^1$.

If we are interested in $SU(k)$ gauge theories with hypermultiplets, then we should first replace (III.29) and (III.30) by the corresponding relations,

$$t \sim v^{k-k_\alpha - 1}, \tag{III.39}$$

and

$$t \sim v^{-k-k_\alpha + 1}, \tag{III.40}$$

for t large and small, respectively. These are, in fact, the relations we get from the beta functions for these theories. If we take $k_{\alpha_1} = 0$, and $N_f = k_{\alpha+1}$, the curve becomes

$$t^2 + B(v)t + C(v) = 0, \tag{III.41}$$

with $C(v)$ a polynomial in v, of degree N_f, parametrized by the masses of the hypermultiplets,

$$C(v) = f \prod_{j=1}^{N_f} (v - m_j), \tag{III.42}$$

with f a complex constant.

Summarizing, we have been able to find a moduli of brane configurations reproducing four dimensional $N = 2$ supersymmetric $SU(k)$ gauge theories. The exact Seiberg-Witten solution is obtained by reduction of the worldvolume fivebrane dynamics on the surface $\Sigma_{\vec{u}}$ defined at (III.38) and (III.40). Obviously, reducing the fivebrane dynamics to $\mathbf{R}^4$ on $\Sigma_{\vec{u}}$ leads to an effective coupling constant in $\mathbf{R}^4$, the $k-1 \times k-1$ Riemann matrix $\tau(\vec{u})$ of $\Sigma_{\vec{u}}$.

Before finishing this section, it is important to stress some peculiarities of the brane construction. First of all, it should be noticed that the definition of the curve Σ, in terms of the brane configuration, requires working with uncompactified x^4 and x^5 directions. This is part of the brane philosophy, where we must start with a particular configuration in flat spacetime. A different approach will consist in directly working with a spacetime $Q \times R^7$, with Q some Calabi-Yau manifold, and consider a fivebrane worldvolume $\Sigma \times \mathbf{R}^4$, with $\mathbf{R}^4 \subset \mathbf{R}^7$, and Σ a lagrangian submanifold of Q. Again, by Mc Lean's theorem, the $N=2$ theory defined on $\mathbf{R}^4$ will have a Coulomb branch with dimension equal to the first Betti number of Σ, and these deformations of Σ in Q will represent scalar fields in the four dimensional theory. Moreover, the holomorphic top form Ω of Q will define the meromorphic λ of the Seiberg-Witten solution. If we start with some Calabi-Yau manifold Q, we should provide some data to determine Σ (this is what we did in the brane case, with Q non compact and flat). If, on the contrary, we want to select Σ directly from Q, we can only do it in some definite cases, which are those related to the *geometric mirror construction*. Let us then recall some facts about the geometric mirror. The data are

- The Calabi-Yau manifold Q.
- A lagrangian submanifold $\Sigma \to Q$.
- A $U(1)$ flat bundle on Σ.

The third requirement is equivalent to interpreting Σ as a D-brane in Q. This is a crucial data, in order to get from the above points the structure of abelian manifold of the Seiberg-Witten solution. Namely, we frist use Mc Lean's theorem to get the moduli of deformations of $\Sigma \to Q$, preserving the condition of lagrangian submanifold. This space is of dimension $b_1(\Sigma)$. Secondly, on each of these points we fiber the jacobian of Σ, which is of dimension g. This family of abelian varieties defines the quantum moduli of a gauge theory, with $N=2$ supersymmetry, with a gauge group of rank equal $b_1(\Sigma)$. Moreover, this family of abelian varieties is the moduli of the set of data of the second and third points above, i. e., the moduli of Σ as a D-2brane. In some particular cases, this moduli is Q itself or, more properly, the geometric mirror of Q. This will be the case for Σ of genus equal one, i. e., for the simple $SU(2)$ case. In this cases, the characterization of Σ in Q is equivalent to describing Q as an elliptic fibration. The relation between geometric mirror and T-duality produces a completely different physical picture. In fact, we can, when Σ is a torus, consider in type IIB a threebrane with classical moduli given by Q. After T-duality or mirror, we get the type IIA description in terms of a fivebrane. In summary, it is an important problem to understand the relation of quantum mirror between type IIA and type IIB string theory, and the M-theory strong coupling description of type IIA strings.

But before ending this section we would like to discuss the "rationale" of the M-theory approach. First of all, we have started with a brane configuration in type IIA string theory which allows us to define a theory that in the infrared behaves as $N=2$ supersymmetric theory, with gauge group $SU(N)$. Preserving the value of the bare coupling constant we can change the separation between the vertical solitonic fivebranes if at the same time we turn on, in appropiate way, the string coupling constant. Through this procedure we can describe the macroscopic gauge theory with $N = 2$ in four dimensions, using strongly coupled type IIA string theory. i. e., using M-theory. Now, in M-theory the brane configuration becomes a single fivebrane. The Coulomb branch of the four dimensional theory is now the moduli of holomorphic curves Σ in $Q = \mathbf{R}^3 \times S^1$, for a fivebrane with worldvolume $\mathbf{R}^4 \times \Sigma$.

B $\;\;N = 2$ Models with Vanishing Beta Function.

Let us come back to brane configurations with $n + 1$ solitonic fivebranes, with k_α Dirichlet fourbranes extending between the αth pair of NS-5branes. The beta function, derived in (III.25), is

$$-2k_\alpha + k_{\alpha+1} + k_{\alpha-1}, \tag{III.43}$$

for each $SU(k_\alpha)$ factor in the gauge group. In this section, we will compactify the x^6 direction to a circle of radius L. Imposing the beta function to vanish in all sectors immediately implies that all k_α are the same. Now, the compactification of the x^6 direction does not allow to eliminate all $U(1)$ factors in the gauge group: one of them can not be removed, so that the gauge group is reduced from $\prod_{\alpha=1}^n U(k_\alpha)$ to $U(1) \times SU(k)^n$. Moreover, using the definition (III.24) of the mass of the hypermultiplets we get, for periodic configurations,

$$\sum_\alpha m_\alpha = 0. \tag{III.44}$$

The hypermultiplets are now in representations of type $k \otimes \bar{k}$, and therefore consists of a copy of the adjoint representation, and a neutral singlet.

Let us consider the simplest case, of $N = 2$ $SU(2) \times U(1)$ four dimensional theory, with one hypermultiplet in the adjoint representation [6]. The corresponding brane configuration contains a single solitonic fivebrane, and two Dirichlet fourbranes. The mass of the hypermultiplet is clearly zero, and the corresponding four dimensional theory has vanishing beta function. A geometric procedure to define masses for the hypermultiplets is a fibering of the v-plane on the x^6 S^1 direction, in a non trivial way, so that the fourbrane positions are identified modulo a shift in v,

$$\begin{aligned} x^6 &\to x^6 + 2\pi L, \\ v &\to v + m, \end{aligned} \tag{III.45}$$

so that now, the mass of the hypermultiplet, is the constant m appearing in (III.45), as $\sum_\alpha m_\alpha = m$.

From the point of view of M-theory, the x^{10} coordinate has also been compactified on a circle, now of radius R. The (x^6, x^{10}) space has the topology of $S^1 \times S^1$. This space can be made non trivial if, when going around x^6, the value of x^{10} is changed as follows:

$$x^6 \to x^6 + 2\pi L,$$
$$x^{10} \to x^{10} + \theta R, \qquad \text{(III.46)}$$

and, in addition, $x^{10} \to x^{10} + 2\pi R$. Relations (III.46) define a Riemann surface of genus one, and moduli depending on L and θ for fixed values of R. θ in (III.46) can be understood as the θ-angle of the four dimensional field theory: the θ-angle can be defined as

$$\frac{x_1^{10} - x_2^{10}}{R}, \qquad \text{(III.47)}$$

with $x_2^{10} = x^{10}(2\pi L)$, and $x_1^{10} = x^{10}(0)$. Using (III.46), we get θ as the value of (III.47). This is the bare θ-angle of the four dimensional theory.

A question inmediately appears concerning the value of the bare coupling constant: the right answer should be

$$\frac{1}{g^2} = \frac{2\pi L}{R}. \qquad \text{(III.48)}$$

It is therefore clear that we can move the bare coupling constant of the theory keeping fixed the value of R, and changing L and θ. Let us now try to solve this model for the massless case. The solution will be given by a Riemann surface Σ, living in the space $E \times C$, where E is the Riemann surface defined by (III.46), and C is the v-plane. Thus, all what we need is defining Σ through an equation of the type

$$F(x, y, z) = 0, \qquad \text{(III.49)}$$

with x and y restricted by the equation of E,

$$y^2 = (x - e_1(\tau))(x - e_2(\tau))(x - e_3(\tau)), \qquad \text{(III.50)}$$

with τ the bare coupling constant defined by (III.47) and (III.48). In case we have a collection of k fourbranes, we will require F to be a polynomial of degree k in v,

$$F(x, y, z) = v^k - f_1(x, y)v^{k-1} + \cdots \qquad \text{(III.51)}$$

The moduli parameters of Σ are, at this point, hidden in the functions $f_i(x, y)$ in (III.51). Let us denote $v_i(x, y)$ the roots of (III.51) at the point (x, y) in E. Notice that (III.51) is a spectral curve defining a branched covering of E, i. e., (III.51) can be interpreted as a spectral curve in the sense of Hitchin's integrable system. If f_i has a pole at some point (x, y), then the same root $v_i(x, y)$ should go to infinity. These poles have the interpretation of locating the position of the solitonic fivebranes. In the simple case we are considering, with a single fivebrane, the Coulomb branch of the theory will be parametrized by meromorphic functions on E with a simple pole at one point, which is the position of the fivebrane. As we have k functions entering (III.51), the dimension of the Coulomb branch will be k, which is the right one for a theory with $U(1) \times SU(k)$ gauge group.

Now, after this discussion of the model with massless hypermultiplets, we will introduce the mass. The space where now we need to define Σ is not $E \times C$, but the non trivial fibration defined through

$$x^6 \to x^6 + 2\pi L,$$
$$x^{10} \to x^{10} + \theta R,$$
$$v \to v + m \tag{III.52}$$

or, equivalently, the space obtained by fibering C non trivially on E. We can flat this bundle over all E, with the exception of one point p_0. Away from this point, the solution is given by (III.51). If we write (III.51) in a factorized form,

$$F(x, y, z) = \prod_{i=1}^{k}(v - v_i(x, y)), \tag{III.53}$$

we can write f_1 in (III.51) as the sum

$$f_1 = \sum_{i=1}^{k} v_i(x, y); \tag{III.54}$$

therefore, f_1 will have poles at the positions of the fivebrane. The mass of the hypermultiplet will be identified with the residue of the differential $f_1\omega$, with ω the abelian differential, $\omega = \frac{dx}{y}$. As the sum of the residues is zero, this means that at the point at infinity, that we identify with p_0, we have a pole with residue m.

IV BRANE DESCRIPTION OF $N = 1$ FOUR DIMENSIONAL FIELD THEORIES.

In order to consider field theories with $N = 1$ supersymmetry, the first thing we will study will be R-symmetry. Let us then recall the way R-symmetries were defined in the case of four dimensional $N = 2$ supersymmetry, and three dimensional $N = 4$ supersymmetry, through compactification of six dimensional $N = 1$ supersymmetric gauge field theories. The $U(1)_R$ in four dimensions, or $SO(3)_R$ in three dimensions, are simply the euclidean group of rotations in two and three dimensions, respectively. Now, we have a four dimensional space $\mathcal{Q}$, parametrized by coordinates t and v, and a Riemann surface Σ, embedded in $\mathcal{Q}$ by equations of the type (III.31). To characterize R-symmetries, we can consider transformations on $\mathcal{Q}$ which transform non trivially its holomorphic top form Ω. The unbroken R-symmetries will then be rotations in $\mathcal{Q}$ preserving the Riemann surface defined by the brane configuration. If we consider only the assymptotic behaviour of type (III.29), or (III.39), we get $U(1)_R$ symmetries of type

$$t \to \lambda^k t,$$
$$v \to \lambda v. \tag{IV.1}$$

This $U(1)$ symmetry is clearly broken by the curve (III.38). This spontaneous breakdown of the $U(1)_R$ symmetry is well understood in field theory as an instanton induced effect. If instead of considering Q, we take the larger space $\hat{Q}$, containing the x^7, x^8 and x^9 coordinates, we see that the $N = 2$ curve is invariant under rotations in the (x^7, x^8, x^9) space.

Let us now consider a brane configuration which reproduces $N = 1$ four dimensional theories [20]. We will again start in type IIA string theory, and locate a solitonic fivebrane at $x^6 = x^7 = x^8 = x^9 = 0$ with, as usual, worldvolume coordinates x^0, x^1, x^2, x^3, x^4 and x^5. At some definite value of x^6, say x_0^6, we locate another solitonic fivebrane, but this time with worldvolume coordinates x^0, x^1, x^2, x^3, x^7 and x^8, and $x^4 = x^5 = x^9 = 0$. As before, we now suspend a set of k D-4branes in between. They will be parametrized by the positions $v = x^4 + ix^5$, and $w = x^7 + ix^8$, on the two solitonic fivebranes. The worldvolume coordinates on this D-4branes are, as in previous cases, x^0, x^1, x^2 and x^3. The effective field theory defined by the set of fourbranes is macroscopically a four dimensional gauge theory, with coupling constant

$$\frac{1}{g^2} = \frac{x_0^6}{gl_s}. \tag{IV.2}$$

Moreover, now we have only $N = 1$ supersymmetry, as no massless bosons can be defined on the four dimensional worldvolume (x^0, x^1, x^2, x^3). In fact, at the line $x^6 = 0$ the only possible massless scalar would be v, since $w = 0$ and $x^9 = 0$, so that we project out x^9 and w. On the other hand, at x_0^6 we have $v = 0$ and $x^9 = 0$ and, therefore, we have projected out all massless scalars. Notice that by the same argument, in the case of two solitonic fivebranes located at different values of x^6 but at $x^7 = x^8 = x^9 = 0$, we have one complex massless scalar that is not projected out, which leads to $N = 2$ supersymmetry in four dimensions. The previous discussion means that v, w and x^9 are projected out as four dimensional scalar fields; however, w and v are still classical moduli parameters of the brane configuration.

Now, we return to a comment already done in previous section: each of the fourbranes we are suspending in between the solitonic fivebranes can be interpreted as a fivebrane wrapped around a surface defined by the eleven dimensional S^1 of M-theory, multiplied by the segment $[0, x_0^6]$. Classically, the four dimensional theory can be defined through dimensional reduction of the fivebrane worldvolume on the surface Σ. The coupling constant will be given by the moduli τ of this surface,

$$\frac{1}{g^2} = \frac{2\pi R}{S}, \tag{IV.3}$$

with S the length of the interval $[0, x_0^6]$, in M-theory units. In $N = 1$ supersymmetric field theories, on the contrary of what takes place in the $N = 2$ case, we have not a classical moduli and, therefore, we can not define a wilsonian coupling constant depending on some mass scale fixed by a vacuum expectation value. This fact can produce some problems, once we take into account the classical dependence of x^6 on v and w. In principle, this dependence should be the same as that in the case studied in previous section,

$$x^6 \sim \tilde{k} \ln v,$$
$$x^6 \sim \tilde{k} \ln w. \tag{IV.4}$$

Using the t coordinate defined in (III.27), equations (IV.4) become

$$t \sim v^k,$$
$$t \sim w^k, \tag{IV.5}$$

for large and small t, respectively, or, equivalently, $t \sim v^k$, $t^{-1} \sim w^k$.

Using the relations in (IV.2) we get, for $k = N_c$,

$$\frac{1}{g^2} \sim -N_c [\ln\left(\frac{v}{l_s}\right) + \ln\left(\frac{w}{l_s}\right)], \tag{IV.6}$$

where we are measuring v and w in string units. Following the same approach as in the $N = 2$ case, we can try read, directly from (IV.6), the one loop beta function of the theory,

$$\mu \frac{\partial g}{\partial \mu} = -\frac{g^3}{16\pi^2} 3 N_c, \tag{IV.7}$$

or, equivalently,

$$\frac{8\pi^2}{g(\mu)^2} = -N_c \ln \left(\frac{\Lambda}{\mu}\right)^3. \tag{IV.8}$$

In order to relate (IV.6) and (IV.8) we can imposse the relation

$$v = \zeta w^{-1}, \tag{IV.9}$$

where ζ is some constant. Using (IV.9), we get

$$\frac{1}{g(\mu)^2} \sim -N_c \ln \left(\frac{\zeta}{l_s^2}\right), \tag{IV.10}$$

which leads to the relation

$$\frac{\zeta}{l_s^2} = \left(\frac{\Lambda}{\mu}\right)^3. \tag{IV.11}$$

Choosing l_s^{-1} for the renormalization point μ, we get

$$\zeta = \Lambda^3 l_s^5. \tag{IV.12}$$

In other words, the constant ζ in (IV.9) is related to the scale of the $N = 1$ theory, Λ. The specific relation will depend on the unit we choose. The previous argument was done in the context of type IIA string theory, without invoquing M-theory. Moreover, (IV.9), together with (IV.29) can be interpreted as defining a fivebrane worldvolume in M-theory of the type $\Sigma \times \mathbf{R}^{3,1}$, where the Riemann surface Σ is defined by

$$t = v^{N_c},$$
$$\zeta^{N_c} t^{-1} = w^{N_c},$$
$$v = \zeta w^{-1}. \qquad (IV.13)$$

This is a holomorphic curve of genus zero, in the space $\mathcal{Q} = \mathbf{R}^5 \times S^1$. As can be seen from (IV.13), the assymptotic behaviour of the curve is determined by the value of ζ^{N_c}. Thus, the possible vacua of the theory are determined by the N_c different roots ζ of ζ^{N_c}, in agreement with the known value for tr $(-1)^F$.

For a given value of ζ, (IV.13) defines a Riemann surface of genus zero, i. e., a rational curve. This curve is now embedded in the space of (t, v, w) coordinates. We will next observe that these curves, (IV.13), are the result of "rotating" [9] the rational curves in the Seiberg-Witten solution, corresponding to the singular points. However, before doing that let us comment on $U(1)_R$ symmetries. As mentioned above, in order to define an R-symmetry we need a transformation on variables (t, v, w) not preserving the holomorphic top form,

$$\Omega = dv \wedge dw \wedge \frac{dt}{t} R. \qquad (IV.14)$$

A rotation in the w-plane, compatible with the assymptotic conditions (IV.5), and defining an R-symmetry, is

$$v \to v,$$
$$t \to t,$$
$$w \to e^{2\pi i/k} w. \qquad (IV.15)$$

Now, it is clear that this symmetry is broken spontaneously by the curve (IV.13). More interesting is an exact $U(1)$ symmetry, that can be defined for the curve (IV.13):

$$v \to e^{i\delta} v,$$
$$t \to e^{i\delta k} t,$$
$$w \to e^{-i\delta} w. \qquad (IV.16)$$

As can be seen from (IV.14), this is not an R-symmetry, since Ω is invariant. Fields charged with respect to this $U(1)$ symmetry should carry angular momentum in the v or w plane, or linear momentum in the eleventh dimension interval (i. e., zero branes) The fields of $N = 1$ SQCD do not carry any of these charges, so all fields with $U(1)$ charge should be decoupled from the $N = 1$ SQCD degrees of freedom. This is equivalent to the way we have projected out fields in the previous discussion on the definition of the effective $N = 1$ four dimensional field theory.

A Rotation of Branes.

A different way to present the above construction is by performing a rotation of branes. We will now concentrate on this procedure. The classical configuration of NS-5branes with worldvolumes extending along x^0, x^1, x^2, x^3, x^4 and x^5, can be modified to a configuration where one of the solitonic fivebranes has been rotated, from the $v = x^4 + ix^5$ direction, to be also contained in the (x^7, x^8)-plane, so that, by moving it a finite angle μ, it is localized in the (x^4, x^5, x^7, x^8) space. Using the same notation as in previous section, the brane configuration, where a fivebrane has been moved to give rise to an angle μ in the (v, w)-plane, the rotation is equivalent to impossing

$$w = \mu v. \tag{IV.17}$$

In the brane configuration we obtain, points on the rotated fivebrane are parametrized by the (v, w) coordinates in the (x^4, x^5, x^7, x^8) space. We can therefore impose the following assymptotic conditions [19]:

$$\begin{aligned} t &= v^k, & w &= \mu v, \\ t &= v^{-k}, & w &= 0, \end{aligned} \tag{IV.18}$$

respectively for large and small t. Let us now assume that this brane configuration describes a Riemann surface, $\hat{\Sigma}$, embedded in the space $(x^6, x^{10}, x^4, x^5, x^7, x^8)$, and let us denote by Σ the surface in the $N=2$ case, i. e., for $\mu = 0$. In these conditions, $\hat{\Sigma}$ is simply the graph of the function w on Σ. We can interpret (IV.17) as telling us that w on Σ possesses a simple pole at infinity, extending holomorphically over the rest of the Riemann surface. If we impose this condition, we get that the projected surface Σ, i. e., the one describing the $N=2$ theory, is of genus zero. In fact, it is a well known result in the theory of Riemann surfaces that the order of the pole at infinity depends on the genus of the surface in such a way that for genus larger than zero, we will be forced to replace (IV.17) by $w = \mu v^a$ for some power a depending on the genus. A priori, there is no problem in trying to rotate using, instead of $w = \mu v$, some higher pole modification of the type $w = \mu v^a$, for $a > 1$. This would provide Σ surfaces with genus different from zero; however, we would immediately find problems with equation (IV.6), and we will be unable to kill all dependence of the coupling constant on v and w. Therefore, we conclude that the only curves that can be rotated to produce a four dimensional $N=1$ theory are those with zero genus. This is in perfect agreement with the physical picture we get from the Seiberg-Witten solution. Namely, once we add a soft breaking term of the type $\mu \text{tr}\phi^2$, the only points remaining in the moduli space as real vacua of the theory are the singular points, where the Seiberg-Witten curve degenerates.

B QCD Strings and Scales.

The $N=1$ four dimensional field theory we have described contains, in principle, two parameters. One is the constant ζ introduced in equation (IV.9) which, as we have already mentioned, is, because of (IV.6), intimately connected with Λ, and the radius R of the eleven dimensional S^1. Our first task would be to see what kind of four dimensional dynamics is dependent on the particular value of R, and in what way. The best example we can of course use is the computation of gaugino-gaugino condensates. In order to do that, we should try to minimize a four dimensional suerpotential for the $N = 1$ theory. Following Witten, we will define this superpotential W as an holomorphic function of Σ, and with critical points precisely when the surface Σ is a holomorphic curve in $\mathcal{Q}$. The space $\mathcal{Q}$ now is the one with coordinates x^4, x^5, x^6, x^7, x^8 and x^{10} (notice that this second condition was the one used to prove that rotated curves are necesarily of genus equal zero). Moreover, we need to work with a holomorphic curve because of $N = 1$ supersymmetry. A priori, there are two different ways we can think about this superpotential: maybe the simplest one, from a physical point of view, is as a functional defined on the volume of Σ, where this volume is given by

$$\text{Vol}(\Sigma) = J.\Sigma, \qquad (IV.19)$$

with J the Kähler class of $\mathcal{Q}$. The other posibility is defining

$$W(\Sigma) = \int_B \Omega, \qquad (IV.20)$$

with B a 3-surface such that $\Sigma = \partial B$, and Ω the holomorphic top form in $\mathcal{Q}$. Definition (IV.20) automatically satisfies the condition of being stationary, when Σ is a holomorphic curve in $\mathcal{Q}$. Notice that the holomorphy condition on Σ means, in mathematical terms, that Σ is an element of the Picard lattice of $\mathcal{Q}$. i. e., an element in $H_{1,1}(\mathcal{Q}) \cap H_2(\mathcal{Q})$. This is what allows us to use (IV.19), however, and this is the reason for temporarily abandoning the approach based on (IV.19); what we require to W is being stationary for holomorphic curves, but it should, in principle, be defined for arbitrary surfaces Σ, even those which are not part of the Picard group. Equation (IV.20) is only well defined if Σ is contractible, i. e., if the homology class of Σ in $H_2(\mathcal{Q}; \mathbf{Z})$ is trivial. If that is not the case, a reference surface Σ_0 needs to be defined, and (IV.20) is modified to

$$W(\Sigma) - W(\Sigma_0) = \int_B \Omega. \qquad (IV.21)$$

where now $\partial B = \Sigma \cup \Sigma_0$. For simplicity, we will assume $H_3(\mathcal{Q}; \mathbf{Z}) = 0$. From physical arguments we know that the set of zeroes of the superpotential should be related by $\mathbf{Z}_{N_c}$ symmetry, with N_c the number of transversal fourbranes. Therefore, if we choose Σ_0 to be $\mathbf{Z}_k$ invariant, we can write $W(\Sigma_0) = 0$, and $W(\Sigma) = \int_B \Omega$. Let us then take B as the complex plane multiplied by an interval $I = [0, 1]$, and let us first map the complex plane into Σ. Denoting r the coordinate on this complex plane, Σ, as given by (IV.13), is defined by

$$t = r^{N_c},$$
$$v = r,$$
$$w = \zeta r^{-1}. \qquad (IV.22)$$

Writing $r = e^\rho e^{i\theta}$, we can define Σ_0 as

$$t = r^{N_c},$$
$$v = f(\rho) r,$$
$$w = \zeta f(-\rho) r^{-1}, \qquad (IV.23)$$

with $f(\rho) = 1$ for $\rho > 2$, and $f(\rho) = 0$ for $\rho < 1$. The $\mathbf{Z}_k$ transformation $t \to t$, $w \to e^{2\pi i/k} w$ and $v \to v$, is a symmetry of (IV.23) if, at the same time, we perform the reparametrization of the r-plane

$$\rho \to \rho,$$
$$\theta \to \theta + b(\rho), \qquad (IV.24)$$

with $b(\rho) = 0$ for $\rho \geq 1$, and $b(\rho) = -\frac{2\pi}{k}$ for $\rho \leq -1$. Thus, the 3-manifold entering the definition of B, is given by

$$t = r^{N_c},$$
$$v = g(\rho, \sigma) r,$$
$$w = \zeta g(-\rho, \sigma) r^{-1}, \qquad (IV.25)$$

such that for $\sigma = 0$ we have $g = 1$, and for $\sigma = 1$, we get $g(\rho) = f(\rho)$. Now, with

$$\Omega = R dv \wedge dw \wedge \frac{dt}{t}, \qquad (IV.26)$$

we get

$$W(\Sigma) = N_c R \int_B dv \wedge dw \wedge \frac{dr}{r}. \qquad (IV.27)$$

The dependence on R is already clear from (IV.26). In order to get the dependence on ζ we need to use (IV.25),

$$W(\Sigma) = N_c R \zeta \int d\sigma d\theta d\rho \left(\frac{\partial g_+}{\partial \sigma} \frac{\partial g_-}{\partial \rho} - \frac{\partial g_+}{\partial \rho} \frac{\partial g_-}{\partial \sigma} \right), \qquad (IV.28)$$

for $g_\pm = g(\pm \rho, \sigma)$. Thus we get

$$W(\Sigma) \sim N_c R \zeta. \qquad (IV.29)$$

Now we should compare (IV.29) with the known value of the gaugino condensate,

$$W = N_c < \text{tr } \lambda\lambda >, \qquad (IV.30)$$

with

$$< \mathrm{tr}\,\lambda\lambda >_j = N_c \Lambda^3 \exp 2\pi i j/N_c \qquad (\text{IV}.31)$$

the value of the gaugino condensate in the j-vacua. Thus, from (IV.31) we get

$$N_c^2 \Lambda^3 l_s^6 \sim N_c R \zeta, \qquad (\text{IV}.32)$$

where we have used the string length as reference scale. Notice that our definition of W is the volume of a 3-surface. From (IV.32) and (IV.14), we get the relation

$$l_s = \frac{R}{N_c}. \qquad (\text{IV}.33)$$

A different way to connect ζ with Λ is defining, in the M-theory context, the QCD string and computing its tension. Following Witten, we will then try an interpretation of ζ independent of (IV.6), by computing in terms of ζ the tension of the QCD string. We will then, to define the tension, consider the QCD string as a membrane, product of a string in $\mathbf{R}^4$, and a string living in $\mathcal{Q}$. Let us then denote by C a curve in $\mathcal{Q}$, and assume that C ends on Σ in such a way that a membrane wrapped on C defines a string in $\mathbf{R}^4$. Moreover, we can simply think of C as a closed curve in $\mathcal{Q}$, going around the eleven dimensional S^1,

$$\begin{aligned} t &= t_0 \exp(-2\pi i \sigma), \\ v &= t_0^{1/N_c}, \\ w &= \zeta v^{-1}. \end{aligned} \qquad (\text{IV}.34)$$

This curve is a non trivial element in $H_1(\mathcal{Q}; \mathbf{Z})$, and a membrane wrapped on it will produce an ordinary type IIA string; however, we can not think that the QCD string is a type IIA string. If $\mathcal{Q} = \mathbf{R}^3 \times S^1$, then $H_1(\mathcal{Q}; \mathbf{Z}) = \mathbf{Z}$, and curves of type (IV.34) will be the only candidates for non trivial 1-cycles in $\mathcal{Q}$. However, we can define QCD strings using cycles in the relative homology, $H_1(\mathcal{Q}/\Sigma; \mathbf{Z})$. i. e., considering non trivial cycles ending on the surface Σ. To compute $H_1(\mathcal{Q}/\Sigma; \mathbf{Z})$, we can use the exact sequence

$$H_1(\Sigma; \mathbf{Z}) \to H_1(\mathcal{Q}; \mathbf{Z}) \xrightarrow{\imath} H_1(\mathcal{Q}/\Sigma; \mathbf{Z}), \qquad (\text{IV}.35)$$

which implies

$$H_1(\mathcal{Q}/\Sigma; \mathbf{Z}) = H_1(\mathcal{Q}; \mathbf{Z})/\imath H_1(\Sigma; \mathbf{Z}). \qquad (\text{IV}.36)$$

The map $\imath$ is determined by the map defining Σ ($t = v^{N_c}$), and thus we can conclude that, very likely,

$$H_1(\mathcal{Q}/\Sigma; \mathbf{Z}) = \mathbf{Z}_{N_c}. \qquad (\text{IV}.37)$$

A curve in $H_1(\mathcal{Q}/\Sigma; \mathbf{Z})$ can be defined as follows:

$$t = t_0,$$
$$v = t_0^{1/N_c} e^{2\pi i\sigma/N_c},$$
$$w = \zeta v^{-1}, \qquad (IV.38)$$

with t_0^{1/N_c} one of the N_c roots. The tension of (IV.38), by construction, is independent of R, because t is fixed. Using the metric on $\mathcal{Q}$, the length of (IV.38) is given by

$$\left(\frac{\zeta^2 t^{-2/N_c}}{N_c^2} + \frac{t^{2/N_c}}{N_c^2} \right)^{1/2}, \qquad (IV.39)$$

and its minimum is obtained when $t^{2/n} = \zeta$. Thus, the length of the QCD string should be

$$\frac{|\zeta|^{1/2}}{N_c}. \qquad (IV.40)$$

which has the right length units, as ζ behaves as (length)^2. The relation of (IV.40) with the QCD tension implies

$$|\zeta|^{1/2} = \Lambda N_c l_s^2 \qquad (IV.41)$$

or, equivalently,

$$l_s = \frac{N_c^2}{\Lambda}. \qquad (IV.42)$$

Consistency between (IV.33) and (IV.42) requires

$$R = \frac{1}{\Lambda N_c}. \qquad (IV.43)$$

These are not good news, as they imply that the theory we are working with, in order to match QCD, posseses 0-brane modes, with masses of the order of Λ, and therefore we have not decoupled the M-theory modes.

C Domain Walls.

Given the vacuum structure of $N = 1$ supersymmetric QCD, we can define domain walls interpolating between the different vacua. The M-theory picture of these domain walls can be given in terms of a fivebrane with worldvolume $S \times \mathbf{R}^{2,1}$, with S a 3-cycle in $\mathcal{Q} \times \mathbf{R}$, with $\mathbf{R}$ standing for one of the space-coordinates, let us say x_3, and such that S in the region defined by $x_3 = +\infty$ coincide with the Riemann sphere Σ_{j+1}. In order to compute the tension the tension of this domain wall we can use the superpotential deirved above, obtaining

$$\Delta W = W_{j+1} - W_j = N_c R \zeta (1 - e^{2\pi i j/N_c}). \tag{IV.44}$$

Therefore, the tension will behave as

$$T \sim R|\zeta|. \tag{IV.45}$$

Using relation (IV.33) we get

$$T \sim N_c \Lambda^3 l_s^6. \tag{IV.46}$$

A crucial aspect of equation (IV.46) is that when considering, as is the case in the large N limit, that the string coupling constant for the QCD string goes like $\frac{1}{N_c}$, then the tension of the domain wall goes like $\frac{1}{g_{string}^{QCD}}$, which is the typical relation for D-branes. In other words, we get the suggesting picture of a relation betweeen the QCD string and domain walls, in perfect parallel to D-branes and open strings. Moreover, we can expect the QCD string ending on the domain wall in a similar way to fundamental strings ending on D-branes. This, in particular, means that very likely we can interpret the domain walls of $N = 1$ supersymmetric QCD as Chan-Paton factors for the QCD string. If this is the case, we should be able to have quarks in the fundamental representation living on the domain wall.

Acknowledgments

C. G. wishes to thank the organizers of the VI Taller de Partículas y Campos for a nice ambience. This work is partially supported by European Community grant ERBFM-RXCT960012, and by grant AEN-97-1711.

REFERENCES

1. M. J. Duff, R. R. Khuri and J. X. Lu, "String Solitons", Phys. Rep. **259** (1995), 213.
2. For a more complete reference, see Phys. Rep., C. Gómez and R. Hernández, to appear.
3. A. Hanany and E. Witten, "Type IIB Superstrings, BPS Monopoles and Three Dimensional Gauge Dynamics", Nucl. Phys. **B492** (1997), 152.
4. S. Elitzur, A. Giveon and D. Kutasov, "Branes and $N = 1$ Duality in String Theory", Phys. Lett. **B400** (1997), 269.
5. J. de Boer, K. Hori, Y. Oz and Z. Yin, "Branes and Mirror Symmetry in $N = 2$ Supersymmetric Gauge Theories in Three Dimensions", **hep-th/9702154**.
6. E. Witten, "Solutions of Four-Dimensional Fields Theories Via M-Theory", Nucl. Phys. **B500** (1997), 3.
7. N. Evans, C. V. Johnson and A. D. Shapere, "Orientifolds, Branes and Duality of $4D$ Field Theories", **hep-th/9703210**.
8. S. Elitzur, A. Giveon, D. Kutasov, E. Ravinovici and A. Schwimmer, "Brane Dynamics and $N = 1$ Supersymmetric Gauge Theory", **hep-th/9704104**.
9. J. L. F. Barbon, "Rotated Branes and $N = 1$ Duality", Phys. Lett. **B402** (1997), 59.
10. J. Brodie and A. Hanany, "Type IIA Superstrings, Chiral Symmetry, and $N = 1$ $4D$ Gauge Theory Dualities", **hep-th/9704043**.
11. A. Brandhuber, J. Sonnenschein, S. Theisen and S. Yanckielowicz, "Brane Configurations and $4D$ Field Theories", **hep-th/9704044**.
12. O. Aharony and A. Hanany, "Branes, Superpotentials and Superconformal Fixed Points", **hep-th/9704170**.
13. R. Tartar, "Dualities in $4D$ Theories with Product Gauge Groups from Brane Configurations", **hep-th/9704198**.
14. I. Brunner and A. Karch, "Branes and Six Dimensional Fixed Points", **hep-th/9705022**.
15. B. Kol, "5d Field Theories and M Theory", **hep-th/9705031**.
16. A. Marshakov, M. Martellini and A. Morozov, "Insights and Puzzles from Branes: $4d$ SUSY Yang-Mills from $6d$ Models", **hep-th/9706050**.
17. K. Landsteiner, E. López and D. A. Lowe, "$N = 2$ Supersymmetric Gauge Theories, Branes and Orientifolds", **hep-th/9705199**.
18. A. Brandhuber, J. Sonnenschein, S. Theisen and S. Yanckielowicz, **hep-th/9705232**.
19. K. Hori, H. Ooguri and Y. Oz, "Strong Coupling Dynamics of Four Dimensional $N = 1$ Gauge Theory from M theory Fivebrane" **hep-th/9706082**.
20. E. Witten, "Branes and the Dynamics of QCD", **hep-th/9706109**.
21. N. Seiberg and E. Witten, "Electric-Magnetic Duality, Monopole Condensation and Confinement in $N=2$ Supersymmetric Yang-Mills Theory", Nucl. Phys. **B426** (1994), 19.

A Brief Introduction to Duality and D-branes[1]

Norma Quiroz*[2] and Barton Zwiebach†[3]

*Departamento de Física
Centro de Investigación y de Estudios Avanzados del IPN.
Apdo. Postal 14-740, 07000, México D.F., México.

† Center for Theoretical Physics,
LNS, and Department of Physics, MIT,
Cambridge, Massachusetts 02139, U.S.A.

Abstract. A brief introduction to duality symmetries and D-branes is given minimizing the amount of technical details.

INTRODUCTION

At present superstring theory seems to be a consistent quantum theory of gravity and, perhaps, a unified theory of all interactions. In ten spacetime dimensions, the maximal dimension for superstrings, there are only five consistent supersymmetric string theories: Type I, IIA, IIB, heterotic SO(32) and heterotic $E_8 \times E_8$. Until a few years ago, it was thought that these theories were simply different theories, but with the recent discoveries of duality symmetries is now believed that they are different solutions of one single theory called, for a lack of better name, M-theory.[4] While in the past 11 dimensional (11D) supergravity was thought to be only peripherally related to string theory, we now believe that this theory is a low energy limit of M-theory and also a strong coupling limit of some of the 10D string theories. This way all five string theories in 10D and Supergravity in 11D could be connected by string duality symmetries as can be seen in fig.(1).

---

[1] Based on a lectures delivered by B. Zwiebach
[2] E-mail:nquiroz@fis.cinvestav.mx
[3] E-mail:zwiebach@irene.mit.edu.
[4] It has been suggested that M could stand for Mystery, Magic or Membranes.

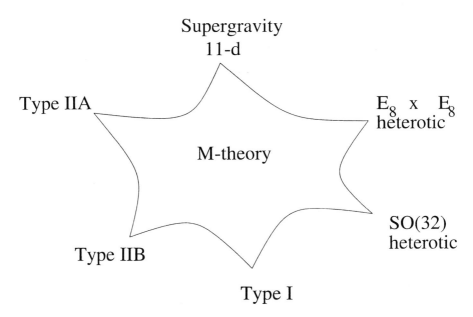

FIGURE 1. The moduli space vacua.

In this course, I will give a brief introduction to the following topics in string theory:

BRANES
STRINS
  Modular Invariance
  T-Duality
D-BRANES
M-THEORY

BRANES

A p-brane is an object with p spatial dimensions which sweeps a $p+1$ dimensional hypersurface when it moves in time. Let us consider the simplest brane: the 0-brane in four dimensions. A 0-brane (particle) is an object with zero dimensions, a point, such that its evolution generates a one-dimensional space, a world-line M^1 as fig.(2).

When the particle is coupled to an electromagnetic field, the action that describes this field and its interaction with the particle is given by:

$$S' = S_{field} + S_{int} \tag{1}$$

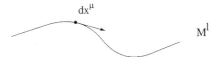

FIGURE 2. Worldline of the particle.

where the field integral is over spacetime

$$S_{field} \sim \int d^D x \, F^{\mu\nu} F_{\mu\nu}$$

and the interaction integral is over the world-line

$$S_{int} \sim \int_{M^1} A_\mu dx^\mu = \int_{M^1} A_\mu(x^\alpha(s)) \frac{dx^\mu}{ds} ds$$

where $F_{\mu\nu}$ is the field-strength tensor defined by $F_{\mu\nu} = \partial_\mu A_\nu - \partial_\nu A_\mu$, and A_μ is the potential that describes the electromagnetic field. The integration is taken along the particle path which is parameterized by one parameter s, and $x^\mu = x^\mu(s)$ describes the position of the particle for each value of the parameter s. The equations of motion for the field are obtained by taking the variation of eq.(1) with respect to the potential:

$$\partial_\mu F^{\mu\nu} = J^\mu$$
$$\partial_\alpha \tilde{F}^{\alpha\beta} = 0,$$

here the four-vector J^ν is the electric current density and $\tilde{F}^{\alpha\beta}$ is the dual field-strength tensor defined as $\tilde{F}^{\alpha\beta} = \frac{1}{2}\epsilon^{\alpha\beta\gamma\delta} F_{\gamma\delta}$.

In the formulation of diferentiable forms, the potential can be regarded as a one-form $A = A_\mu dx^\mu$ and the field-strength tensor as a two-form $F = \frac{1}{2} F_{\mu\nu} dx^\mu \wedge dx^\nu$ (the exterior differential of the potential $F = dA$). For the dual field the two-form associated is $\tilde{F} = \tilde{F}_{\mu\nu} dx^\mu \wedge dx^\nu$. In this formalism the action S' is:

$$S' = \int d^D x \, F^{(2)^2} + \int_{M^1} A^{(1)}$$

where the number between parenthesis denotes the order of the corresponding form. The equations of motion become:

$$d\tilde{F} = J_{elec}$$
$$dF = 0$$

In the first equation J_{elec} is a three-form corresponding to the electric current density. The second equation is a consecuence of the field definition as $F = dA$ and expresses the absence of magnetic charges.

If there were magnetic charges this definition of the field would be modified and the equations of motion would read:

$$d\tilde{F} = J_{elec}$$
$$dF = \tilde{J}_{mag} \quad (2)$$

This time the equations are invariant under

$$F \to \tilde{F}, \quad \tilde{F} \to -F, \quad J_{elec} \to \tilde{J}_{mag}, \quad \tilde{J}_{mag} \to -J_{elec}$$

i.e., equations (2) are invariant when we exchange appropriately electric and magnetic fields, and also electric and magnetic charges. This symmetry, called "electric-magnetic duality", is a symmetry of Maxwell's equations in vacuum ($J_{elec} = \tilde{J}_{mag} = 0$). In the presence of charges it would require the existence of magnetic monopoles in order to hold. This symmetry can be generalized to charges of higher dimensions, as it will be shown later.

In this case, the above mentioned duality between the electric and magnetic charges relates objects with the same dimension as it can be seen briefly in the following diagram:

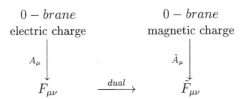

The electric and magnetic fields corresponding to the electric charge q and magnetic charge g are respectively:

$$\vec{E} \sim \frac{q}{r^2}\hat{e}_r$$
$$\vec{B} \sim \frac{g}{r^2}\hat{e}_r$$

and the charges are written in terms of field strengths and dual

$$q \sim \int_{S^2} \tilde{F}$$
$$g \sim \int_{S^2} F$$

since, in four spacetime dimensions $\mathbf{B} = (F_{23}, F_{31}, F_{12}$ and so the two-form F restricted to the spatial sphere can be written as $F|_{S^2} = F_{ij}dx^i \wedge dx^j = \mathbf{B} \cdot d\mathbf{S}$. The integral of this two-form gives the magnetic charge. A similar procedure applies for the electric charge in terms of F.

The previous example was analized in 4D. Now, we will work with the same 0-brane but in 5D. In this case, the dual object to a 0-brane is a 1-brane (string). Indeed

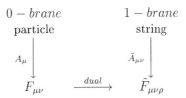

The string is coupled to a potential described by a two-form $\tilde{A}^{(2)}$ and the corresponding action of interaction is proportional to:

$$\int_{M^{(2)}} \tilde{A}^{(2)}$$

here $M^{(2)}$ is the surface swept by the string when it moves in the target space that, in this case, is a 5-dimensional space (see fig. 3).

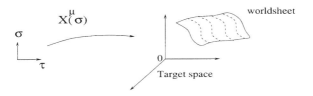

FIGURE 3. The motion of the string in a target space sweeps a worldsheet parameterized by σ and τ.

To understand better the meaning of the charge carried by a string let us return to 4D. The electric charge of any object is given by integration of the corresponding dual field strength. In this case

$$\begin{array}{c} 1-brane \\ A^{(2)} \downarrow \\ F^3 \xrightarrow{dual} \tilde{F}^{(1)} \end{array}$$

the dual field strength is the one-form $\tilde{F}^{(1)}$. We therefore write (see fig.(4)):

$$q_e(1-brane) \sim \int_{S^1} \tilde{F}^{(1)}$$

Notice that this charge is completely different from the usual line charges you have studied in undergraduate electromagnetism. In that case the charge of a wire was

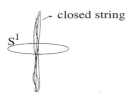

FIGURE 4. The charge of the closed string is found by integrating its dual field along the circle S^1 surrounding the string.

proportional to its length, here the string charge does not depend at all on the length of the string.

Generalizing to any object with p dimensions in a space of D dimensions, an electric p-brane is dual to a magnetic $D-(p+4)$ brane:

$$
\begin{array}{ccc}
p-brane & & D-(p+4)brane \\
A^{(p+1)} \downarrow & & \tilde{A}^{D-(p+3)} \downarrow \\
F^{(p+2)} & \xrightarrow{dual} & \tilde{F}^{D-(p+2)}
\end{array}
\qquad (3)
$$

The electric charge of the p-brane and the magnetic charge of the dual $D-(p+4)$ brane are given by :

$$Q_E \sim \int_{S^{D-p-2}} \tilde{F}^{(D-p-2)}$$

$$Q_M \sim \int_{S^{p+2}} F^{(p+2)}$$

Two important points can be noted in this generalization of the p-branes:
a) The sum of the dimensions of the p-brane and its dual is $D-4$.
b) The number of indices of A plus the number of indices of $\tilde{A}$ is $D-2$.

Now, it is possible to obtain all the potentials and their respective electric and magnetic (dual) objects in any dimensions. For example, in 10 dimensions we have:

form	brane	
	electric	magnetic
A^1	0	6
A^2	1	5
A^3	2	4
A^4	3	3
A^0	-1	7

These electric objects appear in some string theories. The 0-brane can exist in a type IIA theories, the 1-brane is always present in any string theory. The 2-brane (membrane) appears in the IIA string theory and in 11D supergravity as well (since this theory has an $A^{(3)}$ connection, in this case the dual magnetic object is a 5-brane) The type IIB theory has a 3-brane which is selfdual and this object corresponds to dyons since they can have nonvanishing electric and magnetic charges. The (-1)-brane is an instanton, an energy configuration localized in space and time and can appear in the IIB theory. Its magnetic dual is a seven-brane.

STRINGS

The string, a one-brane is a very important object. Its quantization avoids the divergences present in QFT as it will be now sketched. It is known that a Feyman diagram such as that in the fig.(5a) has a divergence when the four interaction points are near each other, but this is an interaction of a point particle and this kind of interaction does not exist in a string theory. The interactions between strings are of the type shown in fig.(5b). In this kind of interaction there is no interaction point where the two strings meet and become a single string. String amplitudes will have contributions in which strings split and join many times and the associated surfaces will have handles. The number of handles on a surface is called the genus.

In string theory when calculating amplitudes one must sum over all inequivalent Riemann surfaces (a topological surface with complex coordinate charts). Two Riemann surfaces are topologically equivalent when they can be smoothly deformed into each other. This is the case when both surfaces have the same genus. So the sum over surfaces is first organized by summing over the possible genus g (see fig.(5c)).

Modular Invariance

Let us study the torus with more details. We can take 1 and τ as the canonical basis of the parallelogram given in fig.(7a), where the *modular parameter* τ is a complex number with $\text{Im}\,\tau \geq 0$. A torus is obtained by gluing the parallel sides of this parallelogram.

While different τ's represent in general inequivalent tori, or tori that cannot be mapped conformally into each other, there is an important exception. When τ and τ' are related as $\tau' = \tau + 1$ and $\tau' = -1/\tau$, or by a composition of such relations, the two tori are actually conformally equivalent.

The first transformation is very trivial. A torus is built by taking a cylinder having two ends and gluing their boundaries with a twist angle θ. If we glue the boundaries using a twist angle $\theta + 2\pi$ the same torus is obtained. This corresponds

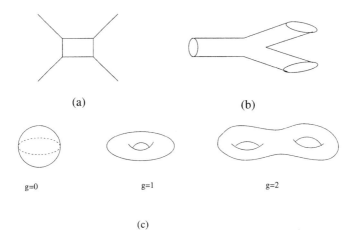

FIGURE 5. (a) Feyman diagram of a four legs particle which has not an equivalent in string theory. (b) Diagram of three legs of a closed string. (c) Different Riemann surfaces caracterized by their genus g.

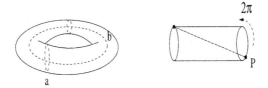

FIGURE 6. a) A torus with its cycles a and b. b) The cut of the torus along of the cycle a and the rotation by 2π of one point p.

to cutting the torus along the cycle a (see fig.6) and taking a fixed point in both opposite edges, to rotate one of them by 2π and after that glue them back. The transformation $\tau \to -1/\tau$ corresponds to an analogue procedure (without rotation) along the cycle b of the torus.

These transformations $\tau \to \tau + 1$ and $\tau \to -1/\tau$ are generators of the modular group $SL(2,Z)$ and the general transformation of the τ parameter is given by:

$$\tau \to \frac{a\tau + b}{c\tau + d} \qquad a,b,c,d \in \mathbf{Z} \qquad ad - cb = 1 \qquad (4)$$

where

$$\begin{pmatrix} a & b \\ c & d \end{pmatrix} \in SL(2,\mathbf{Z})$$

This transformation is independent of the overall sign of the matrices, that is, we can take the elements of the matrices to be positive or negative. So τ and τ' giving

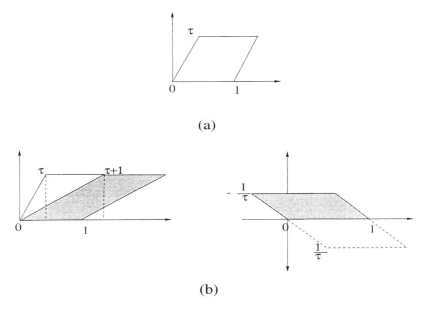

FIGURE 7. (a) The modular parameter τ and 1 are the canonical base. (b) The modular transformations: $\tau \to \tau + 1$ and $\tau \to -1/\tau$.

conformally equivalent tori are related by $PSL(2,\mathbf{Z}) \equiv SL(2,\mathbf{Z})/\mathbf{Z}_2$. Therefore not all of τ are different tori. There are equivalence classes of all of modular parameters related by the transformation (4). These equivalence classes form the modular space $M = H/PSL(2,\mathbf{Z})$ where $H = \{\tau | Im\tau > 0\}$. This is the space of the inequivalent tori (fig.8).

T-Duality

Let us analyze the case of the closed string in more details. The hamiltonian of this string is:

$$H \sim \int d\sigma \left[(\partial_\tau X^\mu)^2 + (\partial_\sigma X)^2\right]$$

The evolution of a string in the target space sweeps a surface named worldsheet (in analogy to the evolution of a particle which sweeps a worldline). The general solution for the wave equation of the closed string which satisfies the periodicity condition $X^\mu(\sigma, \tau) = X^\mu(\sigma + 2\pi, \tau)$ is given in its mode expansion as:

$$X^\mu(\sigma, \tau) = x^\mu + p^\mu \tau + \frac{1}{n} \sum \left(\alpha_n e^{n(\sigma+\tau)} + \tilde{\alpha} e^{-n(\sigma+\tau)}\right)$$

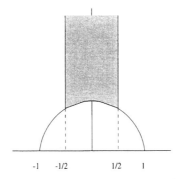

FIGURE 8. The shadowed part corresponds to the space of inequivalent tori.

Now consider the compactification of the X^I coordinate on a circle of radius R. This means that X^I and $X^I+2\pi R$ represent the same spacetime point. It is necessary to reformulate the periodicity condition. Consider a string wrapping m times around that circle. The new periodicity condition is $X^I(\sigma + 2\pi, \tau) = X^I(\sigma, \tau) + 2\pi Rm$. Since there is a circle, momentum is quantized as $p = n/R$. We define a new momentum $w = Rm$ where m is the winding number introduced above. We must have $n, m \in \mathbf{Z}$. The oscillator expansion of this coordinate is:

$$X^I(\sigma, \tau) = x^I + p\tau + w\sigma + \frac{1}{n}\sum \left(\alpha_n^I e^{n(\sigma+\tau)} + \tilde{\alpha}_n^I e^{-n(\sigma+\tau)}\right)$$

and the energy of these oscillations in terms of the quantum numbers is:

$$H = p^2 + w^2 + \cdots = \frac{n^2}{R^2} + R^2 m^2 + \cdots$$

It can be seen from this equation a duality symmetry in the theory. When we change

$$R \longleftrightarrow \frac{1}{R}; \qquad n \leftrightarrow m \tag{5}$$

the theory has the same states. The symmetry is called T-duality of the string theory. When $R \to \infty$, all states with $m \neq 0$ become infinitely massive, but with $m = 0$ there is a continuum for all values of n. Otherwise, when $R \to 0$ only $n = 0$ states are relevant and a continuum appears for m.

D-BRANES

For an open string, the oscillator expansion of the compactified coordinate is:

$$X^I(\sigma, \tau) = x^I + 2p^I \tau + i\sum \frac{\alpha_n^I}{n} \cos(n\sigma) f(\tau)$$

which satisfies the boundary conditions $\partial_\sigma X|_{\sigma=0,\pi} = 0$ and the quantization condition $p = n/R$. Nevertheless, in this case winding numbers do not arise and it seems hard to see how T-duality, as discussed for the closed string, could be a symmetry. As it turns out, the T dual version requires D-branes, and as we will see such objects introduce the possibility of having open string winding. We claim that the dual coordinate must be expanded as

$$X'^I(\sigma,\tau) = X'^I + w^I\sigma + \sum \frac{\alpha_n'^I}{n}\sin(n\sigma)f(\tau)$$

where we see that $X'^I(0) = X'^I(\pi)$ are Dirichlet boundary conditions requiring the string endpoints to be fixed at some point P in the dual circle of radius $1/R$, as in fig.(9), and $w^I \sim n/R$ representing open string winding is dual to the momentum of the original string.

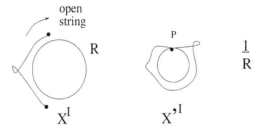

FIGURE 9. a) An open string where one spatial coordinate is a compactified radius R. b) In dual theory with radius of compatification $1/R$, the open string begins and ends on a fixed point P.

Therefore duality is also a symmetry of open strings and involves exchange of Newmann and Dirichlet boundary conditions. The fixed point arising from the Dirichlet conditions represents a topological defect: a Dirichlet-brane or D-brane ! This is a hyperplane in the target space in which the open string is forced to end (the endpoints of the string can move along this brane (see fig.10). Now, taking $R \to 0$ the coordinate X'^I is decompactified and the point P becomes some constant coordinate value where the D-brane is located.

Generalizing for a space of d dimensions and compactifying k spatial coordinates, the dual open string theory will require a D-brane at k fixed coordinates. This represents an object that extends spatially on the other $(D-1) - k$ coordinates and therefore it is a $(D-1) - k$-Dirichlet brane. For finite string coupling a D-brane is a dynamical object of finite mass. To study its evolution it is necessary to introduce functions describing its embedding in spacetime.

Let $(\bar{y}^1, \bar{y}^2, ..., \bar{y}^k)$ denote the coordinates orthogonal to the brane and $(y^1, y^2, ..., y^l)$ with $l = k+1, ..., D-1$ denote the coordinates inside the brane.

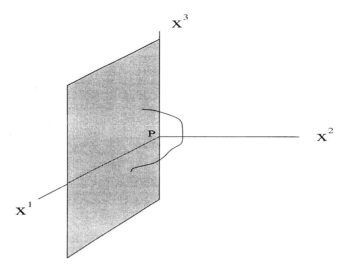

FIGURE 10. A D2-brane in three-dimensional target space. The string satisfies the Dirichlet boundary conditions on a two-dimensional hyperplane in the fixed point $x^3 = 0$.

Functions on the brane $\Phi^i(y)$ with $i = 1, 2, ..., k$, represent the changing position of the brane (in the compactified dimensions) and behave as scalars as far as the brane is concerned. Moreover, it is a universal property of D-branes that they contain (at least) a Maxwell field living on the brane. This is a consequence of the existence of strings in the full spacetime where the brane sits. For a better understanding, recall the diagram (3) where it is shown that the string (1-brane) is electrically coupled to a field of two index $B_{\mu\nu}$. The action of this string is:

$$S \sim \int_M B^{(2)} \tag{6}$$

This action is gauge invariant only for closed strings. For open strings the action is not invariant because the gauge transformation $\delta B^{(2)} = d\lambda^{(1)}$ yields a non null surface term:

$$\int_M d\lambda^{(1)} \sim \int_{\partial M} \lambda^{(1)} \tag{7}$$

To solve this problem we introduce a $U(1)$ gauge field A^μ which is coupled to the edge of the string and lives on the the D-brane where the open string must end. The gauge transformation associated to this potential is $\delta A^{(1)} = -\lambda^{(1)}$. ($A^{(1)}$ could be transformed too in the usual way under the standard gauge transformation: $\delta' A^{(1)} = d\lambda^{(0)}$.) This term will cancel the surface term given by (7). Thus, the action of the string which is gauge invariant is the following:

$$S \sim \int_M B^{(2)} + \int_{\partial M} A^{(1)}$$

We have seen that there are scalars and Maxwell fields living on the D-brane. The action for the brane turns out to be:

$$S_{brane} = -T_p \int d^{p+1}\xi \, e^{-\phi}\sqrt{det[G_{ab} + B_{ab} + F_{ab}]} \qquad (8)$$

where the integration is taken over the p spatial dimensions and the one temporal dimension of the brane world-volumte. Here T is the brane tension and must have units $[T] = M^{p+1}$. In addition, G_{ab} is the metric induced on the world-volume of the brane by the spacetime metric. The field strength F_{ab} is the field strength of the Maxwell field A and the sum $B_{ab} + F_{ab}$ is gauge invariant under both gauge transformation of $A^{(1)}$. So, the brane action is invariant under the same gauge transformation.

Now, let us make some remarks about quantization on a fixed D-brane background. Consider the k fixed coordinates where the brane cannot move. For the open strings ending on this brane we write

$$X^I \sim \bar{X}^I + \sum_n \alpha_n^I sin(n\sigma) \, f(t)$$

with $I = 1, ..., k$. Consider now the $p+1$ other coordinates, where the string endpoints are not fixed. The oscillator expansion of these coordinates is given by:

$$X^a = x^a + p^a\tau + i\sum \frac{\alpha_n^a}{n}\cos(n\sigma)f(\tau) \qquad (9)$$

When the string is quantized, we get physical states of the form:

$$\alpha_1^a |p^a, p^I = 0> \qquad (10)$$
$$\alpha_{-1}^I |p^a, p^I = 0> \qquad (11)$$

They are massless states ($p^2 = 0$) and for each one of them there is an associated field. The first state corresponds to a vector field $A_a(y^b)$ where a is a spacetime index on the brane and the second state correponds to scalar field $\Phi^I(y^a)$.

In the figure (11) two parallel branes are shown with the corresponding Maxwell fields $U(1)$ and oriented strings between them. There are open strings that begin and end on the same brane, these give us two $U(1)$ gauge fields, one from each brane. There are, however, two other types of strings. Strings that begin on brane one and end on brane two, and viceversa. Consider an open string that stretches from the first brane to the second one. For such open string the oscillator expansion along the fixed coordinates would read

$$X^I = \bar{X}_1^I + (\bar{X}_2^I - \bar{X}_1^I)\frac{\sigma}{\pi} + \sum(oscillators)$$

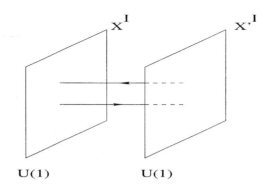

FIGURE 11. Two branes and two oriented strings extending between them.

where $\bar{X}_i^I$ with $i = 1, 2$ are respectively the coordinates of the two D-branes. The oscillations of the string on the free coordinates is given by equation (9). In analogy to the eq. (11) we now get a state:

$$\alpha_{-1}^a |p^a, p^I = 0, d>$$

where d is the separation between the branes. This state has associated a massive vector $A_a(p)$ and its mass depends on the separation between the branes. When d tends to zero the vector mass tends to zero as well. We get two extra gauge vectors (one from each orientation) and the gauge group get enhanced to the nonabelian group $U(2)$. Separating the branes is equivalent to the Higgs mechanism because by separating the branes, some vector fields get mass. Generalizing, when N parallel D-branes approach each other, the group $U(1)^N$ of the separate branes gets enhanced to $U(N)$.

In string theory branes can be classified as:

a) Fundamental branes: are the p-branes which have $T_p \sim (m_s)^{p+1}$

b) Solitonic branes: with $T_p \sim \frac{(m_s)^{p+1}}{g_s^2}$

c) D-branes (Dirichlet): $T_p \sim \frac{(m_s)^{p+1}}{g_s}$

where m_s is the string mass scale and g_s is the dimensionaless string coupling constant given by the expectation value of e^ϕ, with ϕ the dilaton field.

M-THEORY

At the beginning of this paper we have already seen that in 10D there exist only five physically consistent string theories. All of these theories are connected each to the other through the so called string duality. This duality is in general an equivalence relation between the strong coupling region of one string theory and the weak coupling region of a second one or vice versa. For example, the string theory type I and the heterotic $SO(32)$ in 10D are equivalent because the physics described for the strong coupled region of the heterotic is the same that the described for the type I coupled weakly. As a particular case of the string duality we have the T-duality mentioned early in this paper. This duality can relate two different theories in the same weak coupling region but with diferent compactification radii. For example, the type IIA compactified on a circle of radio R is dual to the type IIB compactified on $1/R$.

On the other hand, regarding the fact that in dimensions higher that 11 can exist mass multiplets with spin higher than two, then eleven will be the maximun number of dimensions for any consistent supersymmetric theory. A well known example of such a theory is 11D Supergravity.

At present we believe that 11D Supergravity is a strong coupling limit of some of the 10D string theories. Moreover, all of five 10D string theories and Supergravity could be related through a theory of which 11D Supergravity has been conjectured to be a low energy limit and the five string theories can be constructed as different vacua of the theory. This theory has been called M-theory.

For instance, let us see how the 11D Supergravity and the IIA string theory are connected. In eleven dimensions the bosonic fields content of the theory is the metric $g_{\hat{\mu}\hat{\nu}}$ and the antisymmetric field $A_{\hat{\mu}\hat{\nu}\hat{\rho}}$. When the eleventh spatial dimension is compactified into a circle S^1 of radius R, the massless fields of the IIA string theory are obtained:

$$\begin{array}{cc} Supergravity \longrightarrow & IIATheory \\ 11D & 10D \end{array}$$

$$g_{\hat{\mu}\hat{\nu}} \to g_{\mu\nu}$$
$$g_{11,11} \to \phi$$
$$g_{\mu,11} \to A_\mu$$
$$A_{\hat{\mu}\hat{\nu}\hat{\rho}} \to A_{\mu\nu\rho}$$
$$A_{\hat{\mu}\hat{\nu},11} \to B_{\mu\nu}$$

where $B_{\mu\nu}$ couples to the fundamental string, which happens to have tension $T_p \sim m_s^2$. As we have seen, the string has as dual object a five-brane that in this case is solitonic, and thus with $T_5 \sim m_s^6/g_s^2$. The gauge potential A_μ couples to a D0-brane (a D-particle), having $T_{D0} \sim m_s/g_s$, and the field $A_{\mu\nu\rho}$ couples to a D2-brane with tension $T_{D2} \sim m_s^3/g_s$. So, IIA string theory actually has particles, strings and membranes (and their duals) as it was said at the end of the section 1.1. In 11D Supergravity there is only one coupling constant m_p, the 11D Planck mass. Since, this theory has a potential $A_{\hat{\mu}\hat{\nu}\hat{\rho}}$ then, there are two types of branes: an M2-brane and its dual M5-brane. Their tensions are related with the coupling constant by:

$$T_{M2} \sim m_p^3 \tag{12}$$

$$T_{M5} \sim m_p^6 \tag{13}$$

The compactification on a circle implies that the parameters of 11D supergravity and the parameters of the IIA theory are related. The relation is

$$g_s = Rm_s \tag{14}$$

$$m_s^2 = Rm_p^3 \tag{15}$$

Since the fundamental string tension T is equal to m_s^2 then from eq. (15) it follows that $T = Rm_p^3 = RT_{M2}$ showing that a IIA string is nothing else that an M2 brane wrapped around the compactified dimension. Similarly, for the IIA D2 brane we have $T_{D2} \sim m_s^3/g_s = m_p^3$, and we recognize that this is just the M2 brane, without wrapping. Finally, $T_{D0} = m_s/g_s = 1/R$ showing that D0-branes are simply Kaluza Klein modes from the dimensional reduction.

We have skimmed the relation between IIA superstrings and M-theory, the web of string dualities contains several other interesting relations, but those should be the subject of another set of lectures.

REFERENCES

For more detailed discussiones than the presented here including a fairly complete list of reference see for example:

[1] Joseph Polchinski, "TASI Lectures on D-branes", hep-th/9611050
[2] John H. Schwarz, "The Status of String Theory" hep-th/9711029
[3] P.K. Townsend, "Four Lectures on M-theory", hep-th/9612121
[4] C. Gómez and R. Hernández, "Fields, Strings and Branes", hep-th/9711102
[5] Ashoke Sen, "An Introduction to Non-perturbative String Theory", hep-th/9802051

SEMINARS

Dirac confinement of a heavy quark-light quark system $(Q, \bar{q})$ in high orbital angular momentum states.

M. A. Avila [1]

*Departamento de Física, Facultad de Ciencias, UAEM,
Cuernavaca 62210, Morelos, Mexico.*

Abstract.
The Regge behaviour of the solutions of a Dirac hamiltonian describing a heavy quark-light quark system in high orbital angular momentum states is analyzed. It is found that the solutions of a scalar confining potential are physically admissible while those of a vector confining potential are not. It is concluded that with a Dirac hamiltonian a scalar confining potential is preferred over a vector confining potential for any value of the orbital angular momentum.

Recently there has been given a discussion about the nature of the confinement potential in a heavy quark - light quark $(Q, \bar{q})$ system [1]- [2]. In Ref. [1] an analysis was performed from a phenomenological point of view using diverse techniques. In this work the authors found that with a Dirac-like equation only a Lorentz scalar confinement accounts for the unphysical phenomenon of the mixing of negative with positive energy states also called the Klein paradox, while with the no-pair variant of the Dirac equation only a Lorentz vector confinement potentials leads to a normal Regge behaviour. Concerning to the calculation of the IW function they found that the no-pair equation predicts a value of its slope at zero recoil point in better agreement with the heavy-light data than the Dirac equation. The authors of [1] conclude arguing against scalar confinement.

It is worth it to stress at this stage two points about the work of Ref. [1] that

---

[1]) Electronic address: manuel@servm.fc.uaem.mx

eventually modify the results. The first one is that the analysis of Regge behavior was done partially, since only the the no-pair equation was considered, while the Dirac-like equation was not. The other is that the relativistic corrections to the Dirac hamiltonian were not included. On the basis of the above remarks, the purpose of this work is then to investigate the nature of the confinement when the $(Q, \bar{q})$ system is described by a Dirac equation in the Regge limit of high orbital angular momentum states.

In order to do the above we start with a simple model where we are neglecting the relativistic corrections to the hamiltonian. Consequently, the hamiltonian for the c.m. system is [1], [2]

$$\left[\alpha \cdot \mathbf{p} + m\beta + U(r) + \beta S(r) + V(r) \right] \psi = E\psi, \tag{1}$$

where m (M_Q) is the light (heavy) quark mass, $\mathbf{p}$ is the momentum of the light quark, $U(r) = -\xi/r$ is a color Coulomb-like potential, $S(r) = \kappa_s r$ and $V(r) = \kappa_v r$ are linear increasing Lorentz scalar and vector potentials, respectively. The non-perturbative potentials S and V in Eq. (1) are dynamically responsible for the confinement of the light quark, while the Lorentz vector potential U describes the perturbative color interaction between the quarks.

If one writes (1) as a matrix equation then

$$\begin{pmatrix} m + S + U + V - E & \sigma \cdot \mathbf{p} \\ \sigma \cdot \mathbf{p} & -m - S + U + V - E \end{pmatrix} \begin{pmatrix} G \\ i\sigma \cdot \hat{\mathbf{r}} F \end{pmatrix} \chi_\kappa^m = \begin{pmatrix} 0 \\ 0 \end{pmatrix}. \tag{2}$$

It is not very difficult to show [3] that Eq. (2) leads to the following system of two coupled linear differential equations

$$\frac{dG}{dr} = -\frac{\kappa + 1}{r} G - \left[E + m + S - (U + V) \right] F, \tag{3}$$

$$\frac{dF}{dr} = \left[E - (m + S + U + V) \right] G + \frac{\kappa - 1}{r} F. \tag{4}$$

where $\kappa \simeq l$.

It is possible to show also that the the Regge slopes predicted by the last equations in the limit of very high angular momentum are [3]

$$\alpha' = \frac{|\kappa|}{E^2} = \frac{\sqrt{2\kappa_v \left(\kappa_v + \sqrt{\kappa_v^2 + 8\kappa_s^2} \right) + 4\kappa_s^2}}{5\kappa_v^2 + 3\kappa_v\sqrt{\kappa_v^2 + 8\kappa_s^2} + 4\kappa_s^2}. \tag{5}$$

We must note that these Regge slopes coming from the Dirac equation (1) are the same than those previously found in Ref. [4] which were obtained from a semi-classical confinement model for a $(Q, \bar{q})$ system.

By solving (3)-(4) we obtain the Regge solutions for a $(Q, \bar{q})$ system [3]

$$G = r^{-\kappa} e^{\kappa r} \left[A_1 e^{\sqrt{\kappa_s^2 + 2l\kappa_v - \kappa_v^2}\, r} + A_2 e^{-\sqrt{\kappa_s^2 + 2l\kappa_v - \kappa_v^2}\, r} \right], \tag{6}$$

$$F = -\frac{r^{-\kappa} e^{\kappa r}}{(l - \kappa_v) + \kappa_s} \times$$

$$\left[A_1 \left(\kappa + \sqrt{\kappa_s^2 + 2l\kappa_v - \kappa_v^2} \right) e^{\sqrt{\kappa_s^2 + 2l\kappa_v - \kappa_v^2}\, r} + \right. \tag{7}$$

$$\left. A_2 \left(\kappa - \sqrt{\kappa_s^2 + 2l\kappa_v - \kappa_v^2} \right) e^{-\sqrt{\kappa_s^2 + 2l\kappa_v - \kappa_v^2}\, r} \right].$$

To avoid unphysical sinusoidal behavior we are assuming $\kappa_s^2 + 2l\kappa_v - \kappa_v^2 > 0$ and taking $A_1 = 0$. Another observation concerning Eq. (7) is that the strength of the potentials must be such that $l + \kappa_s \neq \kappa_v$ in order to avoid an unphysical divergence in the lower component of the solution.

Let us consider first the situation where the confinement potential is strictly scalar ($\kappa_v = 0$). In this case the solutions are

$$G = r^{-\kappa} e^{(\kappa - \kappa_s)\, r}, \tag{8}$$

$$F = -\frac{\kappa - \kappa_s}{l + \kappa_s} r^{-\kappa} e^{(\kappa - \kappa_s)\, r}. \tag{9}$$

As we may observe from these equations, if the strength of the scalar potential is strong enough to compensate the intense 'centrifugal forces' it is possible to find physically admissible solutions for $r \to \infty$ for either $\kappa = l$ and $\kappa = -l$.

Let us turn now to consider the case when the confining potential is exclusively vectorial ($\kappa_s = 0$). As can be seen from Eqs. (6) and (7) in this case the solutions are

$$G = r^{-\kappa} e^{[\kappa - \sqrt{(2l/\kappa_v - 1)}\, \kappa_v]\, r}, \tag{10}$$

$$F = -\frac{\kappa - \kappa_v \sqrt{2l/\kappa_v - 1}}{l - \kappa_v} r^{-\kappa} e^{[\kappa - \sqrt{(2l/\kappa_v - 1)}\, \kappa_v]\, r}. \tag{11}$$

We note from these equation that the condition needed to avoid that the exponentials in (6) and (7) become oscillatory is that $\kappa_v < 2l$ which restricts the value of κ_v. On the other hand, if we ask for values of κ_v in the Regge region, $\kappa_v \sim l$ to compensate for the intense 'centrifugal force', then the lower component of the wave function ψ would diverge strongly. Indeed, if we take the limit $\kappa_v \to l$ in Eq. (11) we find

$$\lim_{\kappa_v \to l} F = \begin{cases} l\, r^{-l}\, e^{2lr} & \kappa = l \\ l\, r^{l}\, e^{-2lr} \lim_{\kappa_v \to l} \frac{1}{l - \kappa_v} & \kappa = -l \end{cases} \quad (12)$$

From Eq. (12) we conclude that for values of the strength of the vector potential in the Regge regions it is not possible to find physically admisible solutions. In the case of a very weak vector potential $\kappa_v \ll l$, Eqs. (26) and (27) yield

$$-F \sim G \sim r^{-\kappa} e^{\kappa r}, \quad (13)$$

which shows that the solutions also diverge in this case for $\kappa = l$.

Equations (12) and (13) indicate that the norm of the wavefunctions of a $(Q, \bar{q})$ system in high orbital angular momentum states confined by a vector potential and described by a Dirac equation is not finite.

As a result of all of the above discussed we can conclude in general that if a $(Q, \bar{q})$ system is described by a Dirac equation a scalar confinement is prefered over a vector confinement at states of high orbital angular momentum.

Acknowledgement I thank to N. A. and to N. H. with whom part of this work was done. I acknowledge support from CONACyT grant 3135.

REFERENCES

1. M. G. Olsson, S. Veseli, and K. Williams, Phys. Rev. **D 51**, (1995) 5079.
2. M. Avila, preprint **hep-ph/9710241**.
3. M. Avila, preprint **hep-ph/9711500**.
4. M. G. Olsson, Phys. Rev. **D 55** (1997) 5479.

The equations of motion for this action in the flat F.R.W. metric

$$ds^2 = dt^2 - a^2(t)[dr^2 + r^2 d\theta^2 + r^2 \sin^2\theta d\varphi] \tag{2}$$

are invariant under a stringy symmetry known as "scale factor duality":

$$t \Rightarrow -t \qquad a(t) \Rightarrow \tilde{a}(-t) = a^{-1}(-t) \qquad \phi \Rightarrow \phi - \ln\sqrt{|g|} \tag{3}$$

We simulate matter with a perfect fluid stress-energy tensor $T^\nu_\mu = \text{diag}(\rho, -p\delta^j_i)$ where ρ is the energy density and p is the pressure. Furthermore, we set $\phi = \text{cte} = \phi_0$ and $V(\phi) = 0$. For radiation we get for $t > 0$ our universe in its radiation dominated epoch:

$$a = \left(\frac{t}{t_0}\right)^{\frac{1}{2}} \qquad \phi = \phi_0 \tag{4}$$

$$\rho = 3p = \rho_0 \left(\frac{t}{t_0}\right)^{-2} \qquad G \sim \alpha' e^{\phi_0} = \text{constant} \tag{5}$$

Applying the duality transformation, we find for $t < 0$:

$$a = \left(-\frac{t}{t_0}\right)^{-1/2} \qquad \phi = \phi_0 - 3\ln\left(-\frac{t}{t_0}\right) \tag{6}$$

$$\rho = -3p = \rho_0 \left(-\frac{t}{t_0}\right) \qquad G \sim \alpha' e^{\phi_0} \left(-\frac{t}{t_0}\right)^{-3} \neq \text{constant} \tag{7}$$

The universe starts at $t = -\infty$ with flat empty Minkowski space with zero coupling; Newton's coupling then starts growing and the universe inflates non-exponentially $t << 0$. For $t >> 0$ we recover flat Minkowski space with a weak coupling and a decelerated expansion (which corresponds to our universe). This is a description of the evolution of our universe at times well before and after the big bang ($t = 0$), but what happens during the high curvature regime is, of course, unknown. We may try to simulate it with a time dependent equation of state such as

$$p = \gamma(t)\rho \tag{8}$$

where $-1/3 < \gamma(t) < 1/3$. There is an expanding solution [2] ($H > 0$) for all t: the universe dominated by string matter ($p = -\rho/3$), starts from flat space ($H \to 0$), an unstable solution, with weak coupling ($e^\phi \to 0$) regime evolving through an inflationary phase [$a(t) \sim (-t)^{-1/2}$] phase with gravitational coupling ($e^\phi = \text{const.}$). An analogous solution exists, in which the universe is always contracting ($H < 0$) in correspondence with the dual equation of state $\gamma(-t)$.

Pre-Big-Bang in String Cosmology

Mónica Borunda and M. Ruiz Altaba

Departamento de Física Teórica
Instituto de Física
Universidad Nacional Autónoma de México
Apartado Postal 20-364
01000 México, D.F.

Abstract.
We compute the amount of inflation required to solve the horizon problem of cosmology in the pre-big-bang scenario. First we give a quick overview of string cosmology as developed by Veneziano and collaborators. Then we show that the amount of inflation in this background solves the horizon problem. We discuss fine-tuning.

The standard cosmological model works well at late times, explaining the red shift, the cosmic microwave background and the cosmic primordial nucleosynthesis, but it has problems associated with the initial singularity, the homogeneity, the isotropy, the flatness, and the large-scale structure. Inflation solve these problems except for the initial singularity. Inflationary models are constrained by demanding a graceful exit, the right amount of reheating and the right amount of large-scale inhomogeneities. This requires fine-tuned initial conditions and inflation potentials. But inflation does not even attempt to solve the initial singularity problem.

A few years ago a stringy cosmology was built with the very basic postulate that the universe did indeed start near its trivial vacuum, solving the initial singularity problem [1].

String theory is the only consistent theory containing quantum gravity. Each of the normal modes of vibration of a quantum string is a conventional particle. At large distances ($\lambda \gg 10^{-33}$cm), strings appear as particles. In string theory there are symmetries known as dualities [1] [2] [3] [4] which allow us to find two equivalent solutions to the problem of the initial singularity.

The low-energy effective action for bosonic closed strings is:

$$S = \frac{1}{4\pi\alpha'} \int d^4x \sqrt{\det g}\; e^{-\phi}(R + \partial_\mu \phi \partial^\mu \phi + ...) \tag{1}$$

Where R is the Ricci scalar, $(\alpha')^{-1}$ is the string tension, $g_{\mu\nu}$ is the metric and ϕ is the so-called dilaton, a scalar massless particle which may play the role of inflaton.

What about the other dimensions that string theory allows?

Consider a background in which, during the pre-big-bang phase ($t < 0$), d dimensions expand with scale factor $a(t)$ while n dimensions shrink with scale factor $a^{-1}(t)$ with an equation of state $p = -q = -\rho/(d+n)$.

$$g_{\mu\nu} = \text{diag}(1, -a^2(t)\delta_{ij}, -a^{-1}(t)\delta_{ab}) \tag{9}$$

The solution is $a(t) \sim (-t)^{-2/(d+n+1)}$, $t < 0$ [2]. This background evolves into a phase of maximal, finite curvature, after which it approaches the dual, decelerated regime ($t > 0$) in which the internal dimensions are not frozen, but keep contracting like $a^{-1}(t) \sim t^{-2/(d+n+1)}$ for $t \to +\infty$. The dilaton vacuum expectation value does not settle down to a finite constant value after the big-bang, but tends to decrease during the phase of decreasing curvature. Such a decrease of ϕ is driven by the decelerated shrinking of the internal dimensions which are not frozen.

CONSTRAINTS ON INITIAL CONDITIONS

The condition to solve the horizon problem (in the Einstein frame, thus the tildes) is

$$d_{\text{HOR}}(t_f) = \tilde{a}(t_f) \int_{t_i}^{t_f} dt'/\tilde{a}(t') > \tilde{a}(t_f) H_0^{-1}/\tilde{a}_0 \tag{10}$$

where H_0^{-1} is the size of the observed Universe ($H_0^{-1} \sim 10^{28}$cm), t_f is the time by which the horizon problem is solved and t_i the time when inflation begins. For us, t_i and t_f determines the time range when the pre-big bang description remains valid. Obviously, letting $t_i \to -\infty$ we see that the horizon problem is solved both for $k = 0$ and for $k = -1$. Still, it is of some interest to ask how long did the universe have to behave stringily before the big bang in order for it to come out free of flatness and horizon problems from the high curvature epoch (around $t = 0$) [5] [6].

The amount of expansion required to solve the horizon problem is given by the ratio

$$Z = \frac{H(t_f)a(t_f)}{H(t_i)a(t_i)} \tag{11}$$

Experimentally (or rather, observationally), we need

$$Z > e^{60} \tag{12}$$

in order to solve the horizon problem for our big universe.

Since our effective actions stops being valid when gravity becomes strongly coupled, we expect the pre big bang inflationary epoch to be over by the time t_f when

$$e^{-\phi(t_f)} \gg 1 \tag{13}$$

Similarly, the same effective actions remain valid only while the curvature is not too big:

$$H^{-1}(t_f) \sim (-t_f) \gg l_{st} \tag{14}$$

When $k = 0$, the amount of inflation is thus

$$Z = \left(\frac{-t_i}{l_{st}}\right)^{3/2} \tag{15}$$

This give us that $t_i < -10^{17} l_{st}$ in order to get the amount of inflation for succesfully solve the horizon problem.

We conclude from string cosmology that: (a) inflation comes naturally, without ad-hoc fields, (b) initial conditions are natural, (c) the kinematical problems of the standard cosmological model are solved and (d) a hot big bang could be a natural outcome of our inflationary scenario. Furthermore, it can be shown that (e) perturbations do not grow too fast to spoil homogeneity, (f) our understanding of the high curvature (stringy) phase is still poor and (g) the amount of inflation required needs some fin e-tuning of initial conditions for $k = 0$ and for $k = -1$ in order to solve the horizon problem.

This work is supported in part by the project DGAPA IN103997.

REFERENCES

1. G. Veneziano, *Phys. Lett.* **B265**, 287 (1991)
2. M. Gasperini and G. Veneziano, *Astropart. Phys.*1,317(1993)
3. M. Gasperini and G. Veneziano, *Phys. Rev.***D**50, 2519 (1994)
4. R. Brustein and G. Veneziano, *Phys. Lett.***B**329 429 (1994)
5. M. Turner and E. Weinberg, *Phys. Rev* **D**56: 4604 (1997)
6. N. Kaloper, A. Linde and R. Bousso, SU-ITP-97-46 (1998) hep-th/9801073

Braided Spin groups

A. Criscuolo, M. Rosenbaum and J. D. Vergara

Instituto de Ciencias Nucleares, U.N.A.M.
A. Postal 70-543, México D.F. C.P. 04510, México

Abstract. Using as a starting point a deformed Clifford algebra with an involutive braid we introduce a natural deformation of the Cartan procedure to construct $Spin(4-h, h)$ groups. The method presented produces braided groups instead of quantum groups.

INTRODUCTION

Since the discovery of Quantum Groups, many authors have considered the possibility of q deforming the Lorentz and Poincaré groups as part of a more general program of studying quantum deformed field theories. The purpose of this paper is to show that the natural deformation of the classical Cartan approach to construct spin groups [1], produces braided spin groups. In consequence the q-deformed Lorentz group constructed from the braided spin group is also a braided group.

Braided categories arose naturally from knot theory. They provide a formalism for generalizing supersymmetry and quantum groups, with the hope of achieving a systematic approach to q-deforming structures in physics [2]. In the formalism of braided geometry, vector spaces, linear maps and tensor products of linear algebras, on which most of the mathematics used in physics is based, are replaced by a new category which mimics, by axiomatization, most of the properties of the $Vect$ category. The main axiom that is changed is that in a general category with product, one postulates the existence of a natural transformation Ψ, between the two functors $(C, B) \rightsquigarrow C \otimes B$ and $B \otimes C$. Thus one has a braided tensor product algebra $(\cdot_B \otimes \cdot_C)\Psi_{C,B}$. Here the *braiding* $\Psi_{C,B} : C \otimes B \to B \otimes C$ is a collection of isomorphisms that expresses the degree of commutativity of the algebra structure on the tensor product, i.e., we deform $(a \otimes c)(b \otimes c) = ab \otimes cd$ to

$$(a \otimes c) \cdot_\Psi (b \otimes d) = a\Psi(c \otimes b)d. \tag{1}$$

where, the braiding Ψ satisfies the braid relation

$$(id \otimes \Psi)(\Psi \otimes id)(id \otimes \Psi) = (\Psi \otimes id)(id \otimes \Psi)(\Psi \otimes id). \tag{2}$$

Let $\mathcal{U}$ be an associative algebra with unit $1 \in \mathcal{U}$, and with multiplication map $m : \mathcal{U} \otimes \mathcal{U} \to \mathcal{U}$. We assume that $\mathcal{U}$ is endowed with a structure of a coalgebra, given by a coproduct $\phi : \mathcal{U} \to \mathcal{U} \otimes \mathcal{U}$ and a counit $\epsilon : \mathcal{U} \to$. Furthermore there exists a bijective linear map $\kappa : \mathcal{U} \to \mathcal{U}$. Then, a braided group is a Hopf algebra where the braiding Ψ defines an associative algebra structure on $\mathcal{U} \otimes \mathcal{U}$, such that $1 \otimes 1$ is the unit element.

CLIFFORD, SPECE-TIME, AND SPINOR ALGEBRAS

A general theory of deformed Clifford algebras was developed in [3], based on a quantum generalization of Cartan's theory of spinors. By particularizing this theory to the case of multiparametric involutive braids, we obtained in [4] a deformed Clifford algebra $Cl(\tau, W)$ and requiring that the fundamental property of spinor transformations be preserved in the quantum case, a non-commutative algebra $\mathcal{A}$ was obtained for the *coordinates* of the underlying pseudo-Euclidean spaces given by

$$x^i x^j = m\bar{\tau}(x^i \otimes x^j) = \mu_{ij} x^j x^i \quad x'^i x^j = m\tau(x'^i \otimes x^j) = \mu_{ij}^{-1} x^j x'^i$$
$$x^i x'^j = m\tau(x^i \otimes x'^j) = \mu_{ij}^{-1} x'^j x^i \quad x'^i x'^j = m\bar{\tau}(x'^i \otimes x'^j) = \mu_{ij} x'^j x'^i. \qquad (3)$$

Based on the fact that the total number, $2^{2\nu}$, of products of the components of two spinors equals the sum of the degrees of irreducible tensors found, a classical theorem in spinor calculus(cf. [1]) states that a spinor bilinear is completely reducible with respect to the group of rotations and reversals and decomposes into a scalar, a vector, a bivector,...,an n-vector.

We shall make use of this theorem to obtain commutation relations for the free algebra $\mathcal{S}$ of q-spinors. Thus, guided by the fact that in the limit $\mu \to 1$ we must obtain the classical expression for the components of a vector,

$$x^1 = \frac{1}{2}(\psi^1 \tilde{\varphi}^2 + \tilde{\psi}^1 \varphi^2), \quad x'^1 = -\frac{1}{2}(\psi^2 \tilde{\varphi}^1 + \tilde{\psi}^2 \varphi^1), \qquad (4)$$

$$x^2 = \frac{1}{2}(\psi^2 \tilde{\varphi}^2 + \tilde{\psi}^2 \varphi^2), \quad x'^2 = \frac{\mu}{2}(\psi^1 \tilde{\varphi}^1 + \tilde{\psi}^1 \varphi^1). \qquad (5)$$

Denoting by Ψ the braid operator between the coordinates and the spinor components, we have $(x^\alpha \xi^\beta)^T = m\Psi(x^\alpha \otimes \xi^\beta)$. We now require that $(H(\mathbf{x})\xi, H(\mathbf{x})\tilde{\xi}) = (\xi, \tilde{\xi})$, as in the classical case, together with the fundamental property of spinor transformations $H(\mathbf{x})H(\mathbf{x}) = \langle \mathbf{x}, \mathbf{x} \rangle E$, where $\langle \mathbf{x}, \mathbf{x} \rangle = x'^1 x^1 + x'^2 x^2$. It follows that

$$\varphi^1 \varphi^2 = \varphi^2 \varphi^1, \quad \varphi^1 \psi^1 = \sqrt{\mu} \psi^1 \varphi^1, \quad \varphi^1 \psi^2 = \frac{1}{\sqrt{\mu}} \psi^2 \varphi^1,$$

$$\psi^1 \psi^2 = \psi^2 \psi^1, \quad \varphi^2 \psi^2 = \sqrt{\mu} \psi^2 \varphi^2, \quad \varphi^2 \psi^1 = \frac{1}{\sqrt{\mu}} \psi^1 \varphi^2. \qquad (6)$$

In this way the spinor algebra $\mathcal{S}$ becomes a factor algebra $\mathcal{S}_B = \mathcal{S}/I_B$ with I_B being the 2-sided ideal generated by (6). In the following section we shall show that the generators of this factor algebra acquire the structure of a comodule vector space for the braided spin group.

BRAIDED SPIN GROUPS

In the classical Cartan spinor theory, the action of the operator $H(\mathbf{x}) = \sum_{i=1}^{\nu}(x^i H_i + x'^i H'_i)$ on spinors, with $\mathbf{x}$ a unit vector, corresponds to a reflection in the hyperplane perpendicular to $\mathbf{x}$. A proper rotation on spinors then corresponds to an even product of Clifford operators.

Consider now the element $s = H(x_1)...H(x_{2k})$. Since the Clifford algebra is associative, we can group the above product in pairs so that $s = (H(\mathbf{x}_1)H(\mathbf{x}_2))...(H(\mathbf{x}_{2k-1})H(\mathbf{x}_{2k}))$. But each pair $B(\mathbf{x}_i, \mathbf{x}_{i+1})$, with i odd, has matrix representation $\begin{pmatrix} B_1^i & 0 \\ 0 & B_2^i \end{pmatrix} \in SL(2^{\nu-1},) \times SL(2^{\nu-1},)$. Hence s has the matrix representation $s = \begin{pmatrix} \prod_i B_1^i & 0 \\ 0 & \prod_i B_2^i \end{pmatrix} \in SL(2^{\nu-1},) \times SL(2^{\nu-1},)$. This is due to the fact that $\det(B_j^1...B_j^k) = \prod_i \det(B_j^i) = 1$, for $j=1,2$.

Consequently in the classical case, the quadratic algebra of the Clifford generators is the essential building block for the *Spin* groups, and is the basis of Cartan's construction of the double covering of the (pseudo)-orthogonal groups.

It is the purpose of this section to show that a similar approach may be followed to obtain the braided $Spin(4-h,h)$ groups, as a braided Hopf algebra of polynomial functions in the generators made out of the non-commutative entries of matrix representations of the dyadic operators $B(\mathbf{x}_i, \mathbf{x}_{i+1})$. We thus begin by considering the operator $B(\mathbf{x}, \mathbf{y}) = H(\mathbf{x})H(\mathbf{y})$, where the components $\{x^1, x^2, x'^1, x'^2\}$ and $\{y^1, y^2, y'^1, y'^2\}$ (in an isotropic basis) are no longer commutative, *i.e.*, we shall assume that the *coordinates* of the unit vectors $\mathbf{x}$ and $\mathbf{y}$ are required to satisfy the commutation relations (3), extended to apply to *coordinates* of different vectors also. Note that we can still assume $|\mathbf{x}| = |\mathbf{y}| = 1$, because these products are central to $\mathcal{A}_{\hat{R}}$. We shall denote by $\mathcal{B}$ the algebra generated by the matrix elements of the operator $B(\mathbf{x}, \mathbf{y})$.

We can give to the vector space $\mathcal{S}$ of the generators of the factor algebra $\mathcal{S}_B$ a comodule structure by introducing the coaction map $\delta: \mathcal{S} \to \mathcal{B} \otimes \mathcal{S}$, where $B(\mathbf{x}, \mathbf{y})$ acts on $\mathcal{S}$ through Clifford multiplication. Thus,

$$B(\mathbf{x},\mathbf{y})\dot{\otimes}\xi = \sum_{p=0}^{2}\sum_{i,j=1}^{2}\sum_{k_1<...<k_p}[x^i y^j \otimes \xi^{k_1...k_p}H_i H_j H_{k_1}...H_{k_p}\cdot 1$$
$$+ x'^i y^j \otimes \xi^{k_1...k_p}H'_i H_j H_{k_1}...H_{k_p}\cdot 1$$
$$+ x^i y'^j \otimes \xi^{k_1...k_p}H_i H'_j H_{k_1}...H_{k_p}\cdot 1$$
$$+ x'^i y'^j \otimes \xi^{k_1...k_p}H'_i H'_j H_{k_1}...H_{k_p}\cdot 1] \qquad (7)$$

Taking now as a basis the 4 elements $\{1, H_{k_1} H_{k_2} | k_1 < k_2\}$, and applying the Cartan ordering procedure in terms of semi-spinors of the first type followed by semi-spinors of the second type, we can recombine the coefficients in (7) (making use of our q-Clifford algebra) as new factors in such a basis. Thus we can rewrite (7) as $\sum_{\beta=1}^{4} b^{\alpha}{}_{\beta} \otimes \xi^{\beta}$, $\alpha, \beta = 1, \ldots, 4$, where the first 2 entries in the column ξ^{β} correspond to a semi-spinor of the first type, while the last 2 entries correspond to a semi-spinor of the second type. The rearranged coefficients in (7) yield the entries $b^{\alpha}{}_{\beta}$ of the block-diagonal matrix representation of $B(\mathbf{x}, \mathbf{y})$, from which the free algebra $\mathcal{B}$ of non-commutative polynomials is generated. Furthermore, the algebra (3) of the "coordinates" which occur in $b^{\alpha}{}_{\beta}$, determines the commutation relations for the latter. Thus, $B \otimes \xi = (b^{\alpha}{}_{\beta}) \otimes \xi^{\beta}$, and the entries of the block-diagonal matrix $(b^{\alpha}{}_{\beta})$, are given by :

$$b^1{}_1 = x'^1 y^1 + x'^2 y^2, \quad b^1{}_2 = x'^2 y^1 - \mu^{-1} x'^1 y^2, \quad b^2{}_1 = x^1 y^2 - \mu x^2 y^1,$$
$$b^2{}_2 = x^1 y^1 + x^2 y^2, \quad b^3{}_3 = x^1 y'^1 + x'^2 y^2, \quad b^3{}_4 = x^1 y'^2 - \mu^{-1} x'^2 y^1, \quad (8)$$
$$b^4{}_3 = x^2 y'^1 - \mu x'^1 y^2, \quad b^4{}_4 = x'^1 y^1 + x^2 y'^2, \quad b^i{}_{j+2} = 0, \quad b^{i+2}{}_j = 0, \quad i, j = 1, 2.$$

Moreover, making use of (3), it immediately follows that

$$b^1{}_2 b^3{}_4 = \mu^2 b^3{}_4 b^1{}_2, \quad b^1{}_2 b^4{}_3 = \mu^{-2} b^4{}_3 b^1{}_2, \quad b^2{}_1 b^3{}_4 = \mu^{-2} b^3{}_4 b^2{}_1,$$
$$b^2{}_1 b^4{}_3 = \mu^2 b^4{}_3 b^2{}_1, \quad \left[b^i{}_j, b^l{}_m\right] = 0, \quad [b^{i+2}{}_{j+2}, b^{l+2}{}_{m+2}] = 0, \quad i, j, l, m = 1, 2. \quad (9)$$

Note from the above that elements in the same 2×2 block matrix commute with each other. As a next step we need to show that the algebra $\mathcal{B}$, generated by the matrix elements $b^{\alpha}{}_{\beta}$, has a natural braided bialgebra structure. With this purpose let I_Ψ be the two-sided ideal of $\mathcal{B}$ generated by (9), (note that the braid operator Ψ is induced by the previously obtained braid operator for the *coordinates*). The quotient algebra $\mathcal{B}_\Psi = \mathcal{B}/I_\Psi$ has the structure of a braided group, as defined in Sec.2. It is easy to verify that $\Psi = \Psi^{-1}$, and that the braid relation (2) is satisfied. Thus Ψ is an involutive braid operator. Finally, the antipode is clearly given by $B(\mathbf{y}, \mathbf{x})$. An extended version of the present work has been report elsewhere [5].

REFERENCES

1. Cartan E.,*The Theory of Spinors*, New York: Dover, 1966.
2. Majid S., *J. Pure and Applied Algebra* **86**, 187 (1993); Majid S., *J. Math. Phys.* **32**, 3246 (1991).
3. Bautista R., Criscuolo A., Durdevic M., Rosenbaum M., and Vergara J.D., *J. Math. Phys.* **37**, 5747 (1996).
4. Criscuolo A., Durdevic M., Rosenbaum M., and Vergara J.D., *J. Phys. A: Math. Gen.* **30**, 6451 (1997).
5. Criscuolo A., Rosenbaum M., and Vergara J.D., *J. Geom. Phys.* (1998) to be published.

Quark mass ratios from mixing angles and flavor symmetry considerations

Maritza de Coss and Rodrigo Huerta

Depto. de Física Aplicada
CINVESTAV-IPN, Unidad Mérida
A. P. 73 Cordemex, 97310 Mérida, Yucatán, México

Abstract. Using the flavor symmetry, we "visualize" the dependence of the constituent quark mass on the mixing angles. We further show that a simple relation can be proposed to express its functional dependence, relating the bare and the constituent quark masses.

Several relations are given in the literature for the mixing angles in terms of the quark masses. In the case of the Cabibbo angle, which relates the first and second family of quarks, the equation

$$\theta_{12} = \theta_c = \arctan\sqrt{\frac{m_d}{m_s}} - \arctan\sqrt{\frac{m_u}{m_c}} \tag{1}$$

was obtained by Fritzsch [1] long time ago.

Later, with the introduction of the third family, the angles θ_{13} and θ_{23} were given in terms of quark masses including the heavy ones. However in the latter case, there are several relations depending on the model or texture assumed for the quark mass matrix.

In any case, the mixing angles, being dimensionless are expressed in terms of quark masses ratios. No other physical parameters are involved as is clear from Eq. (1). Other condition is that it should be a mass ratio of equal charge quarks: therefore we have only the mass ratios m_d/m_s, m_d/m_b, m_s/m_b, m_u/m_c, m_u/m_t and m_c/m_t. Two of them are very small, due to the large mass of the top quark.

In the above, all the masses are current or Lagrangian quark masses, in contrast with the constituent masses. For the heavy quarks b and t, the difference between current and constituent masses is negligible, but for the light quarks, u and d, this difference is overwhelming. For the s and c quarks there is a sort of intermediate case.

In the present work we obtain a relation between bare and constituent quark masses using the mixing angle to connect them. Our basic assumption is: we are allowed to draw a diagram in which the quark masses are considered as a dynamical entities, and one can change their values to study flavor symmetry limits [2].

In consequence we give in Fig. 1 a plot in which one of the axis is the ratio m_d/m_s and in the other we have the tangent of the mixing angle θ_{12}. Similar plots are possible which involve m_d/m_b, m_s/m_b and $\tan\theta_{13}$ and $\tan\theta_{23}$ respectively. In order to visualize this connection we have neglected the ratios m_u/m_c, m_c/m_t and m_u/m_t which are smaller than the corresponding ratios, m_d/m_s, m_s/m_b and m_d/m_b.

In Fig. 1, the point A is where flavor symmetry is totally broken, i. e., where the mass of the s quark, compared with the d quark mass, is infinitely large. The point B is obtained when the bare quark mass ratio m_d/m_s crosses the θ_{12} experimental value, indicated by a dotted horizontal line.

Point C is similarly obtain, only that in this case we use constituent quark masses. We use the bare and constituent quark masses given by PDG [3]. Finally point D is the symmetry limit where we have $m_d = m_s$ and the ideal mixing of 45° between the d and s quarks. Notice that C is closer to D as compared to B.

It is clear from Fig. 1, that a curve that passes through points A, B and D has to have a power law smaller than one. As we can expect this power law is equal to one half. In the other hand, to connect points A, C, and D we need a power law larger than one. It happens that a cubic fitts well the points shown in Fig. 1. Then we arrive to our main result,

$$\tan\theta_{12} \simeq \left(\frac{m_d^o}{m_s^o}\right)^{1/2} \simeq \left(\frac{m_d^c}{m_s^c}\right)^3 \tag{2}$$

Here, we have written the bare quark masses with a zero and with a c the constituent quarks masses.

Similar relations are obtained for $\tan\theta_{13}$ and $\tan\theta_{23}$, with different exponents. Also the quark mass ratios m_u/m_c, m_u/m_t and m_c/m_t can be introduced.

However, the main conclusión is not changed: we can relate the current and the constituent quark masses using the mixing angles and the flavor symmetry limits.

Whether this relation is meaningfull, is something that we shall explore in the future, when we compare it with several other relations obtained using different arguments. In particular there is a work [4] in which a relation between constituent, dynamical and current quark masses is given using hyperfine interactions of constituent quarks. It is possible that a connection between the approach in Ref. 4 and the one presented here exists.

One of us (M. C.) wants to thank the support from Fondo Yucatán, for the presentation of this work.

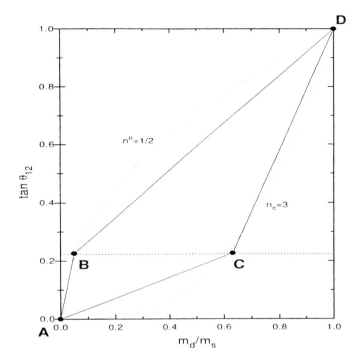

FIGURE 1. The plot of m_d/m_s versus $\tan\theta_{12}$ gives that point A is where the symmetry is fully broken; whereas point D is where we have ideal mixing between d and s. Intermediate cases B and C are obtained using the current and the constituent quark masses, respectively. Two power laws fit the points, shown in the figure. The horizontal dotted line corresponds to the experimental value for $\tan\theta_{12}$.

REFERENCES

1. S. Weinberg, *I.I. Raby Festschrift* (1977); F. Wilczek and A. Zee, *Phys. Lett.* **70B**, 418 (1977); H. Fritzsch, *Phys. Lett.* **70B**, 436 (1977); **73B**, 317 (1978).
2. H. Fritzsch and Z. Xing, *Phys. Lett. B* **13**, 396-404 (1997).
3. R. M. Barnett et al, Particle Data Group, *Phys. Rev.* **D54**, 1 (1996).
4. V. Elias, M. Tong and M. D. Scadron, *Phys. Rev.* **D40**, 3670 (1989).

A Model for a Dynamically Induced Large CP-Violating Phase[1]

D. Delépine

*Institut de Physique Théorique, Université catholique de Louvain,
B-1348 Louvain-la-Neuve, Belgium*

Abstract. Assuming a new interaction with a θ term for the third generation of quarks, a value of θ different from zero dynamically induces the top-bottom mass splitting and a large CP-violating phase.

In the electroweak standard model, there are 2 independent sources of CP violation: one is the phase of the quark mixing matrix (the Cabibbo-Kobayashi-Maskawa matrix (V_{CKM})) [5] and the second is the strong θ angle of QCD [6].

Nowadays, the mechanism for CP violation is still not yet understood and the CP puzzles have been turned into questions concerning tiny quark masses and a large CP-violating phase. The fact that the top quark is very heavy compared to the other quarks, $m_t = 180 \pm 12$ GeV [1], suggests that the third generation may be playing a special role in the dynamics at the electroweak scale. So we assume the existence of a new interaction playing only with the third generation and strong enough to lead to the formation of quark-antiquark bound states which trigger dynamically the breaking of the electroweak symmetry [2–4]. This new interaction is conserving isospin symmetry between top and bottom quarks and generates a θ term.

We show that in this model, the θ term breaks the symmetry between top and bottom as expected from general theorems [11] and induces naturally a large CP-violating phase in $V_{CKM}(\delta_{CKM})$ due to the smallness of the m_b/m_t mass ratio.

I THE MODEL

We consider a standard model Higgs sector in combination with an effective new strong interaction acting on the third generation of quarks and characterized by a θ term [8,7].

The total effective Lagrangian of our model is thus given by

$$L = L_H + L_\Sigma + L_\theta \,, \tag{1}$$

---

[1] This work was done in collaboration with J.M Gerard, R.Gonzalez Felipe and J.Weyers.

with L_H, L_Σ and L_θ defined as follows:

$$L_H = D_\mu H^\dagger D^\mu H - m_H^2 H^\dagger H + \left(h_t \bar{\psi}_L t_R H + h_b \bar{\psi}_L b_R \hat{H} + \text{h.c.}\right), \tag{2}$$

where $H = \begin{pmatrix} H^0 \\ H^- \end{pmatrix}$, $\hat{H} = \begin{pmatrix} H^+ \\ -H^{0*} \end{pmatrix}$ and $\psi_L = \begin{pmatrix} t_L \\ b_L \end{pmatrix}$; h_t and h_b are the Yukawa couplings and D_μ is the usual covariant derivative of the standard model. L_Σ parametrizes the effects of the new interaction on the top and bottom quark.

$$L_\Sigma = D_\mu \Sigma_t^\dagger D^\mu \Sigma_t + D_\mu \Sigma_b^\dagger D^\mu \Sigma_b - m^2(\Sigma_t^\dagger \Sigma_t + \Sigma_b^\dagger \Sigma_b) + g(\bar{\psi}_L t_R \Sigma_t + \bar{\psi}_L b_R \hat{\Sigma}_b + \text{h.c.}). \tag{3}$$

where Σ_t and Σ_b are 2 complex doublet scalar fields describing the $q\bar{q}$ bound states.

$$\Sigma_t = \begin{pmatrix} \Sigma_t^0 \\ \Sigma_t^- \end{pmatrix} \sim t_R \bar{\psi}_L, \qquad \hat{\Sigma}_b = \begin{pmatrix} \Sigma_b^+ \\ -\Sigma_b^{0*} \end{pmatrix} \sim b_R \bar{\psi}_L \tag{4}$$

For the θ term, we shall take, in analogy with QCD, the lagrangian form

$$L_\theta = -\frac{\alpha}{4}\left[i\text{Tr}\left(\ln U - \ln U^\dagger\right) + 2\theta\right]^2, \tag{5}$$

with

$$U = \begin{pmatrix} \Sigma_t^0 & \Sigma_b^- \\ \Sigma_t^+ & -\Sigma_b^{0*} \end{pmatrix}. \tag{6}$$

This term typically arises as a leading term in a $1/N$ - expansion.

From Eqs.(2) and (3) it follows that the top and bottom field-dependent masses are given at the tree level by the linear combinations

$$M_t = h_t H^0 + g\Sigma_t^0, \qquad M_b = h_b \hat{H}^0 + g\hat{\Sigma}_b^0, \tag{7}$$

II ELECTROWEAK AND ISOSPIN SYMMETRY BREAKINGS

Without loss of generality, we take the phase of the neutral Higgs field H^0 to be zero. This can always be achieved by performing a suitable electroweak gauge transformation. We write the VEVs of the neutral components of the fields in the form

$$\langle H^0 \rangle = \frac{v}{\sqrt{2}}, \quad \langle \Sigma_t^0 \rangle = \frac{\sigma_t}{\sqrt{2}} e^{i\varphi_t}, \quad \langle \Sigma_b^0 \rangle = \frac{\sigma_b}{\sqrt{2}} e^{i\varphi_b}. \tag{8}$$

Including the radiative corrections (induced by top and bottom quark loops) [9], the effective potential in terms of these VEVs reads

$$V = m_H^2 \frac{v^2}{2} + \frac{m^2}{2}(\sigma_t^2 + \sigma_b^2) - \beta\left(\mu_t^2 + \mu_b^2\right) + \lambda\left(\mu_t^4 + \mu_b^4\right) + \alpha\left(\theta - \varphi_t + \varphi_b\right)^2 , \quad (9)$$

where

$$\mu_{t,b}^2 = |\langle M_{t,b}\rangle|^2 = \frac{1}{2}\left(h_{t,b}^2 v^2 + g^2 \sigma_{t,b}^2 + 2 h_{t,b} v g \sigma_{t,b} \cos\varphi_{t,b}\right), \quad (10)$$

while β and λ are some effective quadratic and quartic couplings. In what follows we shall assume all couplings and parameters in the potential to be real and positive.

Note that the potential which leads to Eq.(9) can be viewed either as an effective renormalizable interaction or as an expansion up to quartic terms in a cut-off theory.

The extrema conditions $\frac{\partial V}{\partial v} = \frac{\partial V}{\partial \sigma_t} = \frac{\partial V}{\partial \sigma_b} = \frac{\partial V}{\partial \varphi_t} = \frac{\partial V}{\partial \varphi_b} = 0$ imply a system of equations which can be solved in a simple analytical way assuming $h_t = h_b$, $\alpha \gg \beta m_t^2$ with m_t the physical mass of the top quark, $\beta \gg 2\lambda m_t^2$. In that case, the presence of a phase θ close to $\frac{\pi}{2}$ induces both isospin breaking and CP violation with [2]

$$\sigma_b \ll \sigma_t \neq 0 , \; v \neq 0 , \; \varphi_t \simeq \sigma_b/\sigma_t , \; \varphi_b \simeq -\pi/2 + \sigma_b/\sigma_t . \quad (11)$$

III CP VIOLATION

Let us now investigate whether this new source of CP violation can be responsible for what is observed in the $K^0 - \bar{K}^0$ system. Let us consider the 3×3 quark mass matrices

$$M_{u,d} = (h)_{u,d} v + \begin{pmatrix} 0 & & \\ & 0 & \\ & & 1 \end{pmatrix} g\sigma_{t,b} e^{\pm i\varphi_{t,b}}, \quad (12)$$

with $(h)_{u,d}$ arbitrary real matrices.

If we neglect $O(h^2 v^2)$ terms, $(MM^\dagger)_{u,d}$ are diagonalized by the following unitary matrices:

$$U_u \simeq R_u \begin{pmatrix} 1 & & \\ & 1 & \\ & & e^{-i\varphi_t} \end{pmatrix} , \; U_d \simeq R_d \begin{pmatrix} 1 & & \\ & 1 & \\ & & e^{i\varphi_b} \end{pmatrix} , \quad (13)$$

with $R_{u,d}$ orthogonal. In this approximation and using the Cabibbo-Kobayashi-Maskawa parametrization for the quark mixing matrix [5], we get, last but not

---

[2] the question of the eigenvalues of the scalar mass matrix is discussed in Ref. [7]

least, $\delta_{KM} \simeq -(\varphi_t + \varphi_b) \simeq \pi/2$. In the nowadays standard parametrization of the Cabibbo-Kobayashi-Maskawa mixing matrix [1], the phenomenology in $K^0 - \bar{K}^0$ physics requires that the CP-violating phase (δ_{13}) be around $\frac{\pi}{2}$. The CP-violating phase in the 2 parametrizations are related by $\sin \delta_{13} \simeq \frac{\vartheta_2}{\vartheta_{23}} \sin \delta_{KM}$, in the small mixing angles approximation [13]. From $|V_{ij}|_{KM} = |V_{ij}|_{standard}$, the experimental constraint on the ratio

$$\frac{V_{ub}}{V_{cb}} = 0.08 \pm 0.02 \tag{14}$$

require a texture for the h-matrices such that the mixing angles $\vartheta_2 > \vartheta_3$.

Therefore, we conclude that our solution leads indeed to a sizeable CP-violating phase

$$\delta_{13} \simeq \pi/2 \,, \tag{15}$$

which is welcome by phenomenology in $K^0 - \bar{K}^0$ physics [14].

REFERENCES

1. *Review of Particle Physics*, R.M. Barnett *et al.*, Phys. Rev. **D 54** (1996) 1.
2. Y. Nambu and G. Jona-Lasinio, Phys. Rev. **122** (1961) 345.
3. A. Miransky, M. Tanabashi and K. Yamawaki, Mod. Phys. Lett. **A 4** (1989) 1043; Phys. Lett. **B 221** (1989) 177.
4. W.A. Bardeen, C.T. Hill and M. Lindner, Phys. Rev. **D 41** (1990) 1647.
5. M. Kobayashi and K. Maskawa, Prog. Theor. Phys. **49** (1973) 652.
6. For a recent review, see e.g. R.D. Peccei, *QCD, Strong CP and Axions*, hep-ph/9606475.
7. D.Delepine, J.-M. Gerard, R.Gonzalez Felipe, J.Weyers, Phys. Lett.**B411** (1997) 167.
8. G. Buchalla, G. Burdman, C.T. Hill and D. Kominis, Phys. Rev. **D 53** (1996) 5185.
9. J.P. Fatelo, J.-M. Gérard, T. Hambye and J. Weyers, Phys. Rev. Lett. **74** (1995) 492.
10. E. Witten, Ann. Phys. **128** (1980) 363.
11. C. Vafa and E. Witten, Nucl. Phys. **B234** (1984) 173.
12. D. Delépine, J.-M. Gérard and R. González Felipe, Phys. Lett. **B 372** (1996) 271.
13. H. Fritzsch, Phys. Rev. **D 32** (1985) 3058.
14. A.J. Buras, M. Jamin and M.E. Lautenbacher, Phys. Lett. **B 389** (1996) 749.

Photon Dispersion Relations in a Nuclear Medium

Juan Carlos D'Olivo[*] and José F. Nieves[†]

[*]*Instituto de Ciencias Nucleares*
Universidad Nacional Autónoma de México
Apartado Postal 70-543, 04510 México, D.F., México
[†]*Laboratory of Theoretical Physics,*
Department of Physics, P.O. Box 23343
University of Puerto Rico,
Río Piedras, Puerto Rico 00931-3343

Abstract. We examine the nucleon contribution to the photon self-energy in a plasma, including the effect of the anomalous magnetic moment of the nucleons. Our results are relevant for the study of the photon dispersion relations in the core of a supernova, and has implications with regard to the recent suggestion that the Čerenkov process $\nu \to \nu\gamma$ can take place in such a system.

It has been recently pointed out by Mohanty and Samal [1] that, in a supernova core, the photon dispersion relation receives a contribution from the nucleon magnetic moment which is of the opposite sign to that of the usual electron plasma effect. Under such circumstances, the emission of Čerenkov radiation by neutrinos becomes possible, which can have important implications for the energetics of the system. More recently, Raffelt [2] has noted that the conclusions of Ref. [1] are based on a numerical error in the calculation of the refractive index. These analysis were based in the static paramagnetic susceptibility formula to compute the effect of the nucleons on the photon dispersion relation. However, for a proper account of this effect what is required is the nucleon contribution to photon self-energy for any value of the photon momentum and not just in the zero-frequency limit.

Motivated by the above considerations, in the present work we study the photon dispersion relation in a nuclear medium taking into account the anomalous magnetic moment couplings of the nucleons to the photon. The calculations are based on the application the methods of finite temperature field theory (FTFT) [3], which have proven to be very useful in the study of the matter effects on the properties of the elementary particles. Within the formalism of FTFT, the dispersion relations for the transverse and longitudinal photon modes In a medium are obtained by solving

$$\omega_{T,L}^2 - \mathcal{Q}^2 = \pi_{T,L}(\omega_{T,L}, \mathcal{Q}), \qquad (1)$$

where $\pi_{T,L}$ denote the total background contribution to the transverse and longitudinal components of the photon self-energy. For a background containing electrons and nucleons $\pi_{T,L} = \pi_{T,L}^{(e)} + \pi_{T,L}^{(p)} + \pi_{T,L}^{(n)}$. In general, $\omega_{T,L}$ are complex functions of $\mathcal{Q}$ and can be written in the form $\omega_{T,L} = \Omega_{T,L} - i\gamma_{T,L}/2$.

The quantities $\Omega_{T,L}$ and $\gamma_{T,L}$ are real and have the interpretation of being the dispersion relation and damping rate of the propagating mode, respectively. We focus our attention on the dispersion relation. Retaining terms that are at most linear in $\gamma_{T,L}$, from Eq.(1) we have

$$\Omega_{T,L}^2 - \mathcal{Q}^2 = \mathrm{Re}\,\pi_{T,L}(\Omega_{T,L}, \mathcal{Q}). \tag{2}$$

In what follows, for simplicity we neglect the contribution from the protons, and assume that the electrons can be represented by a relativistic gas and the neutrons by a degenerate non-relativistic gas. The dispersion relations for the transverse modes under such conditions are obtained by solving the equation [4]

$$q^2\left(1 - \chi_0^{(n)} f_T(z_n)\right) = 3e^2 \omega_{0e}^2 \frac{\Omega_T}{\mathcal{Q}}\left[\frac{2\Omega_T}{\mathcal{Q}} - \frac{q^2}{\mathcal{Q}^2}\log\left|\frac{\Omega_T + \mathcal{Q}}{\Omega_T - \mathcal{Q}}\right|\right], \tag{3}$$

where $q^2 = \Omega_T^2 - \mathcal{Q}^2$ and we have defined

$$\chi_0^{(n)} = \frac{\kappa_n^2 m_n \mathcal{P}_{Fn}}{\pi^2},$$

$$z_n = \frac{\Omega}{v_{Fn}\mathcal{Q}}, \tag{4}$$

with $v_{Fn} = \mathcal{P}_{Fn}/\mathcal{E}_{Fn}$ being the Fermi velocity of the neutron gas. Here, κ_n is the neutron magnetic moment given by $\kappa_n = -1.91\,(|e|/2m_n)$ in terms of the the electron charge e. In Eq. (3),

$$f_I(z) = \frac{1}{6}v_F^2 + \left[1 - \frac{1}{2}v_F^2 z^2\right]\left[1 - \frac{1}{2}z\log\left|\frac{1+z}{1-z}\right|\right] \tag{5}$$

and

$$\omega_{0e}^2 = \frac{1}{6\pi^2}\int_0^\infty d\mathcal{P}\,\mathcal{P}(f_e + f_{\bar{e}}), \tag{6}$$

where f_e and $f_{\bar{e}}$ denote the electron and positron number density distributions. The integral in the last equation cannot be performed without specifying f_e; in particular, for a degenerated (relativistic) gas the integration over $\mathcal{P}$ yields

$$\omega_{0e}^2 = \frac{1}{12}\left(\frac{3n_e}{\pi}\right)^{2/3}. \tag{7}$$

For the longitudinal modes the dispersion relations are given by the solutions of

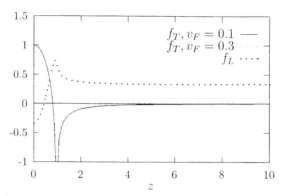

FIGURE 1. Plot of the functions f_T and f_L defined in Eq. (5) and Eq. (8), respectively At $z = 1$, f_T becomes infinitely negative.

$$q^2 \left(1 - \chi_0^{(n)} v_{Fn}^2 f_L(z_n)\right) = -12 e^2 \omega_{0e}^2 \frac{q^2}{Q^2} \left(1 - \frac{\Omega_L}{2Q} \ln \left|\frac{\Omega_L + Q}{\Omega_L - Q}\right|\right), \tag{8}$$

with $q^2 = \Omega_L^2 - Q^2$ and

$$f_L(z) = -\frac{1}{3} + z^2 + \frac{1}{2} z(1 - z^2) \log \left|\frac{1+z}{1-z}\right|. \tag{9}$$

The functions $f_{T,L}$ are plotted in Fig. 1. As we see from it, both functions satisfy $f_{T,L} \leq 1$. If the neutrons are nonrelativistic, then $\chi_0^{(n)} \simeq 3.6\, \alpha v_{Fn} < 1$ and therefore $\chi_0^{(n)} f_{T,L} < 1$.

While it is not possible to find the general solution to Eqs.(3) and (8), some useful conclusions can be drawn from them. For $\Omega_T < Q$ the right hand side of Eq. (3) is positive. On the contrary, the left hand side of this equation is negative since, as we have seen, $\chi_0^{(n)} f_{T,L} < 1$. As a consequence, the solution to Eq. (3) is such that $\Omega_T > Q$. This implies, in particular, that a neutrino propagating in such a medium cannot emit transverse photons in the form of Čerenkov radiation. Furthermore, since z_n is considerably greater than unity for $\Omega_T > Q$, the value of $f_T(z_n)$ is negligible for that range and therefore the solution to Eq. (3) is well approximated by the corresponding solution for the electron background only. Notice that a possibly different, and erroneous, conclusion would have been obtained if we were to approximate the neutron contribution by the value in the static limit $\operatorname{Re} \pi_T^{(n)}(0, Q) = -Q^2 \chi_0^{(n)}$ Then, instead of Eq. (3), such an approach leads to the equation

$$q^2 = -Q^2 \chi_0^{(n)} + \operatorname{Re} \pi_T^{(e)} \tag{10}$$

for the dispersion relation, which can have a solution such that $\Omega_T < Q$ depending on the relative size of the two competing terms in the right hand side.

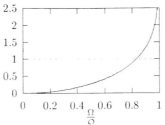

FIGURE 2. The factor $\frac{\Omega}{2\mathcal{Q}} \log \left| \frac{\Omega + \mathcal{Q}}{\Omega - \mathcal{Q}} \right|$ is plotted as a function of $\frac{\Omega}{\mathcal{Q}}$.

On the other hand, the quantity

$$\frac{\Omega}{2\mathcal{Q}} \log \left| \frac{\Omega + \mathcal{Q}}{\Omega - \mathcal{Q}} \right|, \quad (11)$$

which is plotted in Fig. 2, is larger than unity for values of $\frac{\Omega}{\mathcal{Q}}$ larger than about 0.84. Therefore, there is a small range

$$0.84 \lesssim \frac{\Omega_L}{\mathcal{Q}} < 1, \quad (12)$$

for which the solution to Eq. (7) is such that $\Omega_L < \mathcal{Q}$. Hence, the Cerenkov radiation of longitudinal photons (plamons) is, in principle, allowed [5]. For those values of $\Omega_L/\mathcal{Q}$, the value of z_n corresponds to the region in Fig. 1 where the function f_L is negligible. Therefore, the neutron component of the background does not play a significant role in the photon dispersion relation in this case either. The previous results are presented in much more detail in Ref. [4], where the proton effects are also taken into account.

ACKNOWLEDGEMENTS

This work has been partially supported by the U.S. National Science Foundation Grant PHY-9600924 (JFN) and by Grant No. DGAPA-IN100397 at the Universidad Nacional Autónoma de México (JCD).

REFERENCES

1. S. Mohanty and M. K. Samal, Phys. Rev. Lett. **77**, 806 (1996).
2. Georg G. Raffelt, Phys. Rev. Lett. **79**, 773 (1997).
3. N. P. Landsman and Ch. G. van Weert, Phys. Rep. **145**, 141 (1987); M. Le Bellac, *Thermal Field Theory*, Cambridge Univ. Press , 1996.
4. J. C. D'Olivo and José F. Nieves Phys. Rev. **D57**, 3119 (1998).
5. S. Sahu, Phys. Rev. **D56**, 1688 (1997).

Study of Λ^0 Polarization in Exclusive pp Reactions at 27.5 GeV/c

J. Félix[1], C. Avilez[1,‡], D.C. Christian[3], M.D. Church[4b], M. Forbush[5g], E.E. Gottschalk[4d], G. Gutierrez[3], E.P. Hartouni[2a], S.D. Holmes[3], F.R. Huson[5], D.A. Jensen[2b], B.C. Knapp[4], M.N. Kreisler[2a], G. Moreno[1], J. Uribe[2c], B.J. Stern[4d], M.H.L.S. Wang[2a], A. Wehmann[3], L.R. Wiencke[4f], J.T. White[5]

[1] *Universidad de Guanajuato, León, Guanajuato, México,* [2] *University of Massachusetts, Amherst, Massachusetts, USA,* [3] *Fermilab, Batavia, Illinois, USA,* [4] *Columbia University, Nevis Labs, New York, USA* [5] *Department of Physics, Texas A&M University, College Station, Texas 77843, USA.*

Abstract. We report a measurement of the x_F and P_T Λ^0 polarization dependence; Λ^0's were produced in the specific channels: $pp \to p\Lambda^0 K^+\pi^+\pi^-$, $pp \to p\Lambda^0 K^+\pi^+\pi^-\pi^+\pi^-$, $pp \to p\Lambda^0 K^+\pi^+\pi^-\pi^+\pi^-\pi^+\pi^-$, and $pp \to p\Lambda^0 K^+\pi^+\pi^-\pi^+\pi^-\pi^+\pi^-\pi^+\pi^-$ with 27.5 GeV/c protons incident on a liquid hydrogen target. We find that Λ^0 polarization is independent of the final state.

INTRODUCTION

The discovery that Λ^0 hyperons are polarized when produced in high energy pp collisions [1] has posed an interesting puzzle to theories of particle production. That discovery for Λ^0's, and subsequent observations that other hyperons were polarized as well [2], has called into question the generally accepted assumption that spin plays no role in high energy multi-particle production.

Despite many works on this phenomenon, an understanding of the source of the polarization remains elusive. Models proposed to date do not fit all of the data well and tend not to have predictive power [3].

We report a study of Λ^0 polarization, performed in a high statistics exclusive sample, of the particular reactions:

$$pp \to p\Lambda^0 K^+\pi^+\pi^-, \tag{1}$$

$$pp \to p\Lambda^0 K^+\pi^+\pi^-\pi^+\pi^-, \tag{2}$$

$$pp \to p\Lambda^0 K^+\pi^+\pi^-\pi^+\pi^-\pi^+\pi^-, \tag{3}$$

$$pp \to p\Lambda^0 K^+ \pi^+ \pi^- \pi^+ \pi^- \pi^+ \pi^- \pi^+ \pi^-. \tag{4}$$

We investigated the kinematic dependence of the polarization in each reaction.

Λ^0 POLARIZATION

The data for this study come from the experiment BNL E766. Details of the experiment and analysis procedures can be found elsewhere [5-8,10,11]. The numbers of exclusive events selected for these measurements are 5421, 51195, 48195, 14582 for the Reaction (1), (2), (3), and (4), respectively. This study of Λ^0 polarization explores its dependence on the specific final state (1), (2), (3) and (4), and on the kinematic variables: P_T, the transverse momentum of Λ^0 with respect to the incident proton beam, and x_F, defined by $x_F = \frac{P_Z}{P_{Zmax}}$, where P_Z is the longitudinal Λ^0 momentum with respect to the beam proton momentum, in the event's center of mass, and P_{Zmax} is the maximum value of P_Z in this frame.

We have used two different methods, Methods A and B, to determine Λ^0 polarization, $\mathcal{P}$. In both methods, the decay angular distribution is assumed to be described by the expression:

$$dN/d\Omega = N_0(1 + \alpha \mathcal{P} cos\theta), \tag{5}$$

where $dN/d\Omega$ is the angular distribution of the proton from the Λ^0 decay in the Λ^0 rest frame, N_0 is a normalization constant, α is the asymmetry parameter (0.642±0.013) [9] and $\mathcal{P}$ is the polarization. θ is the angle between the direction of the proton from the decay of the Λ^0 and the normal to the production plane, $\hat{n} \equiv \frac{\vec{P}_\Lambda \times \vec{P}_{beam}}{|\vec{P}_\Lambda \times \vec{P}_{beam}|}$ where $\vec{P}_{beam}$ and $\vec{P}_\Lambda$ are the momentum vectors of the Λ^0 and the incident beam proton respectively.

RESULTS

Λ^0 polarization has been observed to be odd in x_F in Reaction (2) [4]. We have observed that to be the case as well in Reactions (1), (3), and (4). Thus in order to improve the statistical power of this measurement, we have combined the data from $x_F > 0$ and $x_F < 0$ by multiplying $cos\theta$ by the sign of x_F. In what follows, we present our discussion in terms of $|x_F|$.

In Method A, the data are separated into x_F and P_T bins and histogrammed in $cos\theta$. These histograms are then fit to Eq. (5) with N_0 and $\mathcal{P}$ as free parameters. This method is described in detail in Ref. [10]. In Method B, $\mathcal{P}$ is determined by the maximum likelihood method [9] using Eq. (5) as the probability distribution for having dN protons in $d\Omega$. The likelihood function has been calculated with two different polarization parameterizations:

$$f_1(x_F, P_T) = -a_1 x_F P_T, \tag{6}$$

$$f_2(x_F, P_T) = -a_2 x_F P_T^2, \tag{7}$$

Eq. (6) represents the simplest bi-linear combination of x_F and P_T. Eq. (7) is the small P_T limit of the empirical parameterization of polarization data presented in Ref. [12]. Maximizing the likelihood function over all of the data obtains the best estimate of parameters a_1 and a_2. The maximum likelihood technique works well with small statistics samples and gives the maximum possible information for any statistics [9].

With the results of Method B, we calculate $\mathcal{P}_{ijk}$, the average polarization for the x_{Fi} bin centered at x_{Fi} with width Δx_{Fi} and the P_{Tj} bin centered at P_{Tj} with width ΔP_{Tj} for each polarization parameterization ($k = 1$ for Eq. (6) and $k = 2$ for Eq. (7)). The average was calculated by the expression:

$$\mathcal{P}_{ijk} = \frac{\int_{\Delta x_{Fi}} dx_F \int_{\Delta P_{Tj}} f_k(x_F, P_T) I(x_{Fi}, P_{Tj}) dP_T}{\int_{\Delta x_{Fi}} dx_F \int_{\Delta P_{Tj}} I(x_{Fi}, P_{Tj}) dP_T} \tag{8}$$

where $I(x_{Fi}, P_{Tj})$ is the number of events in the x_{Fi} bin times the number of events in the P_{Tj} bin.

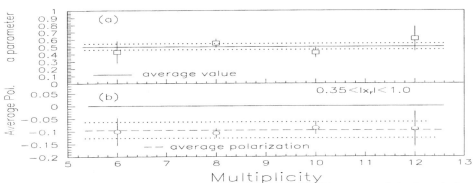

FIGURE 1. (a) constant parameters, calculated using Eq. (6); (b) average polarization calculated using Eq. (6).

The parameters determined by a maximum likelihood fit using Eq. (6) are shown in Fig. 1(a); and in Fig. 1(b), the corresponding average polarizations in the region $0.35 < |x_F| < 1.0$. Those values, both the constant parameters and average polarization, are consistent with that obtained using Eq. (7). We conclude that the polarization is independent of the final state.

The parameters determined by a maximum likelihood fit using Eq. (6) are shown in Fig. 1(a); and in Fig. 1(b), the corresponding average polarizations in the region $0.35 < |x_F| < 1.0$. Those values, both the constant parameters and average polarization, are consistent with that obtained using Eq. (7). We conclude that the polarization is independent of the final state.

Method A and Method B give the same results. Figure 2 and Figure 3 show the results of polarization for the combined sample, as a function of x_F and P_T, -using Method B, Eq. (6)- respectively.

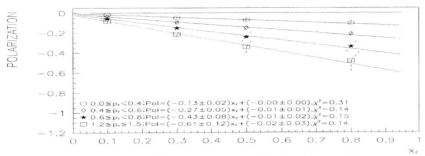

FIGURE 2. Λ^0 polarization vs. x_F, for the entire sample.

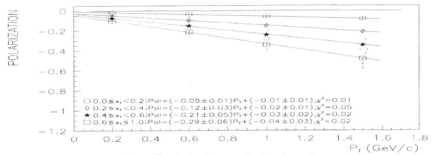

FIGURE 3. Λ^0 polarization vs. P_T, for the entire sample.

CONCLUSIONS

We found that the Λ^0 polarization can be described as either $\mathcal{P} = (-0.443 \pm 0.037)x_F P_T$ or $\mathcal{P} = (-0.485 \pm 0.041)x_F P_T^2$. The agreement of these parameterizations between specific final states and inclusive final states [12] strongly suggests that the mechanism which produces the polarization is independent of a particular final state. Finally, since the mechanism producing the final state particles changes significantly from Reaction (1) through Reaction (4) [11], our results suggest that the Λ^0 polarizing mechanism is independent of the specific Λ^0 associated production mechanism.

ACKNOWLEDGMENTS

This work was supported in part by National Science Foundation Grants No. PHY90-14879 and No. PHY89-21320, by the Department of Energy Contracts No. DE-AC02-76 CHO3000, No. DE-AS05-87ER40356 and No. W-7405-ENG-48, and by CoNaCyT of México under Grant 458100-5-4009PE.

REFERENCES

1. Lesnik A. et al., Phys. Rev. Lett. **35**, 770 (1975); Bunce G. et al., Phys. Rev. Lett. **36**, 1113 (1976); Heller K. et al., Phys. Lett. **68B**, 480 (1977).
2. Bunce G.et al., Phys. Lett. **86B**, 386 (1979); Duryea J. et al., Phys. Rev. Lett. **67**, 1193 (1991); Rameika R.et al., Phys. Rev. **D 33**, 3172 (1986); Wilkinson C. et al., Phys. Rev. Lett. **58**, 855 (1987).
3. DeGrand T. A. et al., Phys. Rev. **D 24**, 2419 (1981); Andersson B. et al., Phys. Lett. **85B**, 417 (1979); Szweed J. et al., Phys. Lett. **105B**, 403 (1981).
4. Félix J. et al., Phys. Rev. Lett. **76**, 22 (1996).
5. Uribe J. et al., Phys. Rev. **D 49**, 4373 (1994) and Ref. 10 and 12 therein.
6. Hartouni E. P. et al., Phys. Rev. Lett. **72**, 1322 (1994).
7. Christian D. C. et al., Nucl. Instr. and Meth. **A345**, 62 (1994).
8. Knapp B. C. and Sippach W. , IEEE Trans. on Nucl. Sci. **NS-27**, 578 (1980); Hartouni E. P. et al., ibid. **NS-36**, 1480 (1989); Knapp B. C., Nucl. Instrum. Methods A **289**, 561 (1990).
9. Particle Data Group, Phys. Rev. **D 50**, 1 (1994).
10. Félix J., Ph.D. thesis, Universidad de Guanajuato, México, 1994.
11. Gottschalk E. E. et al., Phys. Rev. **D 53**, 4756 (1996).
12. Pondrom L. G., Phys. Rep. **122**, 57 (1985).

[‡] Deceased.
[a] Present address: L. Livermore National Laboratory. Livermore CA 94550.
[b] Present address: Fermilab, Batavia, IL 60510.
[c] Present address: U. of Texas, M.D. Anderson Cancer C., Houston, TX 77030.
[d] Present address: University of Illinois, Urbana, Illinois.
[e] Present address: AT&T Research Laboratories, Murray Hill, NJ 07974.
[f] Present address: University of Utah, Salt Lake City, UT 84112.
[g] Present address: University of California, Davis, CA 95616.

String effective potential with massive ends

G. Germán

Instituto de Física, Laboratorio de Cuernavaca,
Universidad Nacional Autónoma de México,
Apartado Postal 48-3, 62251 Cuernavaca, Morelos, México

Yu Jiang

Facultad de Ciencias, Universidad Autónoma del Estado de Morelos,
Av. Universidad 1001, Col. Chamilpa,
62210 Cuernavaca, Morelos, México

Abstract. Following recent work by Lambiase and Nesterenko we study in detail the interquark potential for a Nambu-Goto string with point masses attached to its ends. We obtain accurate solutions to the gap equations for the Lagrange multipliers and metric components and determine the potential without simplifying assumptions. We also discuss Lüscher term and argue that it remains universal.

I THE MODEL AND GAP EQUATIONS

There has been considerable effort in trying to understand the forces between quarks in terms of strings and several models [1-15] have been proposed with different degrees of success. In all of these models one important problem is to determine the potential between two sources i.e., the so called interquark static potential. This potential has been calculated by various perturbative and non-perturbative methods. The common feature has been, however, the assumption of infinitely massive quarks at the ends of the string which is equivalent to impossing fixed ends boundary conditions. In a recent series of papers a consistent method has been proposed to study the effects of finite point masses attached to the ends of the string [16]. In particular a variational estimation of the Nambu-Goto string potential has been worked out although with some simplifying assumptions[17]. Here we reconsider this problem and present the solutions to the gap equations and determine the interquark potential without simplifying assumptions. We also provide a discussion of Lüscher term and argue that it remains universal with no mass contributions coming from the point particles attached to the ends of the string.

At the quantum level the Nambu-Goto model is given by the following functional integral

$$Z = \int [Dx^\mu] e^{-S}, \tag{1}$$

in Euclidean space, the action S is

$$S = M_0^2 \int d^2\xi \sqrt{g} + \sum_{a=1}^{2} m_a \int_{C_i} ds_a, \tag{2}$$

where M_0^2 is the string tension, $C_i (i = 1, 2)$ are the world trajectories of the string massive ends and g is the determinant of the metric

$$g_{ij} = \partial_i x^\mu(\xi_i) \partial_j x^\nu(\xi_i) \eta_{\mu\nu}, \qquad i = 0, 1. \tag{3}$$

The x^μ, $\mu = 0, 1, ..., d-1$ are the string coordinates and $\eta_{\mu\nu}$ is the embedding Euclidean metric of the space where the string evolves, g^{ij} is thus the induced metric on the world sheet swept out by the string. To study the model further it is convenient to specify a gauge, we choose the "physical gauge" or Monge parametrization

$$x^\mu(\xi_i) = (t, r, u^a(t, r)), \tag{4}$$

where the $\vec{u}^a(t, r)$, $a = 2, ..., d-1$ are the $(d-2)$ transverse oscillations of the string. We further introduce composite fields σ_{ij} given by

$$\sigma_{ij} = \partial_i \vec{u} \cdot \partial_j \vec{u}. \tag{5}$$

The metric g_{ij} and string coordinates $\vec{u}$ become independent fields by introducing Eq. 3 as a constraint. This requires the use of Lagrange multipliers α^{ij} which also become independent variables. The functional integral Eq. (2.1) then becomes

$$Z = \int [D\vec{u}][D\alpha][D\sigma] e^{-S(\vec{u}, \alpha, \sigma)}, \tag{6}$$

where the action Eq. 2 is now given by

$$S = M_0^2 \int_0^\beta dt \int_0^R dr [\sqrt{\det(\delta_{ij} + \sigma_{ij})} + \frac{1}{2} \alpha^{ij} (\partial_i \vec{u} \cdot \partial_j \vec{u} - \sigma_{ij})]$$
$$+ \sum_{a=1}^{2} m_a \int dt \sqrt{1 + \dot{\vec{u}}^2(t, r_a)} \qquad r_1 = 0, \qquad r_2 = R. \tag{7}$$

It has been shown by Alvarez that, at the saddle point, the Lagrange parameters α^{ij} as well as the metric components σ_{ij} become symmetric constant matrices with no dependence on t and r. Thus while $\vec{u} = \vec{u}(t, r)$ is in general a function of t and r, $\dot{\vec{u}}^2 = \sigma_0$ becomes, at the saddle point, a constant. This fact simplifies the problem considerably. Since the action is quadratic in the string oscillations $\vec{u}^a$ we can do the gaussian integral inmediately. The resulting action, in the particular case where $m_1 = m_2 = m$, can be written as

$$S(\alpha, \sigma) = M_0^2 \beta R [\sqrt{(1+\sigma_0)(1+\sigma_1)} - \frac{1}{2}(\alpha_0 \sigma_0 + \alpha_1 \sigma_1) - \sqrt{\frac{\alpha_1}{\alpha_0}} \lambda] + 2m\beta. \quad (8)$$

Here λ is related to the Casimir energy $E_c = \frac{1}{2}\sum_{k=1}^{\infty} \omega_k$ as follows

$$\lambda = -\frac{(D-2)}{M_0^2 R} E_c, \quad (9)$$

and the last term in Eq. 8 is the contribution to the action due to the point masses at the ends of the string. This term can be set to zero with an appropriate redefinition of S. Thus we ignore this term in what follows. The Casimir energy E_c depends on the eigenmomenta ω_k which on its turn depend on the boundary conditions imposed on the system. For a string with infinitely heavy quarks attached to its ends we impose fixed ends boundary conditions, in this case

$$\omega_k = \frac{n\pi}{R} \quad n = 1, 2, ..., \quad (10)$$

and the Casimir energy is

$$E_c = \frac{1}{2}\sum_{k=1}^{\infty} \omega_k = \frac{\pi}{2R}\sum_{n=1}^{\infty} n = -\frac{\pi}{24R}, \quad (11)$$

where the last term was obtained by the use of Riemann's ζ-function i.e., $\sum_{n=1}^{\infty} n = [\sum_{n=1}^{\infty} \frac{1}{n^\nu}]_{\nu=-1} = \zeta(-1) = -\frac{1}{12}$. In the case of finite quark masses the problem becomes increasingly difficult to dealt with even when $m_1 = m_2$. It can be shown that in this case ($m_1 = m_2 = m$) the Casimir energy is given by [17]

$$E_c = \frac{1}{2\pi R} \int_0^\infty dx \ln[1 - (\frac{x-s}{x+s})^2 e^{-2x}], \quad (12)$$

where

$$s = \frac{\rho}{\mu}\alpha_0\sqrt{1+\sigma_0}, \quad (13)$$

and

$$\rho = M_0 R, \quad \mu = \frac{m}{M_0}, \quad (14)$$

are dimensionless quantities corresponding to the (extrinsic) length and point masses attached to the ends of the string, respectively. The equation for λ Eq. 9 becomes

$$\lambda = -\frac{(D-2)}{2\pi\rho^2}\eta(s), \tag{15}$$

where

$$\eta(s) = \int_0^\infty dx \ln[1 - (\frac{x-s}{x+s})^2 e^{-2x}]. \tag{16}$$

It is also convenient to write λ in the form

$$\lambda = \frac{(D-2)\pi}{24\rho^2} - \frac{(D-2)}{2\pi\rho^2}\int_0^\infty dx \ln[1 + \frac{4sx}{(x+s)^2}\frac{1}{e^{2x}-1}]. \tag{17}$$

Note that λ is a function of α_0 and σ_0 through s, Eq. 13. Thus when writing the equations for the Lagrange multipliers and metric components derivatives of λ with respect to σ_0 and α_0 should appear. These are given by

$$\alpha_0 = \sqrt{\frac{1+\sigma_1}{1+\sigma_0}} - \frac{\sqrt{\alpha_0\alpha_1}}{1+\sigma_0}\frac{\partial\lambda}{\partial\alpha_0}, \tag{18a}$$

$$\alpha_1 = \sqrt{\frac{1+\sigma_0}{1+\sigma_1}}, \tag{18b}$$

$$\sigma_0 = \frac{1}{\alpha_0}\sqrt{\frac{\alpha_1}{\alpha_0}}\lambda - 2\sqrt{\frac{\alpha_1}{\alpha_0}}\frac{\partial\lambda}{\partial\alpha_0}, \tag{18c}$$

$$\sigma_1 = -\frac{1}{\sqrt{\alpha_0\alpha_1}}\lambda, \tag{18d}$$

where, in Eqs. (1.18), $\frac{\partial\lambda}{\partial\sigma_0}$ has been replaced by

$$\frac{\partial\lambda}{\partial\sigma_0} = \frac{\alpha_0}{2(1+\sigma_0)}\frac{\partial\lambda}{\partial\alpha_0}. \tag{19}$$

The potential $V(\rho)$ is obtained in the usual way $e^{-\beta V(\rho)} \sim Z$, $\beta \to \infty$ and is given by the simple looking formula

$$\overline{V}(\rho) = \rho\alpha_0, \tag{20}$$

which follows from Eq. 8 and the gap equations (1.18). The potential $\overline{V}(\rho)$ is also a dimensionless quantity, $\overline{V}(\rho) = M_0^{-1}V(\rho)$. Of course there is no way to solve Eqs. (1.18) analytically thus Eq. 20 is only a formal expression for $\overline{V}(\rho)$. One can play with Eqs. (1.18) and write down an expression for α_0

$$\alpha_0 = \sqrt{1 - \frac{1+\alpha_0\alpha_1}{\sqrt{\alpha_0\alpha_1}}\lambda - (1-\alpha_0\alpha_1)\sqrt{\frac{\alpha_0}{\alpha_1}}\frac{\partial\lambda}{\partial\alpha_0}}, \tag{21}$$

which will be useful for discussing some limiting situations in the next section.

The behaviour of the potential $\overline{V}(\rho)$ for several values of the mass μ is shown in Ref.[18]. For big and small values of μ the curves come close together in agreement with Eq. 12 approaching the Nambu-Goto result for $\mu = 0, \infty$. The small bump in Fig.2 of [17] for $\mu \approx 0.3$ is not present. This being probably a numerical artefact. The close similarity of the results we obtain with the accurate solution of the exact problem validates the approximations made in [17]. For a full discussion of the solutions and several interesting figures of the different aspects of the problem see Ref.[18]

II DISCUSSION OF LÜSCHER TERM AND CONCLUSIONS

We have obtained exact results to the problem of quark mass corrections to the string potential for the Nambu-Goto model in the case where the masses attached to the ends of the string are equal. These results are similar to those presented by Lambiase and Nesterenko [17] obtained under some symplifying assumptions. There is, however, a subtle point concerning the Lüscher term which we would like to discuss. For a string with fixed ends Lüscher term has a contribution to the potential of the form

$$\overline{V}_L(\rho) = -\frac{(D-2)\pi}{24\rho}. \tag{22}$$

The importance of this term is that it is universal i.e., independent of the details of a whole class of models, in particular, independent of the parameters of the model under consideration. In the one-loop approximation to the problem discussed above the potential becomes

$$\overline{V}(\rho) = \rho + (D-2)E_c, \tag{23}$$

where E_c is given by

$$E_c = \frac{\eta(s)}{2\pi\rho}, \tag{24}$$

and E_c depends on the mass μ through s (see Eq. 13 thus apparently giving Lüscher term a mass dependence. It is important to notice, however, that this Coulomb-like term arises as a long distance (large-ρ) effect. Thus strictly speaking corrections to Lüscher term, if any, should be obtained after expanding Eq. 23 for large ρ. From our numerical results we can see that large-ρ is equivalent to large-s for a given finite value of μ. Thus for large ρ, α_0 and α_1 are essentially one and from Eq. 20 and Eq. 21 the potential becomes

$$\overline{V}(\rho) \approx \rho\sqrt{1-2\lambda} \approx \rho(1-\lambda+...). \tag{25}$$

For large-s we can approximate the integral involved in the definition of λ Eq. 17 with the result

$$\lambda \approx \frac{(D-2)\pi}{24\rho^2} - \frac{(D-2)\pi}{12\rho^3}\mu, \qquad s \to \infty. \tag{26}$$

Thus the potential becomes

$$\overline{V}(\rho) = \rho - \frac{(D-2)\pi}{24\rho} + \frac{(D-2)\pi}{12\rho^2}\mu + ... \tag{27}$$

leaving Lüscher term universal.

In conclusion the study of the interquark potential for string models with masses attached to its ends is of undoubted interest by itself as a mathematical problem and certainly for the possible physical applications to the low energy regime of QCD. Here we have presented exact solutions to the gap equations and the interquark potential has been obtained for several values of μ. We see that having finite point masses at the ends of the string has considerable effects on the potential. Also the deconfinement radious become a function of μ and its value could be fixed phenomenologically. We also discuss the universality of Lüscher term and argue that it remains universal if we understand it strictly as a long distance effect with mass corrections coming up at higher orders in ρ^{-1}. Finally the tachyon problem of string theories remains unresolved although recently [19] there has been some discussion on how one can possibly avoid it.

REFERENCES

1. Y. Nambu, in *Symmetries and Quark Models*, edited by R. Chand (Gordon and Breach, New York, 1970); T. Goto, Prog. Theor. Phys. **46**, 1560(1971).
2. T. Eguchi, Phys. Rev. Lett. **44**, 126(1980).
3. A. Schild, Phys. Rev. D**16**, 1722(1977).
4. M. Lüscher, K. Symanzik, and P. Weisz, Nucl. Phys. **B173**, 365(1980),
5. O. Alvarez, Phys. Rev. D**24**, 440(1981).
6. J.F. Arvis, Phys. Lett **127B**, 106 (1983).
7. R.D. Pisarski and O. Alvarez, Phys. Rev. D**26**, 3735(1982); A. Antillón and G. Germán, Phys. Rev. D**47**, 4567(1993).
8. M. Flensburg and C. Peterson, Nucl. Phys. **B283**, 141(1987); F. Karsch and E. Laermann, Rep. Prog. Phys. **56**, 1347(1993).
9. J. Polchinski, Phys. Rev. Lett. **68**, 1267(1992); Phys. Rev. D**46**, 3667(1992).
10. G. Germán, H. Kleinert, and M. Lynker, Phys. Rev. D**46**, 1699 (1992); G. Germán, M. Lynker, and A. Macías, Phys. Rev. D**46**, 3640(1992); G. Germán and M. Lynker, Phys. Rev. D**46**, 5678(1992); A. Antillón and G. Germán, Phys. Rev. D**49**, 1966(1994).
11. M. Natsuume, Phys. Rev. D**48**, 835(1993).

12. A.M. Polyakov, Nucl. Phys. **B268**, 406(1986); H. Kleinert, Phys. Lett. **174B** 335(1986); for a review see A. M. Polyakov, *Gauge Fields and Strings* (Harwood Academic Publishers, Chur, 1987).
13. G. Germán, Mod. Phys. Lett. A**6**, 1815(1991).
14. M. Awada, Phys. Lett. **351B**, 468(1995).
15. A.M. Polyakov, PUPT-1632, hep-th/9607049; M.C. Diamantini, F. Quevedo, and C.A. Trugenberger, Phys. Lett. **396B**, 115(1997).
16. V.V. Nesterenko, Z. Phys. C**51**, 643(1991); V.V. Nesterenko and N.R. Shvetz, Z. Phys. C**55**, 265(1992); B.M. Barbashov and V.V. Nesterenko, *Introduction to the Relativistic String Theory* (Worl Scientific, Singapore, 1990).
17. G. Lambiase and V.V. Nesterenko, Phys. Rev. D**54**, 6387(1996).
18. G. Germán and Yu Jiang, "On quark mass correction to the string potential", Preprint IFUNAM, (1997). hep-th/9707037.
19. H. Kleinert, G. Lambiase, and V.V. Nesterenko, Phys. Lett. **384B**, 213(1996).

τ WEAK MAGNETIC DIPOLE MOMENT

Gabriel González–Sprinberg[1]

Facultad de Ciencias, Universidad de la Rep'ublica
Montevideo, URUGUAY

Abstract. The weak magnetic dipole moment of the τ-lepton is reviewed. Standard Model predictions and the last experimental results are presented. These may result in a stringent test for both their point-like structure and also for new physics.

The dipole moments of the electron and muon provide very precise tests of quantum field theories. The prediction first obtained by Schwinger [1] for the electron anomalous magnetic moment was one of the most spectacular achievements of quantum field theory, while CP-violation through electric and weak-electric dipole moments is exhaustively investigated, both for quarks and leptons [2].

Since then, low energy experiments, LEPI and II and SLC accelerators resulted in an enormous variety of measurements that lead, up to now, to the confirmation of the quantum corrections given by the Standard Model (SM).

The dipole moments (DM) have the same chirality flip structure as the mass terms. They receive contributions from electroweak radiative corrections in the SM, but also new physics contributions may show up in them. In particular, they may provide insight into the origin of mass.

The theoretical and experimental situation for the electro-magnetic (*i.e.* the ones related to the γ–coupling) DM of light fermions is firmly established [3]: experiments are sensitive to several orders of magnitude and theoretical predictions of higher orders have been computed.

For heavy fermions (τ, b, t) the magnetic DM are much poorly tested, and also their theoretical significance is more involved [4]. The weak magnetic dipole moment (WMDM) has been tested at LEPI in recent years for the τ, by means of the angular distribution of the decay products acting as spin analyzers. Precise experiments may provide bounds for compositeness and also for the scale of new physics [5] (possibly) hidden in the experimental errors.

In this contribution we will concentrate on the experiments and theoretical predictions related to the τ-WMDM. In what follows we define the WMDM. Next we

---

[1] e-mail: gabrielg@fing.edu.uy

analyze in some detail the Standard Model predictions and experimental bounds and finally we conclude with some comments and prospects for the near future in this issue.

In the context of the Standard Model the definition of the WMDM is as follows. Consider the neutral current coupled to the Z, having vector and axial vector parts. The matrix elements of this current has a Lorentz structure dictated by the symmetries of the theory. In the case of the vector part this is:

$$< f(p_-)\bar{f}(p_+)|J_Z^{V,\mu}(0)|0> = e\, \bar{u}_f(p_-) \left[\frac{v_f(q^2)}{2\,s_w\,c_w}\gamma^\mu + i\frac{a_f^Z(q^2)}{2\,m_\tau}\sigma^{\mu\eta}q_\eta\right] v_f(p_+) \qquad (1)$$

where $q = p_- + p_+$, e is the proton charge, s_w and c_w are the sine and cosine weak mixing angle respectively, and $\sigma^{\mu\eta} = i/2[\gamma_\mu, \gamma_\eta]$. The first term $v_f(q^2)$ is the Dirac vertex (or charge radius) form factor and it is present at tree level with the value $v_f(q^2) = t_{3L}(f) - 2\,q_f\,s_w^2$, where $t_{3L}(f)$ is the weak isospin of the fermion f and q_f the charge in units of e. The second factor $a_f^Z(q^2)$ is the weak magnetic form factor and appears due to quantum corrections. Notice that we prefer to define the WMDM in terms of the fermion anti-fermion vertex that naturally appears in $e^+e^- \to f\bar{f}$. This definition coincide with the usual one for the magnetic DM with the fermion-fermion vertex, more natural in the case of the interaction with an external field.

The weak-magnetic dipole moment is then $a_f^Z(q^2 = m_Z^2)$. Notice that only the on-shell DM, with fermions as well as the boson on the mass-shell are entitled to be gauge invariant: they are residues of the S-matrix.

With this definition in mind it is clear that if we intend to test the predictions of the Standard Model (or any extended gauge theory prediction), one should measure the on-shell effective vertex that contain the tensor Lorentz structure associated with the DM.

The WMDM appears in the SM at one loop in perturbation theory. It is generated through a chirality flip mechanism, so it is expected to be proportional to the mass of the particle. Thereby, only heavy leptons and quarks are good candidates to have an on-shell sizable DM: τ, **c** and **b** quarks.

The electroweak gauge invariant WMDM for the SM was investigated for the τ in Ref. [6,7], and the one for the b-quark is studied in Ref. [8]. For the τ there are 14 diagrams to calculate in the t'Hooft–Feynman gauge. In Figure 1 a generic diagram is shown: α, β, γ stand for the particles circulating in the loop: $N\tau^+\tau^-, \nu C^+C^-, \nu\nu C^-, \tau^- NN'$, where $N, N' = \gamma, Z, \chi, \Phi$, $N \neq N'$ and $C = W^\pm$, $\sigma^\pm$. We denote by $\sigma^\pm$ the charged non–physical Higgs and by χ and Φ the neutral non–physical and physical Higgs particles.

Six of these diagrams are not present in the electroweak corrections to the usual magnetic DM [9]; they are the ones with the following particles circulating in the loop: $W\nu\nu$, $\sigma^-\nu\nu$, $\tau Z\Phi$, $\tau \Phi Z$, $\tau \Phi \chi$, $\tau \chi \Phi$. One of them (the one with $W\nu\nu$ in the loop) gives the leading contribution: the WMDM is governed by quantum weak effects.

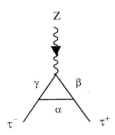

FIGURE 1. Contributing Feynman diagrams to the WMDM of τ in the t'Hooft–Feynman gauge.

In [7] we computed the numerical contribution of the diagrams (each one of order $10^{-6} - 10^{-8}$ approximately).

The SM value for the WMDM is

$$a_\tau^w(M_Z^2) = -(2.10 + 0.61\, i) \times 10^{-6} \qquad (2)$$

The Higgs mass only modifies the real part of this result less than a 1% for $1 < \frac{M_\Phi}{M_Z} < 3$. In Eq.(2) we have chosen $M_\Phi = 2M_Z$. Contrary to the well known electroweak contributions to the anomalous magnetic moment (related to the photon vertex) where it does not appears, there exist a non–vanishing absorptive part in Eq.(2) due to the fact that we compute on the Z mass shell $q^2 = M_Z^2$. In fact, one expects a non–vanishing imaginary part coming from unitarity.

In order to measure the WMDM one should identify observables that vanish when the fermion mass (and the dipole moments) vanishes. In this way most of the unwanted contributions to the observable are eliminated. For example, in $e^+e^- \longrightarrow \tau^+\tau^-$ unpolarized collisions, in the $m_f = 0$ limit both the transverse (within the collision plane) and normal (to the collision plane) polarization components vanish in that limit. The transverse single τ polarization terms were used in Refs. [6,7] to construct asymmetries proportional to the DM. The transverse polarization term in the cross section is proportional to the real part of the WMDM, except for a small helicity–flip suppressed tree level contribution. The normal polarization term is proportional to the absorptive part of the WMDM, except for possible CP-violating weak electric DM or electric DM contributions.

The spin properties of the τ's can be analyzed from the angular distribution of their decay products. In order to have access to the single τ polarization, the τ frame has to be reconstructed. Micro–vertex detectors allow such a reconstruction, as was shown in Ref. [10], for the case in which both τ's decay into (at least one) hadrons and their energies and tracks are reconstructed.

In the processes $e^+e^- \longrightarrow \tau^+\tau^- \longrightarrow h_1^+ X h_2^- \nu_\tau$ and $e^+e^- \longrightarrow \tau^+\tau^- \longrightarrow h_1^+ \bar\nu_\tau h_2^- X$ one can construct angular asymmetries in order to disentangle the dispersive and absorptive parts of the WMDM. After some algebra one finds for these asymmetries:

$$A_{Dis}{}^{\mp} = \mp\alpha_h \frac{s_w c_w}{4\beta} \frac{v^2+a^2}{a^3}\left[-\frac{v}{\gamma s_w c_w} + 2\gamma\, Re(a_\tau^w)\right] \quad (3)$$

$$A_{Abs}{}^{\mp} = \mp\alpha_h \frac{3\pi\gamma}{4} c_w s_w \frac{v}{a^2}\, Im(a_\tau^w) \quad (4)$$

where α_h is a parameter that expresses the sensitivity of each channel to the τ-spin properties. The $\mp$ signs refer to the two processes defined above. These ideas were closely followed by the L3 Collaboration. The following results were obtained collecting data at LEP from 1991 to 1995 and considering $Z\tau\tau$ events from $\rho\rho$, $\rho\pi$ and $\pi\pi$ semileptonic decay channels. L3 has put the following bounds [11,12]:

$$|Re(a_\tau^w)| \leq 4.5 \times 10^{-3} \quad (5)$$

$$|Im(a_\tau^w)| \leq 9.9 \times 10^{-3} \quad (6)$$

at the 95% confidence level. This is the first *direct* measurement of this magnitude. The Standard Model predictions for the real and imaginary parts of the WMDM are not accessible in experiments. Any signal coming from the observables defined above should be related to new physics. Recent papers have also investigated contributions for the WMDM coming from some extended models, specially supersymmetric ones. [13].

Acknowledgements. Jordi Vidal and José Bernabéu are acknowledged for clarifying comments. This work has been supported in part by CONICYT under Grant 1039/94 and the Programa de Cooperación Científica con Iberoamérica from the Agencia Española de Cooperación Internacional.

REFERENCES

1. J.Schwinger, *Phys. Rev.* **73** (1948) 416.
2. CP Violation, C.Jarkslog (cd.), World Scientific, Singapore, 1989.
3. Review of Particle Physics, R.M.Barnett et al., *Phys. Rev.* **D54** (191) 1996.
4. G.A.González—Sprinberg *etal.*, paper in preparation.
5. M.C. González-García, A. Gusso, S.F. Novaes, **hep-ph**/9802254,
   M.C. González-García and S.F. Novaes, *Phys. Lett.* **B389** (1996) 707.
6. J.Bernabéu, G.A.González–Sprinberg and J.Vidal, *Phys. Lett.* **B326** (1994) 168.
7. J.Bernabéu et al., *Nucl. Phys.* **B436** (1995) 474.
8. J.Bernabéu, J. Vidal and G.A.González–Sprinberg, *Phys. Lett.* **B397** (1997) 255.
9. K.Fujikawa, B.W.Lee and A.I.Sanda, *Phys. Rev.* **D6** (1972) 2923.
10. J.H.Kühn, *Phys. Lett.* **B313** (1993) 458.
11. E. Sánchez, Ph.D. Thesis, Universidad Complutense, Madrid (1997).
12. The L3 Coll., CERN-EP/**98-15**, submitted to Phys. Lett. B.
13. W.Hollik et al., *Acta Phys. Polon.* **B28**, 1997, 2389;
    B. de Carlos and J.M.Moreno, hep-ph/**9707487**.

$W^{\pm}$ and Z^0 Production at LEP/LHC Energies

A. Gutiérrez and A. Rosado

Instituto de Física, Universidad Autónoma de Puebla.
Apdo. Postal J-48, Col. San Manuel, Puebla, Pue. 72570, México.

Abstract. We discuss the production of $W^{\pm}$ and Z^0 bosons in deep inelastic e^-p-scattering at LEP/LHC Energies. We present results for the total and differential cross sections as a function of the parameters of the outgoing lepton. We included the contributions of γ-exchange diagrams we have also considered the contributions of $W^{\pm}$- and Z^0- exchange. We also present results for the energy spectrum and the angular distribution of the produced Z^0. Energy and angular distributions for Z^0-production are found to be strongly peaked, which will allow a clear identification of the produced Z^0.

INTRODUCTION

One of the main goals at LEP/LHC is the exploration of physical frontiers beyond the standard model [1]. However, only if $W^{\pm}$-and Z^0-production, which together lay down an important background for new physics, are completely known and characterized can the latter be sucessfully investigated. Therefore our aim in the present work is to discuss in detail the heavy boson production in deep inelastic e^-p-scattering at LEP/LHC Energies ($E_e^{max} = 60\ GeV$, $E_P^{max} = 7\ TeV$). This will be done by using the coupling between fermions and bosons as given by the standard model of the strong and electroweak interactions [2] and the parton model [3] through the parton distribution functions reported by J. Botts et al. [4], which take into account scaling violations and the heavy quarks contribution.

Three types of reaction mechanisms are discussed: $W^{\pm}$- and Z^0-production at the lepton line, at the quark line and through the fusion diagrams with the non-Abelian coupling of the gauge bosons. Our calculations include, besides γ-exchange, Z^0- and $W^{\pm}$-exchange diagram contributions and are thus complete.

The first estimate for $W^{\pm}$-and Z^0-production in e^-p-scattering were given by Llewelyn-Smith and Wiik in 1977 [5]. More recently, other authors have reported more detailed calculations [6–14]. It is not possible to compare directly all of our results with those of the former papers because different cutoffs were applied. In Ref. 8 the total cross section for the reaction $e^-p \to \nu_e W^- X$ was calculated with a cutoff of $O(1)GeV^2$ taking into account the γ-exchange contribution only. For an energy in the center of mass of $1,296\ GeV$ and our cutoffs we get $\sigma_T = 5.5 \times 10^{-37}$ cm^2 to be compared with $\sigma_T = 5.05 \times\ cm^2$ [8]. For the process $e^-p \to eZ^0X$, E. Gabrielli [8] found for the lepton vertex contribution $\sigma_{lep} = 3.3 \times 10^{-37}\ cm^2$. By using the same parameters we get a similar result; namely $\sigma_{lep} = 3.2 \times 10^{-37}$ cm^2. In our calculations we have also included the hadron vertex contribution, $\sigma_{had} = 2.6 \times 10^{-37}\ cm^2$.

One of the aims of LEP/LHC is to explore the physics beyond the standard model. For example, particles predicted by supersymmetry [15], subconstituent models [16] and others. This can be accomplished only if the results of the standard model are known in detail. Our discussion is thus complemented by presenting calculations for the energy spectrum and the angular distribution of the produced Z^0. It is found that for fixed x and y Z^0 is mainly produced in a small, well determined region of the phase space.

This paper is organized as follows: In section II, we describe the calculation of the differential cross section for inclusive $W^\pm$- and Z^0-production and discuss our results for the total cross section and differential cross section as a function of the dimensionless variables x and y, for all the different reactions and reaction mechanisms which contribute to charged vector boson and neutral vector boson production in e^-p-collisions. This section also contains the results for the energy spectrum and the angular distribution of the produced Z^0. Our conclusions are summarized in section III.

$W^\pm$-AND Z^0-PRODUCTION

Five different reactions contribute to $W^\pm$ and Z^0 production at e^-p colliders,

$$e^-p \to e^-W^-X \qquad (1)$$
$$e^-p \to e^-W^+X \qquad (2)$$
$$e^-p \to \nu_e W^-X \qquad (3)$$
$$e^-p \to e^-Z^0X \qquad (4)$$
$$e^-p \to \nu_e Z^0X. \qquad (5)$$

We have discussed in detail Z^0-production and also the kinematics of heavy boson production in deep inelastic e^-p-scattering in the Ref. 17. Notation and formulae as given in the latter to analize the $W^\pm$- and Z^0-production are here adopted, though care is exercised with regard to the different ways in which the bosons are arranged in the non-Abelian coupling diagrams.

In this section numerical results obtained by using the standard model of the electroweak interactions [2] are presented. We take $M_{Z^0} = 91.2\ GeV$ and $M_W = 80.3\ GeV$ for the masses of the neutral and charged bosons, and $sin^2\theta_W = 0.223$ for the electroweak mixing angle. We have included in our computations, in addition to γ-exchange, the Z^0-exchange and the $W^{\pm}$-exchange diagrams. We present results for e^-p-scattering with an electron energy $E = 60\ GeV$ and a proton-energy $E_P = 7\ TeV$, i.e. in the LEP/LHC system. We take cuts of $4\ GeV^2$, $4\ GeV^2$ and $10\ GeV^2$ for Q^2, Q'^2 and the invariant hadronic mass square W, respectively. These values are suitable for the parton distribution functions of J. Botts et al. [4] and the latter are used in our calculations.

Total Cross Sections

Calculations that only take into account the contribution from photon exchange diagrams to the cross section of the processes (1)-(3) have been reported, where different parton distribution functions and cuts for the invariant mass square W are used. For the sake of comparison we give our results in table 1 for the following cases: γ-exchange contribution σ_γ; γ- and Z^0-exchange contribution $\sigma_{\gamma+Z}$; and γ-, Z- and $W^{\pm}$-exchange contribution σ_T. As expected, the dominant contributions to the total cross sections stem from photon exchange.

Processes	$\sigma_\gamma (cm^2)$	$\sigma_{\gamma+Z}(cm^2)$	$\sigma_T(cm^2)$
$e^-p \to e^-W^-X$	5.7×10^{-36}	5.7×10^{-36}	6.1×10^{-36}
$e^-p \to e^-W^+X$	6.8×10^{-36}	6.8×10^{-36}	6.1×10^{-36}
$e^-p \to \nu_e W^-X$	4.6×10^{-37}	6.6×10^{-37}	5.5×10^{-37}
$e^-p \to e^-Z^0X$	5.5×10^{-37}	5.7×10^{-37}	5.8×10^{-37}
$e^-p \to \nu_e Z^0X$	0	0	1.4×10^{-37}

Table 1. Contribution to the total cross sections from the different boson exchange diagrams: $E_e = 60\ GeV$, $E_P = 7\ TeV$.

dσ/dy

In order to illustrate the behaviour of the y-dependence, we show in Fig. 1 the compensation mechanism characteristic of the non-Abelian structure of the electroweak standard model as a gauge theory. In process (3) this compensation reaches two orders of magnitude at LEP/LHC energies. Therefore, even small deviations of the coupling estructure from the standard model, like anomalous moment terms, can be expected to lead to observable deviations on results obtained through the standard model.

Energy and Angular Distribution

Results for the normalized energy distribution $(d\sigma/dE_k dxdy)/(d\sigma/dxdy)$ of the Z^0 are illustrated in Fig. 2 for $x = 0.001$, 0.003, 0.008 and $y = 0.95$. We see that these energy distributions have resonance-like behaviour, the Z^0 bosons being mainly produced in a small E_k interval. Something similar happens with the normalized angular distribution $(d\sigma/dcos\theta_k dE_k dxdy)/(d\sigma/dE_k dxdy)$. Fig. 3 show that the Z^0 is mainly found in the forward direction and that the $cos\theta_k$ distribution develops a sharp peak before reaching its kinematical limit. We also observe that most of the Z^0 bosons are produced in the plane which is spanned by the momenta of the incoming and outgoing leptons *i.e.* $cos\phi_k \approx 0$. The reason for these sharp energy-and angular distributions is that the differential cross section becomes large when $Q'^2 = -(p - p' - k)^2$ approaches its minimal value, $Q'^2 \approx Q'^2_{cut}$.

CONCLUSIONS

We have presented the complete calculation of heavy vector boson production in unpolarized deep inelastic e^-p-scattering in the context of the standard model in conjunction with the parton model, at LEP/LHC energies. By taking $M_W = 80.3$ GeV, $M_{Z^0} = 91.2\ GeV$, $\sin^2\theta_W = 0.223$, $\sqrt{s} = 1,296\ GeV$ and by using the parton distributions reported by J. Botts *et al.*, a total production of about 290 Z^0, 3030 W^+, 3310 W^- bosons for an integrated luminosity de 500 pb^{-1}, for LEP/LHC was found. We also analyzed in the LEP/LHC system the energy E_k and $cos\theta_k$ distributions of the produced Z^0 for a given x and y (*i.e.* for a given energy and polar angle of the final electron). Considering the region where the Z^0-production is almost purely leptonic, we found that the E_k and θ_k distributions are strongly peaked. It is shown that if the energy and polar angle of the scattered electron are given by scaling variables x and y, the boson Z^0 will be mostly produced with energy $E_k = Ey + E_p x(1 - y) + E_p M_Z^2/(sy)$, polar angle $cos\theta_k \approx (2Ey - E_k)/\sqrt{E_k^2 - M_Z^2}$ and azimuthal angle $cos\phi_k \approx 0$.

Finally, it is worth mentioning that the resonance-like behaviour of the E_k and θ_k distributions will help discriminate between the production of a standard Z^0 and the production of other particles, such as those predicted by theories dealing with other reaction mechanisms, for example supersymmetry [15] and subconstituent models [16], through experiments which will be carried at the collider LEP/LHC. The cross section and consequently the even rate is presumably not large enough to allow for detailed investigation of the $W^\pm$ and Z^0 properties. However, it is important to perform complete and detailed calculations of $W^\pm$- and Z^0-production since new physical features can be observed only if standard background events are correctly distinguished. In this sense the results of the present report may help clear the way toward new physics at LEP/LHC.

ACKNOWLEDGMENTS

A scholarship provided by *Consejo Nacional de Ciencia y Tecnología* (CONA-CyT) and financial support from *Sistema Nacional de Investigadores* (SNI) are gratefully acknowledged.

REFERENCES

1. See e.g. R. Rückl, preprint MPI-PAE/PTh 76/90 and references therein.
2. S. Weinberg, Phys. Rev. Lett. **19**, 1264 (1967); A. Salam: Proc. 8th Nobel Symposium, S. 367 (N. Svartholm, Ed., Stockholm 1968); S.L. Glashow, J. Illiopoulos and L. Maiani, Phys. Rev. **D2**, 1285 (1970).
3. R.P. Feynman, Photon-Hadron Interactions, Benjamin, Reading, Massachusetts, 1972.
4. J. Botts, J. G. Morfin, J. F. Owens, J. Qiu, W. K. Tung and H. Weerts, preprint Fermilab-Pub-92/371.
5. C. H. Llewelyn-Smith and B. H. Wiik, DESY Report 77/36 (1977).
6. P. Salati and J. C. Wallet, Z. Phys. **C16**, 155 (1982).
7. G. Altarelli, G. Martinelli, B. Mele and R. Rückl, Nucl. Phys. **B262**, 204 (1985).
8. E. Gabrielli, Mod. Phys. Lett. **A1**, 465 (1986).
9. D. Atwood, U. Baur, G. Couture and D. Zeppenfeld, Proc. of the Snowmass DPF Summer Study 1988, p. 264.
10. H. Baer, J. Ohnemus and D. Zeppenfeld, Z. Phys. **C43**, 675 (1989).
11. D. Atwood, U. Baur, D. Goddard, S. Godfrey and B. A. Kniehl, Proc. of the Snowmass DPF Summer Study 1990, p. 557.
12. U. Baur, B. A. Kniehl, J. A. M. Vermaseren and D. Zeppenfeld, Proc. of the Aachen ECFA Workshop 1990, p. 956.
13. U. Baur, J. A. M. Vermaseren, D. Zeppenfeld, Nucl. Phys. **B375**, 3 (1992).
14. U. Baur, Proc. of the Rencontres de Moriond on QCD and High Energy Hadronic Interactions, Les Arcs, France 1992, p. 91.
15. H. E. Haber and G.L. Kane, Phys. Rep. **117**, 75 (1985).
16. See e.g. R. J. Cashmore el. al. Phys. Rep. **22**, 277(1985) and references therein.
17. M. Böhm and A. Rosado, Z. Phys. **C34**, 117 (1987).

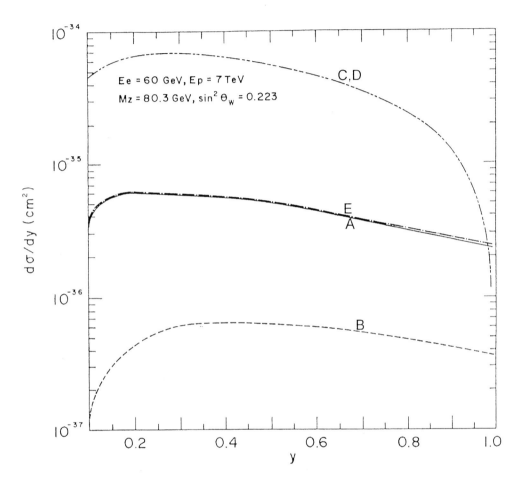

Fig. 1. Comparison of the contribution to $d\sigma/dy$ from the different boson production mechanisms for process (1). Curve A: total contribution; curve B: production at the leptonic vertex; curve C: production at the hadronic vertex; curve D: production through the non-Abelian couplings; curve E: sum of the hadronic vertex and non-Abelian production mechanisms ($\sqrt{s} = 1,296\ GeV$).

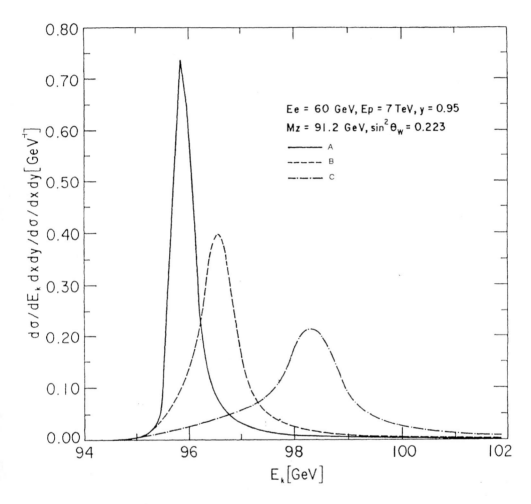

Fig. 2. Normalized energy spectrum of the produced Z^0 bosons for $y = 0.95$ and $x = 0.001$ (curve A), $x = 0.003$ (curve B), $x = 0.008$ (curve C) ($\sqrt{s} = 1,296\ GeV$).

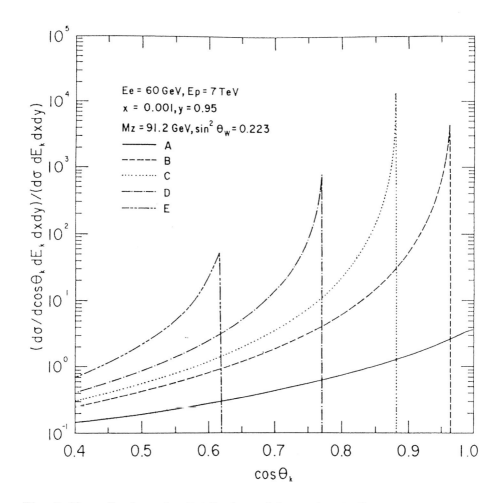

Fig. 3. Normalized angular distributions of the produced Z^0 bosons for $x = 0.001$, $y = 0.95$ and $E_k = 94.5\ GeV$ (curve A), $E_k = 95\ GeV$ (curve B), $E_k = 95.5\ GeV$ (curve C), $E_k = 96.5\ GeV$ (curve D), $E_k = 99\ GeV$ (curve E).

Finite Soft SUSY Breaking Terms[1]

T. Kobayashi[1], J. Kubo[2], M. Mondragón[3] and G. Zoupanos[4,5]

[1] *Inst. of Particle and Nuclear Studies, Tanashi, Tokyo 188, Japan*
[2] *Dept. of Physics, Kanazawa Univ., Kanazawa 920-1192, Japan*
[3] *Inst. de Física, UNAM, Apdo. Postal 20-364, México 01000 D.F., México*
[4] *Physics Dept., Nat. Technical University, GR-157 80 Zografou, Athens, Greece*
[5] *Insitut f. Physik, Humboldt-Universität, D10115 Berlin, Germany*

Abstract. We give a new solution to the requirement of two-loop finiteness of the soft supersymmetry breaking terms (SSB) in Finite-Gauge-Yukawa unified theories, which has the form of a sum rule for the scalar masses. Using this sum rule we investigate two promising models and determine their spectrum in terms of a few parameters. Some characteristic features of the models are that a) the old agreement of the top quark mass prediction with the measured value remains unchanged, b) the lightest Higgs mass is predicted around 120 GeV, c) the s-spectrum starts above 200 GeV.

INTRODUCTION

Finite Unified Theories (FUTs) support the hope that it might be possible to have at some scale a theory without infinity renormalizations. Finiteness is based on the fact that it is possible to find renormalization group invariant (RGI) relations among couplings that keep finiteness in perturbation theory, even to all orders [4].

Consider a chiral, anomaly free, $N = 1$ globally supersymmetric gauge theory based on a group G with gauge coupling constant g. The superpotential of the theory is given by

$$W = \frac{1}{2} m_{ij} \Phi_i \Phi_j + \frac{1}{6} C_{ijk} \Phi_i \Phi_j \Phi_k , \qquad (1)$$

where m_{ij} and C_{ijk} are gauge invariant tensors and the matter field Φ_i transforms according to the irreducible representation R_i of the gauge group G.

The vanishing of $\beta_g^{(1)}$ and $\gamma_{ij}^{(1)}$ at one-loop gives us the following conditions

$$\sum_i \ell(R_i) = 3C_2(G), \qquad C^{ikl}C_{jkl} = 2\delta_j^i g^2 C_2(R_i) . \qquad (2)$$

A natural question to ask is what happens at higher loop orders. The finiteness conditions (2) impose relations between gauge and Yukawa couplings. We would

---

[1]) Presented by M. Mondragón. Supported partly by the projects CONACYT 3275-PE, PAPIIT IN110296, FMBI-CT96-1212, ERBFMRXCT960090, and PENED95/1170;1981.

like to guarantee that such relations leading to a reduction of the couplings hold at any renormalization point. The necessary, but also sufficient, condition for this to happen is to require that such relations are solutions to the reduction equations (REs)

$$\beta_g \frac{dC_{ijk}}{dg} = \beta_{ijk} \tag{3}$$

and hold at all orders. Remarkably the existence of all-order power series solutions to (3) can be decided at the one-loop level. One-loop finiteness implies that the Yukawa couplings C_{ijk} must be functions of the gauge coupling g. To find a similar condition to all orders it is necessary and sufficient for the Yukawa couplings to be a formal power series in g, which is a solution of the REs (3) [2].

FINITE SUPERSYMMETRY BREAKING TERMS

Let us now examine what happens when we add soft supersymmetry breaking terms. Consider the superpotential given by (1) along with the Lagrangian for SSB terms,

$$-\mathcal{L}_{\rm SB} = \frac{1}{6} h^{ijk} \phi_i \phi_j \phi_k + \frac{1}{2} b^{ij} \phi_i \phi_j + \frac{1}{2} (m^2)^j_i \phi^{*i} \phi_j + \frac{1}{2} M \lambda\lambda + {\rm H.c.} \tag{4}$$

where the ϕ's are the scalar parts of the chiral superfields Φ's, λ are the gauginos and M their unified mass. Since we would like to consider only finite theories here, we assume that the gauge group is a simple group and the one-loop β function of the gauge coupling g vanishes. We also assume that the reduction equations (3) admit power series solutions of the form

$$C^{ijk} = g \sum_{n=0} \rho^{ijk}_{(n)} g^{2n} , \tag{5}$$

According to the finiteness theorem of ref. [2], the theory is then finite to all orders in perturbation theory, if, among others, the one-loop anomalous dimensions $\gamma^{(1)j}_i$ vanish. The one- and two-loop finiteness for h^{ijk} can be achieved by

$$h^{ijk} = -MC^{ijk} + \ldots = -M\rho^{ijk}_{(0)} g + O(g^5) . \tag{6}$$

Now, to obtain the two-loop sum rule for soft scalar masses, we assume that the lowest order coefficients $\rho^{ijk}_{(0)}$ and also $(m^2)^i_j$ satisfy the diagonality relations

$$\rho_{ipq(0)}\rho^{jpq}_{(0)} \propto \delta^j_i \text{ for all } p \text{ and } q \text{ and } (m^2)^i_j = m^2_j \delta^i_j , \tag{7}$$

respectively. Then we find the following soft scalar-mass sum rule

$$(m^2_i + m^2_j + m^2_k)/MM^\dagger = 1 + \frac{g^2}{16\pi^2} \Delta^{(1)} + O(g^4) \tag{8}$$

for i, j, k with $\rho^{ijk}_{(0)} \neq 0$, where $\Delta^{(1)}$ is the two-loop correction, which vanishes for the universal choice in accordance with the previous findings of ref. [3].

FINITE UNIFIED THEORIES

A predictive Gauge-Yukawa unified SU(5) model which is finite to all orders, in addition to the requirements mentioned already, should also have the following properties:

(1) The one-loop anomalous dimensions are diagonal, i.e., $\gamma_i^{(1)j} \propto \delta_i^j$, according to the assumption (7).
(2) Three fermion generations, $\overline{\mathbf{5}}_i$ ($i = 1, 2, 3$), which should not couple to the **24**.
(3) The two Higgs doublets of the MSSM should mostly be made out of a pair of Higgs quintet and anti-quintet, which couple to the third generation.

In the following we discuss two versions of the all-order finite model.
A: The model of ref. [1].
B: A slight variation of the model **A**.

The superpotential which describe the two models takes the form [1,5]

$$W = \sum_{i=1}^{3} [\, \frac{1}{2} g_i^u \, \mathbf{10}_i \mathbf{10}_i H_i + g_i^d \, \mathbf{10}_i \overline{\mathbf{5}}_i \, \overline{H}_i \,] + g_{23}^u \, \mathbf{10}_2 \mathbf{10}_3 H_4 \tag{9}$$

$$+ g_{23}^d \, \mathbf{10}_2 \overline{\mathbf{5}}_3 \, \overline{H}_4 + g_{32}^d \, \mathbf{10}_3 \overline{\mathbf{5}}_2 \, \overline{H}_4 + \sum_{a=1}^{4} g_a^f \, H_a \, \mathbf{24} \, \overline{H}_a + \frac{g^\lambda}{3} (\mathbf{24})^3 \,,$$

where H_a and $\overline{H}_a$ ($a = 1, \ldots, 4$) stand for the Higgs quintets and anti-quintets.

The non-degenerate and isolated solutions to $\gamma_i^{(1)} = 0$ for the models {**A**, **B**} are:

$$(g_1^u)^2 = \{\frac{8}{5}, \frac{8}{5}\} g^2 \,, \quad (g_1^d)^2 = \{\frac{6}{5}, \frac{6}{5}\} g^2 \,, \quad (g_2^u)^2 = (g_3^u)^2 = \{\frac{8}{5}, \frac{4}{5}\} g^2 \,, \tag{10}$$

$$(g_2^d)^2 = (g_3^d)^2 = \{\frac{6}{5}, \frac{3}{5}\} g^2 \,, \quad (g_{23}^u)^2 = \{0, \frac{4}{5}\} g^2 \,, \quad (g_{23}^d)^2 = (g_{32}^d)^2 = \{0, \frac{3}{5}\} g^2 \,,$$

$$(g^\lambda)^2 = \frac{15}{7} g^2 \,, \quad (g_2^f)^2 = (g_3^f)^2 = \{0, \frac{1}{2}\} g^2 \,, \quad (g_1^f)^2 = 0 \,, \quad (g_4^f)^2 = \{1, 0\} g^2 \,.$$

Consequently, these models are finite to all orders. After the reduction of couplings the symmetry of W is enhanced [5].

The main difference of the models **A** and **B** is that three pairs of Higgs quintets and anti-quintets couple to the **24** for **B** so that it is not necessary to mix them with H_4 and $\overline{H}_4$ in order to achieve the triplet-doublet splitting after the symmetry breaking of $SU(5)$.

PREDICTIONS OF LOW ENERGY PARAMETERS

Since the gauge symmetry is spontaneously broken below M_{GUT}, the finiteness conditions do not restrict the renormalization property at low energies, and all it remains are boundary conditions on the gauge and Yukawa couplings (10), the $h = -MC$ relation (6), and the soft scalar-mass sum rule (8) at M_{GUT}. So we examine the evolution of these parameters according to their renormalization group equations at two-loop for dimensionless parameters and at one-loop for dimensional ones with these boundary conditions. Below M_{GUT} their evolution is assumed to be governed by the MSSM. We further assume a unique supersymmetry breaking scale M_s so that below M_s the SM is the correct effective theory.

The predictions for the top quark mass M_t are ~ 183 and ~ 174 GeV in models A and B respectively. Comparing these predictions with the most recent experimental value $M_t = (175.6 \pm 5.5)$ GeV, and recalling that the theoretical values for M_t may suffer from a correction of less than $\sim 4\%$ [4], we see that they are consistent with the experimental data.

Turning now to the SSB sector we look for the parameter space in which the lighter s-tau mass squared $m_{\tilde{\tau}}^2$ is larger than the lightest neutralino mass squared m_χ^2 (which is the LSP). Using the sum rule (8) and imposing the conditions a) successful radiative electroweak symmetry breaking b) $m_{\tilde{\tau}^2} > 0$ and c) $m_{\tilde{\tau}^2} > m_{\chi^2}$, we find a comfortable parameter space for both models (although model B requires large $M \sim 1$ TeV). The particle spectrum of models A and B in turn is calculated in terms of 3 and 2 parameters respectively. In both models the lightest Higgs mass is predicted to be of the order of ~ 120 GeV.

CONCLUSIONS

The conditions for finiteness of the SSB parameters have been studied, and a sum rule for the soft scalar masses has been obtained which quarantees their finiteness up to two-loops [5], avoiding at the same time serious phenomenological problems related to the previously known "universal" solution. It was found that this sum rule coincides with that of a certain class of string models in which the massive string modes are organized into $N = 4$ supermultiplets. Using the sum rule determined the spectrum of realistc models in terms of just a few parameters. In addition to the successful prediction of the top quark mass the characteristic features of the spectrum are that 1) the lightest Higgs mass is predicted ~ 120 GeV and 2) the s-spectrum starts above 200 GeV.

REFERENCES

1. D. Kapetanakis, M. Mondragón and G. Zoupanos, Zeit. f. Phys. **C60** (1993) 181; M. Mondragón and G. Zoupanos, Nucl. Phys. **B** (Proc. Suppl) **37C** (1995) 98.
2. C. Lucchesi, O. Piguet and K. Sibold, Helv. Phys. Acta **61** (1988) 321; Phys. Lett. **B201** (1988) 241.
3. I. Jack and D.R.T. Jones, Phys.Lett. **B333** (1994) 372.
4. For an extended discussion and a complete list of references see: J. Kubo, M. Mondragón and G. Zoupanos, Acta Phys. Polon. **B27** (1997) 3911.
5. T. Kobayashi, J. Kubo, M. Mondragón and G. Zoupanos, *Constraints on Finite Soft SUSY Breaking Terms* to be published in Nucl. Phys. B.

Non-Noether Charges in Classical Mechanics[1]

J. L. Lucio M.[2], A. Cabo[3], and V.M. Villanueva[4]

Instituto de Física de la Universidad de Guanajuato, P.O. Box E-143, 37150, León, Gto., México

Abstract. Starting from Noether's charges we derive an expression for new conserved quantities which extend the algebra of the original group of transformations.

INTRODUCTION

Noether's theorem provides a systematic way of analyzing the conserved quantities associated to a physical system. In the Hamiltonian formalism, the Poisson brackets of the conserved (Noether's) charges define an algebra which is isomorphic to the algebra of the global symmetry group G of transformations from which the charges were obtained [1]. There exist however the possibility that the algebra of the charges Q_r is an extension of that of the global symmetry group. Indeed, as a consequence from the passage from configuration to phase space, new terms -as compared to the algebra of the original group of transformation G- may appear in the Poisson brackets of the Q_r charges. According to our results, a necessary condition for this to happen is that the action but not the Lagrangian be invariant under the symmetry transformation. The Galilei group provide an example where such conditions are met and a classical central extensions appear.

THE MODEL

We consider a system with n degrees of freedom, given as functions $q_j(t)$ ($j = 1, 2, \cdots n$) of the "time" variable t. Throughout the work, we make two basic assumptions: first that the dynamics of the system is described by the action functional:

---

[1] Work supported by CONACyT under contract 3979PE-9608
[2] lucio@ifug.ugto.mx.
[3] On leave of absence from ICIMAF, La Habana, Cuba
[4] victor@ifug1.ugto.mx

$$S[q_j] = \int_{t_i}^{t_f} dt \mathcal{L}(q_j, \dot{q}_j), \qquad (1)$$

and second, that the system is regular *i.e.*, free of constraints. We consider a group of infinitesimal transformations with parameters $\delta\alpha_r$:

$$q'_j(t') = q_j(t) + \delta q_j = q_j(t) + (\delta\alpha_r T_r)q_j,$$
$$t' = t + \delta t = t + (\delta\alpha_r S_r)t, \qquad (2)$$

where T_r, S_r, ($r = 1, 2, \cdots, \omega$) are the group "generators" (appropriated algebraic or differential operators).

Finite transformations can be obtained in terms of the T_r, S_r generators by exponentiating (2)

$$q'_i(t') = e^{\alpha_r T_r} q_i(t) \equiv g_c(\alpha)q_i(t),$$
$$t' = e^{\alpha_r S_r} t \equiv g_\tau(\alpha)t. \qquad (3)$$

The group property of the transformations (2) can be used to bring out the algebra of the generators. This is achieved by considering the commutator of two infinitesimal transformations to obtain:

$$[T_r, T_s] = C_{rs}^u T_u, \qquad (4)$$

where the structure constants C_{rs}^u are expressed in terms of the coordinate ($\delta q_j = \delta\alpha_r \delta^r q_j$) and time ($\delta t = \delta\alpha_r \delta^r t$) variations as:

$$C_{rs}^u \delta^u q_i = \left(\delta^r q_j \frac{\partial}{\partial q_j} \delta^s q_i - \delta^s q_j \frac{\partial}{\partial q_j} \delta^r q_i + \delta^r t \frac{\partial}{\partial t} \delta^s q_i - \delta^s t \frac{\partial}{\partial t} \delta^r q_i \right). \qquad (5)$$

Using Noether's theorem [1] we deduce the conservation of the charges

$$Q_r(q, \dot{q}, t) = \frac{\partial \mathcal{L}}{\partial \dot{q}_j} \delta q_j - \left(\frac{\partial \mathcal{L}}{\partial \dot{q}_j} \dot{q}_j - \mathcal{L} \right) \delta t - \Lambda. \qquad (6)$$

It is convenient to consider the phase space version of these charges:

$$Q_r(q, p, t)\delta\alpha_r = p_i \delta q_i - H(q_i p)\delta t - \Lambda. \qquad (7)$$

By using charge conservation, the Poisson bracket between these charges is calculated, and the result is

$$\{Q_r, Q_s\} = C_{rs}^u Q_u + L_{rs}, \qquad (8)$$

where

$$L_{rs} = \{\Lambda_s, p_i\delta^r q_i\} - \{\Lambda_r, p_i\delta^s q_i\} - C^u_{rs}\Lambda_u. \tag{9}$$

Notice that L_{rs} will not depend on the momenta and that, as it should be, it is antisymmetric in the $r - s$ indices. Explicit evaluation of the Poisson bracket taking into account that $\delta^r q$ are p independent leads to:

$$L_{rs} = \left(\frac{\partial \Lambda_s}{\partial q_j}\right)\delta^r q_j - \left(\frac{\partial \Lambda_r}{\partial q_j}\right)\delta^s q_j - C^u_{rs}\Lambda_u. \tag{10}$$

In fact, if Noether charge is conserved, then (9) implies that L_{rs} is also conserved. Furthermore it is possible to show that L_{rs} is q and p independent. Thus, we have shown that the Poisson bracket of Noether's charges can acquire only coordinate and momentum independent central extensions. Therefore, we have derived an expression for a conserved quantity Eq. (10), not following from Noether's theorem, and which extends the algebra of the original symmetry group.

GALILEI GROUP

As an application of this approach, let us consider a free particle and the Galilei symmetry group. It is well known that the mass of the particle is involved in the algebra of the group and it is considered as a central extension [2]. The system under consideration is described by:

$$\mathcal{L} = \frac{M}{2}\sum_{i=1}^{3}\dot{q}_i^2, \qquad \mathcal{H} = \sum_{i=1}^{3}\frac{p_i^2}{2M}. \tag{11}$$

The Galilei transformations lead to the infinitesimal variations

$$\delta q_j = (\delta v_j)t + \delta a_j, \qquad \delta \dot{q}_j = \delta v_j. \tag{12}$$

The δq_j must be considered as the combination of two independent variations. A pure boost characterized by the parameters (δv_j) and pure translations (δa_j)

$$\delta^r q_j(boost) \equiv \frac{\delta q_j}{\delta v_r} = t\delta_{jr},$$

$$\delta^r q_j(trans) \equiv \frac{\delta q_j}{\delta a_r} = \delta_{jr}. \tag{13}$$

For infinitesimal transformations, the variation of the Lagrangian is:

$$\delta \mathcal{L} = \frac{d}{dt}(Mq_i\delta v_i). \tag{14}$$

Thus, in this case, $\Lambda = Mq_i\delta v_i = \Lambda_r^{boost}\delta v_r + \Lambda_r^{trans}\delta a_r$. Clearly $\Lambda_r^{boost} = Mq_r$ and $\Lambda_r^{trans} = 0$. Noether's theorem leads to the independent conserved charges:

$$Q_r = p_r t - Mq_r, \qquad P_r = p_r \qquad r = 1, 2, 3. \tag{15}$$

The Poisson brackets of these charges are:

$$\{Q_r, Q_s\} = 0, \qquad \{P_r, P_s\} = 0, \qquad \{P_r, Q_s\} = M\delta_{rs}. \tag{16}$$

On the other hand, according to our discussion, the central extension -*if it exist*- should be given by (10). It is straightforward to show using (4) that for this example $C_{rs}^u = 0$. Futhermore, if the indices r and s refer both to boost, or both to translations, then $L_{rs} = 0$. So, the only possibility left is:

$$\begin{aligned}L_{rs} &= \left(\frac{\partial \Lambda_s^{boost}}{\partial q_j}\right)\delta^r q_j(trans) - \left(\frac{\partial \Lambda_r^{trans}}{\partial q_j}\right)\delta^r q_j(boost)\\ &= M\delta_{sj}\delta_{jr} = M\delta_{rs},\end{aligned} \tag{17}$$

therefore, we conclude that for the Galilei group the mass is a central extension.

REFERENCES

1. J. Govaerts *"Hamiltonian Quantization and Constrained Dynamics"*, Leuven Notes in Mathematical and Theoretical Physics Vol. 4. Series B: Theoretical Physics, Leuven University Press, Belgium (1991).
2. J.A. de Azcarraga, *" Wess-Zumino Terms, Extended Algebras and Anomalies in Classical Physics"*, proceedings of the AMS-IMS-SIAM. Summer Research Conference on Mathematical Aspects of Classical Field Theory, University of Washington at Seatle July 1991.

Concerning CP violation in 331 models[1]

J. C. Montero, V. Pleitez and O. Ravinez

Instituto de Física Teórica
Universidade Estadual Paulista
Rua Pamplona 145
01405-900- São Paulo, SP
Brazil

Abstract. We consider the implementation of CP violation in the context of 331 models. In particular we treat a model where only three scalar triplets are needed in order to give all fermions a mass while keeping neutrino massless. In this case all CP violation is provided by the scalar sector.

In spite of the great efforts of theoreticians and experimentalists, the origin and the smallness of CP violation remains an open question. In the context of the standard electroweak model [1] the CP symmetry is violated in the complex Yukawa couplings [2]. Although this is an interesting feature of the model it leaves open the question of why CP is so feebly violated. Since the works of Lee and Weinberg it has been known that in renormalizable gauge theories the violation of CP has the right strength if it occurs through the exchange of a Higgs boson of mass M_H [3], *i. e.*, it is proportional to $G_F m_f^2/M_H^2$, where m_f is the fermion mass. Since then, there have been many realizations of that mechanism in extensions of the electroweak standard model [4]. Recently, it has been proposed models with the electroweak gauge group being $SU(3)_L \otimes U(1)_N$ instead of the usual $SU(2)_L \otimes U(1)_Y$ [5,6]. One interesting feature of these sort of models is that the anomalies cancel only when all three families are taken together. Although neutrinos remain massless there is lepton-flavor violation in the interactions with doubly charged scalar and vector bosons which are present in the model. Hence, it is possible to have CP violation in that mixing matrix [7]. Another possibility is purely spontaneous CP violation through complex value for the vacuum expectation values (VEVs) for the neutral scalars. This however, only happens in the model with three triplets and one sextet [8].

Let us consider a model with 331 symmetry with exotic heavy leptons. The three leptons generations belong to $(\mathbf{1}, \mathbf{3}, 0)$ representation. It means $(\nu_l, l^-, E_l^+)^T$, for $l = e, \mu, \tau$. The scalar content of the model necessary to give masses to all fermions

---

[1] Presented by O. Ravinez

is

$$\chi = \begin{pmatrix} \chi^- \\ \chi^{--} \\ \chi^0 \end{pmatrix} \sim (\mathbf{3}, -\mathbf{1}), \quad \rho = \begin{pmatrix} \rho^+ \\ \rho^0 \\ \rho^{++} \end{pmatrix} \sim (\mathbf{3}, \mathbf{1}), \quad \eta = \begin{pmatrix} \eta^0 \\ \eta_1^- \\ \eta_2^+ \end{pmatrix} \sim (\mathbf{3}, \mathbf{0}). \quad (1)$$

As we said before, the neutrinos are optionally massless if we do not introduce the right-handed components.

We allow VEVs being complex numbers i.e., $v_a = |v_a|\exp(i\theta_a)$, where $a = \eta, \rho$ and χ. However, it is not enough to implement CP violation. The minimization of the potential implies the conditions $\text{Im}(v_\eta v_\rho v_\chi) = 0$ and, since we can choose always two VEVs being real because of the $SU(3)$ symmetry, it means that no phase at all survive in the potential minimum [8]. However, if we allow beside the complex VEVs, the trilinear term in the potential $\alpha\epsilon_{ijk}\eta_i\rho_k\chi_k + H.c.$ (where i,j,k are $SU(3)$ indices) with the complex constant $\alpha = |\alpha|e^{i\theta_\alpha}$, the minimization of the potential in this case implies $\text{Im}(\alpha v_\eta v_\rho v_\chi) = 0$ and the relative phase, say, among α and v_χ will survive in the Lagrangian density. Hence, there are explicit CP violation in the Lagrangian. (Notice however, that CP violation also requires complex VEV's. We will assume real Yukawa couplings too.)

The lepton are assigned to the following representations:

$$\Psi_{aL} = \begin{pmatrix} \nu_{l_a} \\ l'^{-}_a \\ E'^{+}_a \end{pmatrix}_L \sim (\mathbf{3}, \mathbf{0}); \quad l'^{-}_{aR} \sim (\mathbf{1}, -\mathbf{1}), \; E'^{-}_{aR} \sim (\mathbf{1}, +\mathbf{1}), \quad a = e, \mu, \tau. \quad (2)$$

It is possible to absorb all phases in the leptonic mass matrix so that the symmetry eigenstates (primed fields) are related to the mass eigenstates (unprimed fields) thorough orthogonal matrices [9]

$$l'_{aL} = \mathcal{O}^e_{Lai}l_{iL}, \quad l'_{aR} = \mathcal{O}^e_{Rai}l_{iR}, \quad E'_{aL} = \mathcal{O}^E_{Lai}l_{iL}, \quad l'_{aR} = \mathcal{O}^E_{Rai}l_{iR}, \quad i = 1,2,3. \quad (3)$$

It is always possible to choose the CP violation to occur only through the exchange of singly and doubly charged scalar fields. (The CP violation in the neutral Higgs sector is transformed away by a redefinition of the right-handed components of the lepton fields, say $e_{iR} \to \exp(i\theta_\rho)e_{iR}$ and $E_{iR} \to \exp(i\theta_\chi)E_{iR}$.)

The Yukawa couplings of leptons with the doubly charged scalars are

$$\mathcal{L}_Y = \frac{\sqrt{2}}{|v_\rho|}\bar{E}_L(\mathcal{O}^E_L)^T\mathcal{O}^e_L M^e e_L \rho^{++} + \frac{\sqrt{2}}{|v_\chi|}e^{-i\theta_\chi}\bar{e}_L(\mathcal{O}^e_L)^T\mathcal{O}^E_L M^E E_L L\chi^{++} + H.c., \quad (4)$$

where we have used $\Gamma^e = \mathcal{O}^e_L M^e \mathcal{O}^{eT}_R$, $\Gamma^E = \mathcal{O}^E_L M^E \mathcal{O}^{ET}_R$ with $M^e = \text{diag}(m_e, m_\mu, m_\tau)$; $M^E = \text{diag}(m_{E_1}, m_{E_2}, m_{E_3})$ (where $\sqrt{2}\Gamma^e/v_\rho$ and $\sqrt{2}\Gamma^e/v_\chi$ are the arbitrary Yukawa dimensionless couplings). In Eq. (4) the scalar fields are still symmetry eigenstates. In fact, there are one Goldstone boson, G^{++} and a physical one, X^{++}, we denote its mass by m_X. (In this model there is not lepton-number violation in the interactions with the neutral scalars.) We have verified that it is not possible to absorb all phases in the complete Lagrangian density. We have then CP violation through the exchange of physical scalars.

The standard model prediction for the EDM of the electron is rather small, of the order of magnitude of 2×10^{-38} e cm [10]. On the other hand, the experimental upper limit is $\leq 4 \times 10^{-27}$ e cm [11]. Hence, it is interesting that if a large value for the electron EDM (and other elementary particles) is found, it would indicate new physics beyond the standard model. If neutrino remains massless in the standard model the contribution to electric dipole moment will arise at the three loop level [10] (or, at the two level in other models with massless neutrinos [12]). In our case we have

$$d_e = -\frac{em_e}{64\pi^2 m_X^2} \sqrt{2} M_W^2 G_F \, \mathbf{O}_{ee} \, \sin(2\theta_\alpha), \qquad (5)$$

where we have defined

$$\mathbf{O}_{ee} = \sum_j \left[(\mathcal{O}_L^e)^T \mathcal{O}_L^E\right]^2_{ej} \frac{4m_{E_j}^2}{M_U^2} [F_+(m_{E_j}) + F_-(m_{E_j})], \qquad (6)$$

with

$$F_\pm(m_{E_j}) = -\frac{m_X^2}{2m_e^2} \ln \frac{m_X^2}{m_{E_j}^2} + \frac{m_X^2}{2m_e^2}(m_X^2 \pm m_e^2 - m_{E_j}^2)\Delta^{-1} \ln \left[\frac{m_{E_j}^2 + m_X^2 - m_e^2 + \Delta}{m_{E_j}^2 + m_X^2 - m_e^2 - \Delta}\right], \qquad (7)$$

and

$$\Delta^2 = (m_X^2 + m_{E_i}^2 - m_e^2)(m_X^2 - m_{E_i}^2 - m_e^2). \qquad (8)$$

In writing Eq. (5) we have used $M_U^2 = (g^2/4)(|v_\chi|^2 + |v_\rho|^2)$ and $G_F/\sqrt{2} = g^2/8M_W^2$ where M_U is the mass of the doubly charged vector boson that is present in the model. We have chosen $\theta_\eta = \theta_\rho = 0$, and $\theta_\chi = -\theta_\alpha$.

For nondegenerate heavy leptons the mixing angles remain in Eq. (6). For instance, the contribution of E_1, using $m_{E_1} = 50$ GeV, $m_\chi = 100$ GeV [13], we obtain

$$d_e \approx -\left(\frac{M_W^2}{M_U^2}\right) \left[(\mathcal{O}_L^e)^T \mathcal{O}_L^E\right]^2_{e1} \sin(2\theta_\alpha) \times 10^{-17} \text{ e cm}. \qquad (9)$$

Assuming $M_U = 300$ GeV and that the factor with the mixing angles (including $\sin(2\theta_\alpha)$ is $\approx 10^{-8}$ we obtain $d_e \approx 10^{-27}$ e cm, which is compatible with the experimental upper limit of 10^{-27} e cm [11].

For the muon the experimental EDM's upper limit are of the order of $< 10^{-19}$ e cm [14]. It means a constraint in $\left[(\mathcal{O}_L^e)^T \mathcal{O}_L^E\right]^2_{\mu 2} \leq 1$. For the tau lepton a limit of 10^{-17} e cm is derived from $\Gamma(Z \to \tau^+\tau^-)$ [15]. In the present model the EDM of the tau lepton is at least of the order of 10^{-19} e cm.

In this model there is not rare decays such as $\mu \to 3e$ at tree level. However, the same loop diagrams that contribute for the EDM of the leptons imply also magnetic and electric-moment transitions, like $\mu \to e\gamma$. This transitions will constrain the matrix elements $\left[(\mathcal{O}_L^e)^T \mathcal{O}_L^E \right]^2_{\mu 1}$ only.

In the quark sector we have contributions involving the exchange of one simple charged and one doubly charged scalars in the box diagrams that contribute to the ε and ε' parameters of the neutral Kaon system. There are also contributions to the electric dipole moment of the neutron. These issues will be published elsewhere [16].

With three triplets and one sextet which are needed in the model of Ref. [5] it is possible to have truly spontaneous violation of the CP symmetry. In this case, the minimization condition of the scalar potential implies $\text{Im}(v_S v_\eta v_\rho v_\chi) = 0$ [8], with v_S the VEV of one of the neutral component of the sextet which gives mass to the charged leptons. The phenomenology of this model has been studied in Ref. [17].

REFERENCES

1. S. L. Glashow, Nucl. Phys. **22**, 579 (1967); S. Weinberg, Phys. Rev. Lett. **19**, 1264 (1967); A. Salam, *in* Elementary Particle Theory, Ed. by N. Svartholm (Almqviat and Woksell, 1968); S. L. Glashow, J. Iliopoulos and L. Maiani, Phys. Rev. D**D2**, 1285(1970).
2. M. Kobayashi and T. Maskawa, Prog. Theor. Phys. **49**, 652 (1974).
3. T. D. Lee, Phys. Rev. D **8**, 1226 (1973); Phys. Rep. **9**, 143 (1974); S. Weinberg, Phys. Rev. Lett. **37**, 675 (1976).
4. Y-L. Wu and L. Wolfenstein, Phys. Rev. Lett. **73**, 1762 (1994) and references therein.
5. F. Pisano and V. Pleitez, Phys. Rev. D **46**, 410 (1992); R. Foot, O. F. Hernández, F. Pisano and V. Pleitez, Phys. Rev. D **47**, 4158 (1993). See also P. Frampton, Phys. Rev. Lett. **69**, 2889 (1992).
6. V. Pleitez and M. D. Tonasse, Phys. Rev. D **48**, 2353 (1993).
7. J. T. Liu and D. Ng, Phys. Rev. D **50**, 548 (1994).
8. L. Epele, H. Fanchiotti, C. García Canal, D. Gómez Dumm, Phys. Lett. **B343**, 291 (1995).
9. G. C. Branco, Phys. Rev. Lett. **44**, 504 (1980).
10. F. Hoogeven, Nucl. Phys. **B341**, 322 (1990).
11. E. Commins *el al.*, Phys. Rev. A **50**, 2960 (1994); K. Abdullah —it et al., Phys. Rev. Lett. **65**, 2347 (1990).
12. D. Bowser-Chao. D. Chang and W-Y. Keung, Phys. Lett. **79**, 1988 (1997).
13. R. M. Barnett *et al.*, (Review of Particle Physics), Phys. Rev. D **54**, 1 (1996).
14. J. Bailey *et al.*, J. Phys. G **4**, 345 (1978); Nucl. Phys. **B150**, 1 (1979).
15. R. Escribano and E. Massó, Phys. Lett. **301**, 419 (1993).
16. J. C. Montero, V. Pleitez and O. Ravinez, in preparation.
17. D. Gómez Dumm, Int. J. Mod. Phys. **A11**, 887 (1996).

The Dirac Equation as Spontaneously Broken Supersymmetry

Cupatitzio Ramírez

Facultad de Ciencias Físico Matemáticas
Universidad Autónoma de Puebla
A.P. 1364, Puebla 72000, México [1] [2]

Abstract. It is shown that the mass of the quantum mechanical Dirac equation can be interpreted as resulting from spontaneously broken supersymmetry.

The origin of elementary particles mass is an open question whose most popular solution is given by the spontaneous breaking of symmetry. The belief on this mechanism is reflected by the intense search of the associated Higgs particle of the standard model. For field theories on more than two dimensions, Goldstone theorem tells us that when spontaneous symmetry breaking takes place, massless particles, the Goldstone particles, arise in a number equal to the number of broken symmetry directions. However, spontaneous symmetry breaking can be as well observed in classical mechanics. Hence the question of what corresponds to Goldstone theorem in quantum mechanics.

The effects of spontaneous breaking can be studied if we consider effective theories at low energies. These theories can be obtained coupling the theory at low energies to the Goldstone particle [2]. That is, if the lagrangian is an invariant under nonlinear realizations of the symmetry group G, which are linear under the unbroken symmetry subgroup H [2].

In this work, the question of the meaning of the goldstone of supersymmetry, the goldstino, in quantum mechanics. The starting point is the massless spinning point particle, whose quantization delivers the massless Dirac equation. It is well known that this equation posses a local supersymmetric invariance, opposite to the massive case. Consistently coupling this particle to the one dimensional goldstino particle, it is shown that the massive Dirac equation turns out.

The action of the local supersymmetric spinning point particle is

$$L = -ie^{-1}\dot{q}^2 + \frac{1}{2}\chi\dot{\chi} - \frac{1}{2}\psi\dot{\chi}, \qquad (1)$$

---

[1] Partially supported by Conacyt grant 3644E
[2] E-mail: cramirez@fcfm.buap.mx

where χ^μ ($\mu = 0, 1, \ldots, D$), ψ and ζ are anticommutative quantities representing one-dimensional spinors, q^μ are ordinary (commutative) variables, e is the einbein and χ is its supersymmetric partner (gravitino).

This lagrangian is invariant under the transformations

$$\begin{array}{ll} \delta_\zeta q^\mu = \frac{1}{\sqrt{2}} \zeta \chi^\mu & \delta_\zeta \chi^\mu = \zeta e^{-1}(-\dot{q}^\mu + \frac{i}{2\sqrt{2}} \psi \chi^\mu) \\ \delta_\zeta e = -i\zeta\psi & \delta_\zeta \psi = 2\dot{\zeta}. \end{array} \qquad (2)$$

If we set $\delta_\zeta \equiv \zeta \delta_S$, this transformations satisfy the one-dimensional supergravity algebra

$$[\delta_\eta, \delta_\zeta] = -2i\zeta\eta e^{-1}(\partial_t - \frac{1}{2}\psi\delta_S) \qquad (3)$$

Obviously, the variables e and ψ are non dynamical and to them correspond the Hamiltonian $H \equiv \frac{1}{2} p^2 = 0$ and the supersymmetric $S \equiv \chi p = 0$ constraints. where $p^\mu = \dot{q}^\mu$ and the following Dirac brackets are satisfied

$$\{q^\mu, p_\nu\} = \delta^\mu_\nu, \qquad \{\chi^\mu, \chi_\nu\} = \frac{i}{2} \delta^\mu_\nu \qquad (4)$$

from which it turns out that $\{S, S\} = H$.

The quantization of this point particle can be done by the sustitution $\chi^\mu \to \gamma^\mu$, which delivers the massless Dirac equation

$$S = 0 \to \gamma^\mu p_\mu \psi = 0. \qquad (5)$$

If we wish to add a mass term, the usual way is to introduce a χ_5-term [1] of the form

$$L_5 = \frac{k}{2}(i\chi_5\dot{\chi}_5 - i\psi\chi_5 - e) \qquad (6)$$

which is invariant under the local supersymmetry transformations (2), plus

$$\delta_\zeta \chi_5 = \zeta. \qquad (7)$$

With this term, the constraints have the form $H = p^2 + 2k = 0$ and $S = p\chi + \sqrt{2}k\chi_5 = 0$.

Therefore, they can be quantized by the ansatz [1,5]

$$\chi^\mu \to \gamma^5 \gamma^\mu, \qquad \chi_5 \to \gamma^5 \qquad (8)$$

i.e. we get the massive Dirac equation

$$S = 0 \to (\gamma^\mu p_\mu + \sqrt{2}k)\psi = 0. \qquad (9)$$

The transformation law (7) resembles the first term of the one of the goldstino [5]. Apparently, the mass term distroys the local supersymmetry invariance. Indeed, the transformation law (7) satisfies an abelian algebra. However, it turns out that if the χ_5 variable satisfies the nonlinear local supersymmetric transformation law of the goldstino, the lagrangian (6) will be as well invariant.

In order to compute this transformation law, we will follow the superspace language, where this transformation law can be easily computed.

Let us start from the well known supersymmetry quantum mechanical transformations for a superfield $\Phi = A(t) + \theta \phi(t)$. These transformations can be formulated as

$$\delta_\zeta \Phi = \zeta Q \Phi = \zeta(\frac{d}{d\theta} - i\theta\frac{d}{dt})\Phi, \qquad (10)$$

with the covariant derivatives

$$D_\theta = \frac{d}{d\theta} + i\theta\frac{d}{dt} \qquad (11)$$

and which satisfy

$$\{Q, Q\} = -2i\frac{d}{dt}, \qquad \{D, Q\} = 0. \qquad (12)$$

If $\zeta \to \zeta(z) \equiv \zeta(t, \theta)$, we get local supersymmetry transformations $\delta_\zeta \Phi(z) = \zeta^M(z)\partial_M \Phi(z)$ and as usual we introduce a superspace zweibein in order to define covariant derivatives [8]

$$\mathcal{D}_A \Phi(z) = E_A^M(z)\partial_M \Phi(z) \qquad (13)$$

from which a tensor calculus can be developed. In fact, there is a more convenient way to write these transformations, in the "new superspace" [3]. In our case, the supersymmetry transformations can be writen as [4]

$$\delta_\zeta \Phi(z) = \eta_\zeta^M(z)\partial_M \Phi(z) \qquad (14)$$

and the covariant derivatives are

$$\nabla_A \Phi(z) = \nabla_A^M(z)\partial_M \Phi(z) \qquad (15)$$

where $\eta_t^M = \delta_t^M$,

$$\eta_\theta^M = \begin{pmatrix} i\theta e^{-1} \\ 1 - \frac{i}{2}\theta e^{-1}\psi \end{pmatrix} \qquad (16)$$

and

$$\nabla_M^A = \begin{pmatrix} e - i\theta\psi & \frac{1}{2}\psi \\ i\theta & 1 \end{pmatrix}, \qquad (17)$$

where ψ is the superpartner of e and $\nabla_A{}^M \nabla_M{}^B = \delta_A^B$. From these expressions, the invariant density of superspace can be computed to be

$$\mathcal{E} = Sdet(\nabla_M^A) = e - \frac{i}{2}\theta\psi \tag{18}$$

In this way, the superspace massless point particle action can be written as [5]

$$A = \int dt d\theta \mathcal{E} \nabla_\tau \Phi^\mu \nabla_\theta \Phi_\mu. \tag{19}$$

Further, the goldstino superfield can be obtained from its transformation law, which is consistently given by [7,4]

$$\delta_\zeta \Lambda(z) = \zeta(z)[\eta_\theta^\theta(t, -\Lambda) + \eta_\theta^t(t, -\Lambda)\dot{\Lambda}] \tag{20}$$

where $\Lambda(z)|_{\theta=0} = \lambda(t)$ is the goldstino and the corresponding new superspace superfield can be constructed as [8] $\Lambda(z) = exp(\theta\delta_\theta)\lambda(t)$, where $\delta_\theta \lambda(t)$ is given by $\zeta\delta_\theta\lambda(t) \equiv \delta_\zeta \Lambda(z)|_{\theta=0}$. Thus, we obtain

$$\delta_\zeta \lambda(t) = \zeta(1 + \frac{i}{2}e^{-1}\psi\lambda - ie^{-1}\lambda\dot{\lambda} \tag{21}$$

which satisfies the supergravity algebra (3) and from which we get

$$\Lambda(z) = \lambda + \theta(-1 + \frac{i}{2}\lambda e^{-1}\psi + ie^{-1}\lambda\dot{\lambda}) \tag{22}$$

in such a way that a suitable goldstino lagrangean is given by

$$L_\lambda = \frac{1}{2}\int d\theta \mathcal{E}\Lambda = e - i\psi\lambda + i\lambda\dot{\lambda}, \tag{23}$$

which coincides with (6) if we set $\lambda \to \chi_5$.

Therefore, the χ_5 variable corresponds to the goldstino particle and the spinning particle mass term, that is the mass of the spin-1/2 particle, can be interpreted as resulting from the spontaneous breaking of supersymmetry.

REFERENCES

1. F. A. Berezin and M.S. Marinov, *ITEP* 43, Preprint (1976).
2. S. Coleman, J. Wess and B. Zumino, *Phys. Rev.* **177**, 2239 (1969).
3. J. Wess and B. Zumino, *Phys. Lett.* **74B**, 51 (1978).
4. Cupatitzio Ramirez Romero, *Ann. Phys., N.Y.*, **186**, 43 (1988).
5. L. Brink et. al., *Phys. Lett.* **64B**, 435, (1976).
6. J. Grundberg et. al., *Phys. Lett.* **231B**, 61, (1989).
7. E.A. Ivanov and A.A. Kapustnikov, *J. Phys.* **A11**, 2375 (1978), J. Wess in *Quantum Theory of Particles and Fields, Bithday volume dedicated to Jan Lopuszanski* B. Jancewicz and J. Lukierski, Eds., Singapore: World Scientific (1983).
8. J. Wess, J. Bagger, *Supersymmetry and Supergravity*, Princeton, NJ: Princeton Univ. Press (1981).

The CKM matrix from a scheme of flavour symmetry breaking.

E. Rodríguez-Jáuregui and A. Mondragón

Instituto de Física, Universidad Nacional Autónoma de México
Apdo. Postal 20-364, 01000 México, D. F., México.

Abstract. A theoretical $| V^{th}{}_{CKM} |$ mixing matrix which is a function of the four mass ratios and the CP violating phase α is derived from a simple scheme for breaking the flavour permutational symmetry. We assumed that the symmetry breaking pattern is the same in the u and d-sectors, and imposed a phenomenologically motivated constraint on the amount of mixing of singlet and doublet irreducible representations of $S(3)_L \otimes S(3)_R$. A χ^2 fit of the matrix of the absolute values $| V^{th}{}_{CKM} |$ to the experimentally determined $| V^{exp}{}_{CKM} |$ gives the best value for $\alpha = 76.7°$ and the value $J^{th} = -2.18 \times 10^{-5}$ for the Jarlskog invariant in good agreement with the experimental values. The agreement between $| V^{th}{}_{CKM} |$ and $| V^{exp}{}_{CKM} |$ is also very good with $\chi^2 = 0.28$.

Introduction

A number of authors [1] have pointed out that realistic quark mass matrices result from the flavour permutational symmetry $S(3)_L \otimes S(3)_R$ and its spontaneous or explicit breaking. The group $S(3)$ treats three objects symmetrically while the hierarchical nature of the mass matrices is a consequence of the representation structure $\mathbf{1} \oplus \mathbf{2}$ of $S(3)$, which treats the generations differently. Under exact $S(3)_L \otimes S(3)_R$ symmetry the mass matrix is $\mathbf{M_{3q,H}}$ and the mass spectrum for either up or down quark sectors consists of one massive particle in a singlet irreducible representation and a pair of massless particles in a doublet irreducible representation. The third family quarks, t or b are assigned to the $q_{3H}(X)$ invariant singlet representation with mass m_{3q}. The other two families are assigned to $q_{2H}(X)$ and $q_{1H}(X)$, the two components of the doublet representation of $S(3)$ which remain massles. In order to generate masses for the second and third families, we add the terms $\mathbf{M_{2H}}^q$ and $\mathbf{M_{1H}}^q$ to $\mathbf{M_{3H}}^q$. The term $\mathbf{M_2}^q$ breaks the permutational symmetry $S(3)_L \otimes S(3)_R$ down to $S(2)_L \otimes S(2)_R$, it shifts the masses of the second and third family and mixes the singlet and doublet representations of $S(3)$. $\mathbf{M_1}^q$ transforms as the mixed symmetry term in the doublet complex tensorial representation of

$S_{diag}(2) \subset S(2)_L \otimes S(2)_R$. Putting the first family in a complex representation allows us to have a CP violating phase. Then,

$$\mathbf{M}^q = m_{3q} \left(\begin{pmatrix} 0 & A_q e^{-i\phi_q} & 0 \\ A_q e^{+i\phi_q} & 0 & 0 \\ 0 & 0 & 0 \end{pmatrix} + \begin{pmatrix} 0 & 0 & 0 \\ 0 & D_q + \delta_q & B_q \\ 0 & B_q & -\delta_q \end{pmatrix} + \begin{pmatrix} 0 & 0 & 0 \\ 0 & 0 & 0 \\ 0 & 0 & 1 \end{pmatrix} \right) \quad (1)$$

Computing the invariants of $\mathbf{M}^q$, $tr(\mathbf{M}^q)$, $tr(\mathbf{M}^q)^2$ and $det\mathbf{M}^q$ from (1) and comparing with the corresponding expressions in terms of the mass igenvalues $(m_1, -m_2, m_3)$, we get $A_q^2 = \frac{\tilde{m}_{1q}\tilde{m}_{2q}}{1-\delta_q}$, $D_q = \tilde{m}_{1q} - \tilde{m}_{2q} + \delta_q$, $B_q^2 = \delta_q \left((1 - \tilde{m}_{1q} + \tilde{m}_{2q} - \delta_q) - \frac{\tilde{m}_{1q}\tilde{m}_{2q}}{1-\delta_q} \right)$. Where $\tilde{m}_{1q} = m_{1q}/m_{3q}$ and $\tilde{m}_{2q} = m_{2q}/m_{3q}$. If each possible symmetry breaking pattern is now characterized by the ratio $Z_q = (M^q{}_{23}/M^q{}_{22})$, we obtain a cubic equation for δ_q

$$\delta_q \{(1 + \tilde{m}_{2q} - \tilde{m}_{1q} - \delta_q)(1 - \delta_q) - \tilde{m}_{1q}\tilde{m}_{2q}\} - Z_q(-\tilde{m}_{2q} + \tilde{m}_{1q} + \delta_q)^2 = 0. \quad (2)$$

The small parameter δ_q is the solution of (2) which vanishes when Z_q vanishes.

The CKM mixing matrix

The Hermitian mass matrix $\mathbf{M_q}$ may be written in terms of a real symmetric matrix $\bar{\mathbf{M}}_q$ and a diagonal matrix of phases $\mathbf{P}_q$ as $\mathbf{M_q} = \mathbf{P}_q \bar{\mathbf{M}}_q \mathbf{P}_q^\dagger$. Then, the CKM mixing matrix is given by

$$\mathbf{V}_{CKM} = \mathbf{O}_u^T \mathbf{P}^{(u-d)} \mathbf{O}_d \quad (3)$$

where $\mathbf{P}^{(u-d)} = diag[e^{-i(\phi_u - \phi_d)}, 1, 1]$ is the diagonal matrix of the relative phases, and $\mathbf{O}_q$ is the orthogonal matrix that diagonalizes $\mathbf{M_q}$

$$\mathbf{O}^q = \begin{pmatrix} (\tilde{m}_2 f_1/\Delta_1)^{1/2} & -(\tilde{m}_1 f_2/\Delta_2)^{1/2} & (\tilde{m}_1 \tilde{m}_2 f_3/\Delta_3)^{1/2} \\ (C^q \tilde{m}_1 f_1/\Delta_1)^{1/2} & (C^q \tilde{m}_2 f_2/\Delta_2)^{1/2} & (C^q f_3/\Delta_3)^{1/2} \\ -(\tilde{m}_1 f_2 f_3/\Delta_1)^{1/2} & -(\tilde{m}_2 f_1 f_3/\Delta_2)^{1/2} & (f_1 f_2/\Delta_3)^{1/2} \end{pmatrix} \quad (4)$$

where $f_1 = 1 - \tilde{m}_{1q} - \delta_q$, $f_2 = 1 + \tilde{m}_{2q} - \delta_q$, $f_3 = \delta_q$, $\Delta_1 = C_q(1 - \tilde{m}_{1q})(\tilde{m}_{2q} + \tilde{m}_{1q})$, $\Delta_2 = C^q(1 + \tilde{m}_{2q})(\tilde{m}_{2q} + \tilde{m}_{1q})$, and $\Delta_3 = C^q(1 + \tilde{m}_{2q})(1 - \tilde{m}_{1q})$. From eqs. (3)-(4), all entries in the V_{CKM} matrix may be written in terms of four mass ratios: $(\tilde{m}_u, \tilde{m}_c, \tilde{m}_d, \tilde{m}_s)$ and three free real parameters: δ_u, δ_d and $\alpha = \phi_u - \phi_d$. The CP violating phase α measures the mismatch in the $\bar{S}_{diag}(2)$ symmetry breaking in the u and d-sectors. In order to look for a clue about the actual pattern of $S(3)_L \otimes S(3)_R$ symmetry breaking realized in nature, we made a χ^2 fit of the exact expressions for the absolute value of the entries in the mixing matrix, $|V_{CKM}{}^{th}|$, to the experimentally determined values $|V_{CKM}{}^{exp}|$. We left the mass ratios fixed at their central values [5] $\tilde{m}_u = 0.00002$, $\tilde{m}_c = 0.00517$, $\tilde{m}_d = 0.0019$ and $\tilde{m}_s = 0.035$, and we looked for the best values of the three parameters δ_u, δ_d and α, we found the following results

- I.- Excellent fits of similar quality, $\chi^2 \leq 0.3$, were obtained for a continuous family of values of the parameters (δ_u, δ_d).

- II.- In each good quality fit, the best value of α was fixed without ambiguity.

- III.- The best value of α was nearly stable against large changes in the values of (δ_u, δ_d) which produced fits of the same good quality.

- IV.- In all good quality fits, the difference $\sqrt{\delta_d} - \sqrt{\delta_u}$ takes the same value

$$\sqrt{\delta_d} - \sqrt{\delta_u} \simeq 0.040 \tag{5}$$

These results may be understod if we notice that not all entries in $V_{CKM}{}^{th}$ are equally sensitive to variations of the different parameters. Some entries, like V_{us} are very sensitive to changes in α but are almost insensitive to changes in (δ_u, δ_d) while some others, like V_{cb} are almost insensitive to changes in α but depend critically on the parameters δ_u and δ_d. From eqs. (3)-(4), we obtain

$$\begin{aligned} V_{cb}{}^{th} = &-\left\{\frac{\tilde{m}_u(1+\tilde{m}_c-\delta_u)}{(1-\delta_u)(1+\tilde{m}_c)(\tilde{m}_u+\tilde{m}_c)}\right\}^{1/2}\left\{\frac{(\tilde{m}_d\tilde{m}_s\delta_d)}{(1-\delta_d)(1+\tilde{m}_s)(1-\tilde{m}_d)}\right\}^{1/2}e^{i\alpha} \\ &+\left\{\frac{\tilde{m}_c(1+\tilde{m}_c-\delta_u)}{(1-\delta_u)(1+\tilde{m}_c)(\tilde{m}_u+\tilde{m}_c)}\right\}^{1/2}\left\{\frac{\delta_d}{(1+\tilde{m}_s)(1-\tilde{m}_u)}\right\}^{1/2} \\ &-\left\{\frac{\tilde{m}_c\delta_u(1-\tilde{m}_u-\delta_u)}{(1-\delta_u)(1+\tilde{m}_c)(\tilde{m}_u+\tilde{m}_c)}\right\}^{1/2}\left\{\frac{(1-\tilde{m}_d-\delta_d)(1+\tilde{m}_s-\delta_d)}{(1-\delta_d)(1+\tilde{m}_s)(1-\tilde{m}_d)}\right\}^{1/2}. \end{aligned} \tag{6}$$

In the leading order of magnitude, $|V_{cb}|$ is independent of α and given by $|V_{cb}| \approx \sqrt{\delta_d} - \sqrt{\delta_u}$. Hence, good agreement with $|V_{cb}{}^{exp}|$ requires that $\sqrt{\delta_d} - \sqrt{\delta_u} \approx |V_{cb}{}^{exp}| = 0.039$ at least for one pair of values (δ_u, δ_d). In the preliminary χ^2 fit to the data, it was found that (5) is satisfied almost exactly, not just for one pair of values of δ_u and δ_d, but for a continuous range of values of these parameters in which δ_u and δ_d change by more than one order of magnitude. Therefore, equation (5) will be used as a constraining condition on the possible values of (δ_u, δ_d). In this way, we eliminate one free parameter in $V_{CKM}{}^{th}$ without spoiling the good quality of the fit. If instead of taking (δ_u, δ_d) as free parameters, we use the parameters (Z_u, Z_d) to characterize the pattern of symmetry breaking, we should write δ_q as function of Z_q in (5) to restrict the possible values of (Z_u, Z_d). In this way, to each value of Z_u corresponds one value of Z_d. Since Z_u is still a free parameter, to avoid ambiguities, we may further assume that the up and down mass matrices are generated following the same symmetry breaking pattern, that is, $Z_u = Z_d = Z$. The numerical computation of Z was made using the exact numerical solutions of eq. (2). We found $Z_q = \frac{5}{2}$. The corresponding values of (δ_u, δ_d) are $\delta_u = 0.000064$, $\delta_d = 0.002300$.

Once the value of Z is fixed at $5/2$, the theoretical expression for the entries in $V^{th}{}_{CKM}$, obtained from the exact expressions for O_u and O_d, are written in terms of the four mass ratios $(\tilde{m}_u, \tilde{m}_c, \tilde{m}_d, \tilde{m}_s)$ and only one free parameter, namely the CP violating phase α.

We kept the mass ratios fixed at the values given in [5] and made a new χ^2 fit of $|V^{th}{}_{CKM}|$ to the experimental values $|V^{exp}{}_{CKM}|$ [4]. The best value of α was found to be

$$\alpha = 76.77°, \tag{7}$$

with $\chi^2 = 0.28$. The corresponding best value of $|V^{th}{}_{CKM}|$ is

$$|V^{th}{}_{CKM}| = \begin{pmatrix} 0.97531 & 0.22081 & 0.00254 \\ 0.22063 & 0.97457 & 0.03913 \\ 0.00928 & 0.03810 & 0.999231 \end{pmatrix} \tag{8}$$

which is to be compared with [4]

$$|V^{exp}{}_{CKM}| = \begin{pmatrix} 0.9745 - 0.9760 & 0.217 - 0.224 & 0.0018 - 0.0045 \\ 0.217 - 0.224 & 0.9737 - 0.9753 & 0.036 - 0.042 \\ 0.004 - 0.013 & 0.035 - 0.042 & 09991 - 0.9994 \end{pmatrix} \tag{9}$$

we see that the agreement between computed and experimental values of all entries in V_{CKM} is very good [4]. For the Jarlskog invariant we found the value

$$J = -2.128 \times 10^{-5}, \tag{10}$$

in good agreement with current data on CP violation in $K° - \bar{K}°$ mixing system [4].

Aknowledgements

One of us, E. R-J is indebted to Dr. J. R. Soto for useful discussions. This work was partially supported by DGAPA-UNAM under contract No. PAPIIT-IN110296.

REFERENCES

1. H. Harari, H. Haut and J. Weyers, *Phys. Lett. B* **78**, 459 (1978); H. Fritzsch, *Phys. Lett. B* **73**, 317 (1978); H. Fritzsch *Nucl. Phys. B* **155**, 189 (1979); Y. Koide *Phys. Rev. D* **28**, 252, (1983); H. Fritzsch *Proc. Europhysical Topical Conf. on Flavour Mixing in Weak Interactions (Erice)* ed. J. Lemonne, C. vander Velde and F. Verbeure World Scientific. Singapore. 1984. p. 181 ; P. Kaus and S. Meshkov *Phys. Rev. D* **42**, 1563, (1990); H. Fritzsch and D. Holtmannspötter, *Phys. Lett. B* **338**, 290, (1994); H. Fritzsch and Z. Xing *Phys. Lett. B* **353**, 114, (1995)
2. H. Lehmann, C. Newton and T. Wu, *Phys. Lett. B* 384, 249, (1996)
3. Z. Z. Xing J. *Phys. G., Nucl. Part. Phys.* **23**, 1563 (1997)
4. R. M. Barnett et al. *Phys. Rev. D* **1** (1996)
5. H. Fritzsch, *MPI-PhT/96-34, hep-ph/9605388* (1996)

Regge Trajectories and the Renormalization Group

C.R. Stephens, A. Weber, J.C. López Vieyra, S. Dilcher and P.O. Hess[1]

*Instituto de Ciencias Nucleares, UNAM,
Circuito Exterior C.U., A. Postal 70–543, 04510 México D.F., Mexico*

Abstract. The "environmentally friendly" renormalization group is capable of describing the asymptotic behaviour of Green functions in extremely asymmetric kinematical regions. In particular, the Regge limit can be employed to calculate Regge trajectories in perturbation theory containing the information about bound states and resonances. We present results for a bosonic theory with cubic coupling.

Quite apart from the quest for *the* theory of nature underlying the physics accessible in experiments today, there are aspects of already well–established theories which still defy our comprehension or at least our calculational capabilities (a full understanding of which aspects may well turn out to be important for the former quest). Among them is the formation of bound states and the asymmetric kinematical region of large Q^2 and small x in scattering processes (Q^2 and x are the Bjorken scaling variables). These two seemingly unrelated phenomena are beautifully linked by Regge theory.

Let us begin by recapitulating some of the basic ideas in Regge theory (see, for example, ref. [1] for a more detailed exposition). We consider scattering of two particles into two (possibly different) particles, here for simplicity for the case of spinless bosons. The scattering amplitude $A(s,t)$ then depends on the Mandelstam variables s, the square of the center–of–mass energy, and t, the invariant momentum transfer (and the masses of the in– and outgoing particles). Instead of the variable t one may employ the cosine of the c.m.s. scattering angle θ. The fundamental idea of Regge theory now consists in continuing the partial wave amplitudes $a_l(s)$ in the well–known decomposition of the scattering amplitude

$$A(s,t) = A(s,\cos\theta) = \sum_{l=0}^{\infty}(2l+1)\,a_l(s)P_l(\cos\theta) \tag{1}$$

---

[1] e–mail addresses: stephens@nuclecu.unam.mx, axel@nuclecu.unam.mx, vieyra@pythia.nuclecu.unam.mx, dilcher@nuclecu.unam.mx, hess@nuclecu.unam.mx

to arbitrary (complex) values of l. This continuation is made unique by imposing certain natural conditions on the behaviour for asymptotic values of l.

In analogy to the poles of $A(s,t)$ in s corresponding to s–channel bound states, one finds poles in the continued amplitude $a_l(s)$ of the form

$$\frac{\bar{\beta}(s, m_3^2, m_4^2)\, \beta(s, m_1^2, m_2^2)}{l - \alpha(s)} \tag{2}$$

(indices 1 and 2 label the incoming, 3 and 4 the outgoing particles in the s–channel). The (hypothetical) composite particle corresponding to the pole is referred to as a "Reggeon". It is characterized by the *Regge trajectory* $\alpha(s)$ determining the position of the pole in l. For physical values $l = 0, 1, 2, \ldots$, the pole of $a_l(s)$ in s at $\alpha(s) = l$ corresponds to a true bound state (or resonance) with angular momentum l, hence one Reggeon describes a whole family of bound states (and resonances) with different angular momenta. The position of the pole $\alpha(s) = l$ determines the mass M of the Reggeon via $s = M^2$ as a continuous function of l, while the Reggeon's inner structure is encoded in the "vertex" functions $\beta(s, m_1^2, m_2^2)$ continued to off–shell values of m_1^2 and m_2^2.

To determine the function $\alpha(s)$ in a given theory, the most direct way would be to look for the poles of $a_l(s)$. In perturbation theory, to generate such a pole one has to sum an infinite number of Feynman diagrams, a delicate task indeed. A well–known method to systematize such a summation is the Bethe–Salpeter equation, and recently we have proposed to apply the "environmentally friendly" renormalization group directly for the same purpose [2]. Both approaches have their drawbacks. Fortunately, there is an alternative available, to be described in the following.

Let us go back to the scattering amplitude $A(s,t)$ and "cross" the s–channel process $1\,2 \to 3\,4$ to t–channel scattering $1\,\bar{3} \to \bar{2}\,4$ ($\bar{2}$ is the antiparticle of 2) described by the same analytic function $A(s,t)$ with $t > 4m^2$. Relativistic Regge theory predicts, starting from reasonable assumptions about the analytical structure of the amplitude, that $A(s,t)$ is dominated, in the *Regge limit* $t \to \infty$ (corresponding to $x \ll 1$), by Reggeon exchange processes leading to the asymptotic form

$$A(s,t) \sim \bar{\beta}(s, m_3^2, m_4^2)\, \beta(s, m_1^2, m_2^2)\, \frac{t^{\alpha(s)} \pm (-t)^{\alpha(s)}}{\sin(\pi \alpha(s))}, \quad t \to \infty. \tag{3}$$

This opens up the possibility to determine the Regge trajectory (as well as the structure function) from the asymmetric large–t limit of the scattering amplitude.

For $\alpha(s) = 0, 1, 2, \ldots$, the denominator in (3) generates a pole in $A(s,t)$ corresponding to a physical bound state or a resonance in the s–channel as seen before. One of the subtleties of the relativistic theory is the appearance of *signatures*: due to the $\pm$ sign in the numerator in (3) the poles actually appear only for even (signature $+$) or odd (signature $-$) values of $l = \alpha(s)$. In consequence, there will in general be different trajectories for bound states (or resonances) with even and odd angular momenta.

We will show in the following how one can access the Regge limit with the help of "environmentally friendly" renormalization. Let us consider as a toy model a bosonic non–gauge version of QED consisting of a charged and a neutral scalar field with interaction $g_B \phi^\dagger \phi \varphi$. The relevant diagrams in the large–t limit for $\phi\phi \to \phi\phi$ scattering up to one loop are

$$t \to \quad s \uparrow \quad \diagram = \diagram + \diagram + \diagram + \diagram + \ldots$$

$$\xrightarrow{t\to\infty} \quad -\frac{g_B^2}{t}\left(1 + g_B^2 K(s)\ln(-t)\right) + \frac{g_B^2}{t}\left(1 + g_B^2 K(s)\ln t\right), \qquad (4)$$

the other (connected) diagrams being relatively suppressed. The function K corresponds to a *two–dimensional* one–loop diagram (with the horizontal propagators or "d–lines" contracted) owing to the dimensional reduction in the Regge limit. We will denote the expression in the first bracket in (4) as $B_t^B(s,t)$ and the one in the second bracket as $B_u^B(s,t)$.

Clearly, in the limit $t \to \infty$ the perturbative expansions of B_t^B and B_u^B are ill–defined. We renormalize these functions multiplicatively by

$$B_t^{ren}(s,t,\kappa) = Z(s,\kappa)\, B_t^B(s,t) \qquad (5)$$

and similarly for B_u^B (observe that a renormalization of the coupling constant is not necessary in this bosonic theory). Following the philosophy of "environmentally friendly" renormalization [3], we use a fiducial value of t as our renormalization scale κ. The "natural" normalization condition

$$B_t^{ren}(s, t=\kappa, \kappa) = 1 \qquad (6)$$

leads to an anomalous dimension $\gamma(s) = -g_B^2 K(s)$ for B_t^{ren}. Integration of the corresponding renormalization group equation yields

$$A(s,t) = -\frac{g_B^2}{\kappa} Z(s,\kappa)^{-1} \left[\left(\frac{t}{\kappa}\right)^{g_B^2 K(s)-1} + \left(-\frac{t}{\kappa}\right)^{g_B^2 K(s)-1} \right], \qquad (7)$$

where the second term in the bracket results from an analogous renormalization group improvement of B_u^B. Hence we are able to reproduce the form predicted by Regge theory (cf. eq. (3)) and find the explicit expression $\alpha(s) = g_B^2 K(s) - 1$ for the trajectory to one loop together with its signature +.

Let us finally mention some results for the scattering process $\phi^\dagger \phi \to \phi^\dagger \phi$. In this case a mixing of the different effective "contracted" vertices occurs due to processes like $\phi^\dagger \phi \to \varphi\varphi$ and one has to use a matrix renormalization. We find trajectories with both signatures. Fig. 1 shows the real and imaginary parts of α_+ (signature +) for the mass relation $\mu/m = 1/2$ (μ is the mass of φ). Note that in this case one trajectory generates a series of bound states *and* a series of resonances. These results and the underlying theory are presented in much more detail in ref. [4]. The techniques described in this contribution are currently being extended in various directions.

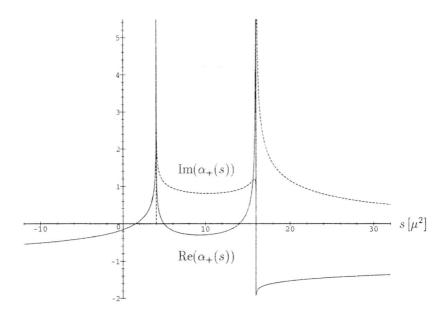

FIGURE 1. The Regge trajectory α_+ in $\phi^\dagger\phi\varphi$ theory. The values of s where $\text{Re}(\alpha_+(s))$ crosses an even integer (with positive slope) give the masses of bound states and resonances.

ACKNOWLEDGEMENTS

This work, as well as A.W.'s participation in the *VI Mexican Workshop*, was supported by CONACyT grant 3298P–E9608. A.W. also acknowledges support by the DAAD and the Mexican Government during part of the work.

REFERENCES

1. Novozhilov, Yu. V., *Introduction to Elementary Particle Theory*, Oxford: Pergamon Press, 1975.
2. López, J. C., Weber, A., Stephens, C. R., and Hess, P. O., in *First Latin American Symposium on High Energy Physics and VII Mexican School of Particles and Fields*, eds. D'Olivo, J. C., Klein–Kreisler, M., and Méndez, H., AIP Conference Proceedings 400, New York: American Institute of Physics, 1997.
3. O'Connor, D., and Stephens, C. R., *Nucl. Phys.* **B360**, 297 (1991); *Int. J. Mod. Phys.* **A9**, 2805 (1994); *Phys. Rev. Lett.* **72**, 506 (1994).
4. Stephens, C. R., Weber, A., López Vieyra, J. C., and Hess, P. O., *Quantum Field Theory in the Limit $x \ll 1$*, UNAM preprint ICN–UNAM–98–01, hep-th/9802031; *Phys. Lett.* **B414**, 333 (1997).

Current status of finite unified models

Liliana E. Velasco-Sevilla

Instituto de Física, UNAM
Apartado Postal 20-364, 01000 México D.F.
México

Abstract. Grand Unified Theories ($GUTs$) can provide predictions for parameters that are free in the Standard Model, but recent experimental data rule them out, favouring supersymmetric $GUTs$. The principle of finiteness in a GUT and the incorporation of supersymmetry can give a robust theoretical frame to make phenomenological reliable models. The purpose of this talk is to comment on the current status of finite unified models and its phenomenological possibilities.

INTRODUCTION

Grand Unified Theories ($GUTs$) can provide predictions for parameters such as $sin^2\theta_W$ and the ratios of the fermion masses, which are free parameters in the Standard Model (SM) of Particles. There exists another principle that certainly points to the direction of further reduction of the free parameters of a gauge theory, namely, the requirement of finiteness. This principle supports the hope that the ultimate theory does not need infinite renormalizations. The predictability of the theory is the most important (phenomenological) motivation for finiteness along with the useful and elegant theoretical idea of finite renormalization.

Supersymmetry (SUSY) plays an important rôle within the Finite Unified Theories ($FUTs$) due to its powerful statements, for instance, within supersymmetric models the absence of quadratic divergences is itself something natural in the model. Besides, there is a strong indication that only SUSY gauge theories can be completely free from ultraviolet divergences. A very interesting fact is that the one-loop finiteness conditions on N=1 SUSY theories automatically ensures also two-loop finiteness [5].

There have been many attempts to classify SUSY theories that satisfy finiteness conditions at one-loop [3] and there is also a theorem which states under which conditions "all-orders finiteness" can be achieved [4], in the sense of vanishing β-functions.

I SUPERSYMMETRIC N=1 FINITE GAUGE THEORIES

The SUSY N=1 Finite Gauge Theories are the simplest supersymmetric finite gauge theories and the most phenomenologically reliable. The models must be chiral and free of anomalies. If we assume the expression of the superpotential to be

$$W = a^i \Phi_i + \frac{1}{2} m^{ij} \Phi_i \Phi_j + \frac{1}{6} C^{ijk} \Phi_i \Phi_j \Phi_k \qquad (1)$$

where a^{ij}, m^{ij} and C^{ij} are invariant tensors and Φ_i are matter fields which transforms under the irreducible representation (irrep.) R_i of the gauge group G. The necessary and sufficient conditions for finiteness to one-loop level are the following:

a) The one-loop finiteness of the gauge fields self-energy, which requires:

$$\Sigma_i l(R_i) = 3 C_2(G) \qquad (2)$$

where $l_i(R_i)$ is the Dynkin index of R_i and $C_2(G)$ is the quadratic Casimir of the operator of the adjoint representation of the gauge group G.

b) The one-loop finiteness of the chiral superfields self-energy. In terms of the cubic coupligs C_{ijk} (Yukawa couplings) appearing in the superpotential given in (1) this condition requires:

$$C^{ikl} C_{jkl} = 2 \delta^i_j g^2 C_2(R_i), \qquad (3)$$

where g is the gauge coupling constant, $C_2(R_i)$ is the quadratic Casimir of the representation R_i, and $C^{ijk} = (C_{ijk})^*$. The condition (3) forbids the presence of singlets with non-zero couplings. Futhermore, it requires that $C^{ikl} C_{jkl}$ is diagonal in its two free indices.

For a theory to be finite at all-loops we must have one-loop finiteness, i.e. the conditions (3) and (2) must be valid, and there must exist an isolated and non degenerate solution to (3). The finiteness conditions at all-loops allows the SUSY breaking probably only with the addition of soft terms [2].

A Finite Unified Models based on SU(5)

From references [3] the difficulties in constructing phenomenologically viable finite unifed theories ($FUTs$) can be studied.

Using $SU(5)$ as a gauge group it has been found that [3], there are only two FUT models which can acomodate three families of fermions. These models contain chiral multiplets 5, $\bar{5}$, 10, $\overline{10}$, 24 with different multiplicities: (4,7,3,0,1) which contains a 24-plet that can be used for the spontaneous symmetry breaking of $SU(5)$ to $SU(3) \times SU(2) \times U(1)$, and (6,9,4,1,0) in which a 24-plet does not exist, and therefore some different mechanism, such as the Wilson flux breaking mechanism, must be incorporated.

The finiteness conditions do not restrict the mass terms, therefore there is freedom in this sector to mix the four Higgs fields and to choose their vacuum expectation values ($vev's$), avoiding problems with too fast proton decay.

Once the model is specified a numerical evolution, with respect to the energy, of the Yukawa and gauge couplings can be performed through the renormalization group equations. The conditions derived from the $FUTs$ at the scale of unification $(1,3) \times 10^{16}$ GeV fix the values of the parameters at the unification point, which are then evolved down to the scale of SUSY breaking, giving the supersymmetric spectrum, to finally evolve them down to the electroweak scale.

If we want to evolve the equations from the electroweak scale up to the unification scale or the SUSY scale we must choose a good "shooting" of the parameters at the electroweak scale, i.e. pick up the best experimental results and then evolve them through the renormalization group equations. Since not all the data are available, for instance the Higgs mass, we have to make a judicious guess according to the experimental bounds, and then evolve them up to the unification scale. The procedure is repeated until we find a convergence in the values of the parameters [6].

B Finite models based on other groups

There have been several propositions on other groups and its representations for constructing finite theories.

The most theoretical appealing seem to be the models based on the $SO(10)$ group, since it is a group of rank 5 with the extra diagonal generator of $SO(10)$ being $B-L$, as in the left-right symmetric groups. Due to its many representantions many different matter contents can be suggested.

The most promising would have the following matter content: 8 10, n 16 and (8-n) $\overline{16}$ with $5 \leq n \leq 8$, but due to the absence of a large representation which can break the gauge group to the SM group, the symmetry breaking must be induced by another mechanism, such as the Wilson line, which would imply that the model comes from a superstring compactification. $SO(10)$ models whose matter content includes the 126 representation have been studied extensibily, however its Dynkin index is larger than the value imposed by the one-loop finiteness condition of the gauge field self-energy (2).

There also exists a model based on E_6 with the matter content n 27 and (12-n) $\overline{27}$ ($7 \leq n \leq 12$), these supermultiplets can acomodate an even number of fermion families.

II DETERMINATION OF THE MINIMUM OF THE POTENTIAL

Once a FUT is specified it is necessary to break the SUSY by an adecuate mechanism in order to give a reliable SUSY spectrum. There has been found that it is also possible to have finiteness in the soft SUSY breaking parameters [2]. Whithin these schemes, in order to find the correct SUSY spectrum, one has to study the low energy effective potential. The central problem here is how to find the minimum of the potential, icluding the radiative corrections, which also corresponds to the physical minimum. There are two dangerous directions in the field-space of the MSSM [7]: the one associated with the existence of color and electric charge breaking minima in the potential deeper than the realistic minimum and the direction in the field-space along which the potential is unbounded from below. A complete study within the finite theories is lacking due to the complexity of the potential.

III CONCLUSIONS

There have been proposed some models, ([1], [2]) which can give good phenomenological predictions, based on the MSSM $SU(5)$. From the studies on the other groups cited above it seems that the $SU(5)$ group could be the most indicated to construct a finite phenomenological reliable model.

For a FUT model to be phenomenological acceptable it is necessary to have an energy spectrum with terms that break SUSY at an appropiate scale, within the experimental bounds, and to give predictions at low energies. For this purpose it is compulsory that the mimimum of the potential includes the radiative corrections and corresponds to a physical minimum, i.e. that does not break the color and electric charge and that it is bounded from below.

REFERENCES

1. D. Kapetanakis, M. Mondragón and G. Zoupanos. Z. Phys. C60:181-186 (1993). M. Mondragón and G.Zoupanos. Nucl. Phys. B (Proc. Suppl.) 37C p. 98-105 (1995).
2. T. Kobayashi, J. Kubo, M. Mondragón and G. Zoupanos. hep-ph/9707425, and see references therein.
3. Hamidi, J. Patera et. al. Phys. Lett. V. 141 B, No. 5,6 (1984). X.D. Jiang and X.J. Zhou (1987). Phys. Lett. B. 216, 1,2 p. 160-166 (1984).
4. C. Lucchesi, O. Piguet and K. Sibold. Helv. Phys. Acta 61 p.321-344 (1988).
5. I. Jack and D.R.T. Jones. Phys. Lett. B 349:294-299 (1995), ibid hep-ph/ 9707278.
6. D.J. Castaño, E.J. Piard, P. Phys. Rev. D V 46, No. 9 p. 3945 (1982), ibid hep-ph/9707425.
7. J.A. Casas, A. Lleyda and C. Muñoz. *Strong Cosntraints on the Parameter Space of the MSSM form Charge and Color Breaking Minima*, hep-ph/9507294.

Classical and path integral analysis of a solvable model with Gribov problems[1]

V. M. Villanueva[2†], Jan Govaerts[3*] and J. L. Lucio M.[4†]

[†]*Instituto de Física de la Universidad de Guanajuato, P.O. Box E-143, 37150, León, Gto., México*
[*]*Institut de Physique Nucléaire, Université catholique de Louvain, B-1348 Louvain-la-Neuve, Belgium*

Abstract. We reanalize a solvable model proposed by Friedberg, Lee, Pang and Ren (FLPR) [1] with special emphasis on the space of gauge orbits and the path integral approach.

INTRODUCTION

As is well kwnow, Gribov [2] obstruction forbids the complete quantization of non-abelian gauge theories. The solution to this problem would be very helpful in the understanding of the non-perturbative regime of QCD for instance. Friedberg *et al.* [1] proposed a quantum mechanical solvable model showing typical characteristics of gauge theories with Gribov problems. According to the authors, in order to get correct results one should include copies belonging to all the gauge orbits in the functional integral. This point of view modifies the present understanding of Gribov amibiguities and the way this problem has to be handled. In Section II we reanalize this model and show that different gauge orbits are picked out depending on the selected gauge fixing condition. In Section III we discuss the path integral formulation of the model using cylindrical coordinates.

I THE MODEL

In cylindrical coordinates, the FLRP model is defined by the fundamental Hamiltonian [3,4]

[1)] Work supported by CONACyT under contract 3979PE-9608
[2)] victor@ifug1.ugto.mx;
[3)] govaerts@fynu.ucl.ac.be;
[4)] lucio@ifug.ugto.mx.

$$H = \frac{1}{2}P_\rho^2 + \frac{1}{2}\frac{P_\varphi^2}{\rho^2} + \frac{1}{2}P_Z^2 + U(\rho) + \xi(P_Z + gP_\varphi), \tag{1}$$

where ρ, φ, Z denote the usual cylindrical coordinates and ξ is a Lagrange multiplier for the first class constraint $\phi = P_Z + gP_\varphi \approx 0$. This Hamiltonian leads to a fundamental first order action that is invariant under the transformations $\rho'(t) = \rho(t), \varphi'(t) = \varphi(t) + \alpha(t), Z'(t) = Z(t) + \frac{1}{g}\alpha(t), \xi'(t) = \xi(t) + \frac{1}{g}\dot\alpha(t)$. Given the gauge invariant boundary conditions $\rho(t_{i,f}) = \rho_{i,f}$, $\varphi(t_{i,f}) = \varphi_{i,f}$, $Z(t_{i,f}) = Z_{i,f}$, and requiring that the gauge parameter $\alpha(t)$ vanishes at the end points, $\alpha(t = t_i) = \alpha(t = t_f) = 0$, the classical solutions to the equations of motion are given by $P_\rho(t) = \dot\rho(t), P_\varphi(t) = L, P_Z(t) = -gL$, as well as,

$$t = t_i + \int_{\rho(t_i)}^{\rho(t)} d\rho \frac{\pm 1}{\sqrt{2(E - U(\rho) - \frac{L^2}{2\rho^2})}},$$

$$\varphi(t) = \varphi_i + L\int_{t_i}^{t} dt' \frac{1}{\rho^2(t')} + g\int_{t_i}^{t} dt'\xi(t'), \tag{2}$$

$$Z(t) = Z_i - gL(t - t_i) + \int_{t_i}^{t} dt'\xi(t'),$$

with the additional condition that $\gamma \equiv \int_{t_i}^{t_f} dt\,\xi(t) = (Z_f - Z_i) + gL(t_f - t_i)$. In these expressions, L and E stand for the constants of motion to be identified with the angular momentum and the energy of the system, respectively.

The parameter defined by $\gamma = \int_{t_i}^{t_f} dt\xi(t)$ is invariant under local gauge tranformations satisfying the boundary conditions specified beneath (1). γ parametrizes the space of gauge orbits in the space of Lagrange multipliers, that is, γ is the Teichmüller parameter of Teichmüller space [3] which in this case is identified with the space of real numbers.

An admissible gauge fixing in Teichmüller space can be achieved with a specification of a section $\xi(t;\gamma)$ which to each possible real value of γ associates one and only one Lagrange multiplier. Recall that in order to get a complete quantization of the model (or any classical model) one must look for gauge fixing conditions that allow the inclusion of all the gauge orbits, this is achieved necessarily by varying the Teichmüller parameter γ over all the real numbers.

II FRIEDBERG ET AL. GAUGES

1) The *time axial* gauge $\xi(t) = 0$. For this $\xi(t)$ it follows that $\gamma = 0$. This implies that the Teichmüller parameter γ takes one single value. Hence, such a condition cannot define an admissible gauge fixing. Indeed, given the boundary conditions, not all possible gauge orbits of the system are accounted for. In other words, there are physically acceptable configurations of the system that are not compatible with the gauge fixing since one must satisfy the condition $\gamma = \Delta Z + gL\Delta t$. Even though

this gauge fixing is complete (no gauge freedom left), it is not global, corresponding to a Gribov problem of the second kind [3].

2) The *axial* gauge $\Omega = Z = 0$. In the conventional approach à la Faddeev [1,3], this is an admissible gauge fixing. Notice that even though it is complete, it is not global. Indeed, the conditions $\{\Omega, \phi\} = 1$ and $\dot{\Omega} = P_Z + \xi = 0$ determine uniquely the Lagrange multiplier $\xi(t) = -P_Z = gL$, leaving no further gauge freedom. This gauge fixing implies a specific time dependence of $\xi(t)$ and also a specific value for the Teichmüller parameter $\gamma = gL\Delta t$. Clearly, this gauge fixing cannot be admissible, since only those configurations for which $Z(t)$ vanishes at all times are included, which is only a subset of all possible configurations.

3) The *lambda* gauge $\Omega = Z - \lambda \rho \cos\varphi = 0$. The analysis of this gauge fixing is performed in analogy to the *axial* gauge. We solve for the gauge degrees of freedom in terms of Dirac brackets. The conclusion is that this is a copy of the *axial* gauge transformation parameter $\alpha(t) = g\lambda\rho(t)\cos\varphi(t)$ as can be seen from the expression that determines $\xi(t)$:

$$\xi(t) = gL + \frac{d}{dt}(\lambda\rho\cos\varphi). \tag{3}$$

Thus, under a gauge transformation with α as specified, we have the following correspondence between these two gauges:

$$Z = 0 \Rightarrow Z = \lambda\rho\cos\varphi. \tag{4}$$
$$\xi = gL \Rightarrow \xi = gL + \frac{d}{dt}(\lambda\rho\cos\varphi).$$

Since the transformation parameter $\alpha = \rho\cos\varphi$ must vanish at the points $t = t_i$ and $t = t_f$, the boundary conditions imply $Z_i = Z_f = 0$. In other words, both gauges describe one and the same single orbit: $\gamma = gL\Delta t$.

4) Gauge fixing conditions of the type $\dot{\xi} = F(\xi)$ were first proposed in Ref. [6], and further investigated in Ref. [3]. In particular the choices $F(\xi) = \beta$ and $F(\xi) = \beta\xi$, β being a constant, are examples of admissible gauge fixing functions leading to one-to-one relations between ξ and the Teichmüller parameter γ [3]. These type of gauge fixing find their best application on the BFV formalism which will be reported elsewhere [10].

The quantum mechanics of the first three gauge fixing conditions discussed in this section were analyzed by FLPR [1]. Notice that our classical analysis shows that these gauge fixings select different sets of gauge orbits, *i.e.* they are not gauge equivalent. The physics that each of these conditions describe is different, and even though they lead to gauge invariant descriptions, they do not provide the correct description of the physical system, since none of them accounts for all physically consistent choices of boundary conditions.

III   PATH INTEGRAL ANALYSIS

Now we consider the path integral formalism for the FLPR model. As in the classical analysis we use cylindrical coordinates which are the most appropriate coordinates for dealing with this model.

Following the Feynman postulate [7] about the path integral, we begin by defining the propagator as

$$K = \sum_{paths} e^{\frac{i}{\hbar} S_{cl}}, \qquad (5)$$

where S_{cl} is the classical action of the system under consideration and the sum is to be considered over all possible classical configurations of the system.

In analogy to the case of spherical coordinates [8,9] we construct a Hamiltonian path integral in cylindrical ones. This path integral has the form

$$K = \int \rho D\rho D\varphi DZ DP_\rho DP_\varphi DP_Z \exp\left\{\frac{i}{\hbar}\int_{t_i}^{t_f} dt\left[P_\rho \dot\rho + P_\varphi \dot\varphi + P_Z \dot Z - H\right]\right\}. \qquad (6)$$

The discrete version of the previos expresion looks like [10]

$$\begin{aligned}
K &= \lim_{N\to\infty} \prod_{j=1}^{N-1} \int_0^\infty \rho_j d\rho_j \int_0^{2\pi} d\varphi_j \int_{-\infty}^{\infty} dZ_j \times \\
&\times \prod_{k=0}^{N-1} (\rho_k \rho_{k+1})^{-\frac{1}{2}} \int_0^\infty \frac{dP_{k+1}}{2\pi\hbar} \sum_{l_{k+1}=-\infty}^{\infty} \frac{1}{2\pi} \int_{-\infty}^{\infty} \frac{dP_{Z_{k+1}}}{2\pi\hbar} \times \qquad (7)\\
&\times \exp\left\{\frac{i\epsilon}{\hbar}\sum_{k=0}^{N-1}\left[P_{k+1}\left(\frac{\Delta\rho_{k+1}}{\epsilon}\right) + l_{k+1}\hbar\left(\frac{\Delta\varphi_{k+1}}{\epsilon}\right) + P_{Z_{k+1}}\left(\frac{\Delta Z_{k+1}}{\epsilon}\right) + \right.\right. \\
&\left.\left. -\frac{1}{2m}P_{k+1}^2 - \frac{(l_{k+1}^2-\frac{1}{4})\hbar^2}{2m\rho_k\rho_{k+1}} - \frac{1}{2m}P_{Z_{k+1}}^2 - U(\mathbf{r}_k)\right]\right\}.
\end{aligned}$$

Now we give the results for the propagators in the *time-axial* and the *axial* gauges, by proceeding in a naive manner, which needs to be justified following the so-called β-limiting procedure [3]. Simply, by imposing by hand the gauge fixing conditions, we implement the *time-axial* gauge on the fundamental Hamiltonian as well as the first class constraint $\phi(P_{Z_{k+1}}, P_{\varphi_{k+1}}) = P_{Z_{k+1}} + gP_{\varphi_{k+1}} = 0$ by including the Dirac delta functions $\delta(P_{Z_{k+1}} + gl_{k+1}\hbar)$ in the path integral for each time slicing (except for the last one). Integrating then over the angular variables φ_j, the path integral reduces to

$$\begin{aligned}
K &= \sum_{l=-\infty}^{\infty} \frac{1}{2\pi} \exp\left\{il(\varphi'' - \varphi') - i\frac{\epsilon}{\hbar}\frac{Ng^2 l^2 \hbar^2}{2m}\right\} \lim_{N\to\infty} \prod_{j=1}^{N-1} \int_0^\infty \rho_j d\rho_j \int_{-\infty}^{\infty} dZ_j \\
&\times \prod_{k=0}^{N-1} (2\pi\hbar)^{-1}(\rho_k\rho_{k+1})^{-\frac{1}{2}} \int_{-\infty}^{\infty} \frac{dP_{\rho_{k+1}}}{2\pi\hbar} \exp\left\{\frac{i\epsilon}{\hbar}\sum_{k=0}^{N-1}\right. \qquad (8)
\end{aligned}$$

$$\times \left[P_{\rho_{k+1}}(\frac{\Delta\rho_{k+1}}{\epsilon}) - gl\hbar(\frac{\Delta Z_{k+1}}{\epsilon}) - \frac{1}{2m}P^2_{\rho_{k+1}} - \frac{(l^2-\frac{1}{4})\hbar^2}{2m\rho_k\rho_{k+1}} - \frac{1}{2}m\omega^2\rho_k^2 \right] \right\}.$$

The next easiest integral to perform is that over the Z variables which lead us to the astonishing result

$$\prod_{j=1}^{N-1}\int_{-\infty}^{\infty} dZ_j \exp\left\{ \frac{i\epsilon}{\hbar}\sum_{k=0}^{N-1}[-gl\hbar(\frac{\Delta Z_{k+1}}{\epsilon})] \right\} \quad (9)$$

$$= \prod_{j=1}^{N-1}\int_{-\infty}^{\infty} dZ_j \exp\left\{ -igl(Z_N - Z_0) \right\} = \infty,$$

from where we conclude that implementing this gauge give us the propagator $K = \infty$, showing the non-admissibility of this gauge.

Now we consider the *axial* gauge $Z = 0$. A long computation [10] results in

$$K_{\Omega=Z} = \frac{m\omega}{\hbar} csc[\omega(t''-t')] \sum_{l=-\infty}^{\infty}\frac{1}{2\pi}\exp\left\{il(\varphi''-\varphi') - \frac{ig^2l^2\hbar(t''-t')}{2m}\right\} \times \quad (10)$$

$$\times \quad I_l\left(-i\frac{m\omega}{\hbar}csc[\omega(t''-t')]\right)\exp\left\{i\frac{m\omega}{2\hbar}ctg[\omega(t''-t')]\Big((\rho')^2 + (\rho'')^2\Big)\right\}.$$

Even when this result is finite, we cannot be sure that it is the correct one until we perform the path integral for gauge fixing in Teichmüller space at the light of the BFV formalism. That result will be reported elsewhere.

REFERENCES

1. Friedberg R., Lee T. D., Pang Y. and Ren H. C., *Ann. Phys.* **246**, 381 (1996).
2. Gribov V. N., *Nucl. Phys.* **B139**, 1 (1978).
3. Govaerts J., *Hamiltonian Quantisation and Constrained Dynamics* (Leuven University Press, Leuven, 1991).
4. Villanueva V. M., Govaerts J. and Lucio J. L. *in Proceedings of the First Latin American Symposium on High Energy Physics and VII Mexican School of Particles and Fields*, Editors: D'Olivo, Klein-Kreisler and Mendez (the American Institute of Physics 1997) p. 563-566.
5. Dirac P. A. M., *Lectures on Quantum Mechanics* (Belfer Graduate School of Science, Yeshiva University, New York, 1964).
6. Teitelboim C., *Phys. Rev.* **D25**, 3159 (1982).
7. R. P. Feynman and A. H. Hibbs, *Quantum Mechanics and Path Integrals* McGraw-Hill Book Company, 1965.
8. D. Peak and A. Inomata, *J. Math. Phys.* **10** (1969) 1422.
9. Ch. Grosche, *DESY 93-141, hep-th/9311001* and Ch. Grosche and F. Steiner *DESY 94-184*.
10. Govaerts J., Lucio J. L. and Villanueva V. M. *in preparation*

Calibration of water Čerenkov detectors for ultraenergetic cosmic rays

L. Villaseñor

Instituto de Física y Matemáticas, Universidad Michoacana de San Nicolás de Hidalgo, Apdo. Postal 2-82, Morelia, Mich., 58040, México

Abstract. A water Čerenkov detector (WCD) prototype of reduced dimensions (cylinder 1.54 m diameter filled with purified water up to a height of 1.2 m) was used to obtain experimental results that validate the concept of remote calibration and monitoring of WCDs based on the use of muons that stop and decay inside the WCD and, in a complementary way, muons that cross the detector. Three clear peaks of PMT charge distributions have been identified which are useful for remote calibration and monitoring of WCDs: one for stopping muons, one for decay electrons and one for crossing muons. This method can be applied to unsegmented as well as segmented detectors.

INTRODUCTION

The Pierre Auger Observatory (PAO) will study cosmic rays reaching the earth with energies above 10^{19} eV [1]. The main purpose of the Auger Project is to study the origin and nature of these cosmic rays by measuring their energy spectra, their arrival direction and their composition. These cosmic rays carry macroscopic energies on microscopic particles and their acceleration mechanism remains as one of the biggest and oldest mysteries in astrophysics. Charged cosmic rays of these extreme energies are constrained to travel distances shorter than about 100 Mpc by the GZK effect [2,3]; as a consequence, their trajectories are deflected less or of the order of a few degrees in the typical intergalactic magnetic fields of a few nanogauss; therefore, their arrival direction provides important information about their source location in the sky. Besides the surface detector, each of the two sites of the PAO will have three fluorescence eyes to measure the longitudinal development of air shower cascades on clear and moonless nights. Both techniques of the hybrid Observatory have been tested extensively in previous detectors: the surface detector system at the Haverah Park [4] detector, and the fluorescence technique at the Fly's Eye [5].

The Auger WCDs will consist of 10 m^2 cross section cylindrical tanks filled with purified water up to a height of 1.2 m; each tank will have with three 8" PMTs looking downwards at the top of the water surface. The detector stations will be instrumented via solar panels. The three PMT's collect the Čerenkov light emitted by particles in the shower front as they cross the tanks with speeds higher than the speed of light in water. Front-end electronics will record the waveforms of each of the three PMT pulses for trigger processing; eventually it will relay the data to a communication system that will send it to a central station for further processing [1].

For a surface array as big as the PAO's, or any other planned in the future, it is indispensable that the initial calibration and the continuous monitoring of each WCD be done in a remote way. In the present paper we present experimental results that validate a novel technique to perform these tasks, it is based on the secondary cosmic ray radiation which has a typical flux of 250 muons/$m^2 s$ at sea level [6]. The low-energy component of these muons, with energies up to about 300 MeV, stop and decay inside the detectors. The corresponding events are characterized by two consecutive PMT pulses separated by a time interval distributed exponentially with a decay constant of about 2 μs. The event trigger requires the coincidence of two consecutive PMT pulses within a time window of 25.6 μs. The signal associated to data collected this way is muon decays, where the first pulse corresponds to the stopping muon, and the second to the decay electron. The background consists of random coincidences of PMT pulses which include PMT noise and PMT after pulses. This muon-decay technique can be complemented with the one based on "lonely muons" crossing the detector, i.e., PMT pulses followed by no further PMT activity in a time window of 25.6 μs.

It is useful to take the Čerenkov signal from muons crossing the detector near vertically as a reference point. The mean values of the charge distributions of decay electrons, as well as stopping and crossing muons are correlated with the mean value of the charge distribution of muons crossing the detector vertically; the latter are selected with a trigger given by the coincidence of signals from two scintillation counters, one placed above and the other below the prototype WCD. A technique based on inclusive muon signals is discussed in [7]. At present it is still debated whether the Auger WCD's should or should not be segmented into three isolated sections each read out by a single PMT to improve the e/μ separation capability of the surface system [8]. One advantage of the methods presented here is that they apply equally well to segmented and unsegmented tanks, with the additional handle in the case of unsegmented tanks of using coincidental pulses in the PMTs to reduce PMT noise.

EXPERIMENTAL SETUP

The WCD we use consists of a reduced-size prototype made of a polyethylene cylinder 1.54 m diameter, white on the inside and black on the outside wall, filled with commercially purified water up to a height of 1.2 m. The inner surface of the tank was covered with a highly reflective tyvek sheet (reflectivity of about 90% in the ultraviolet region of the EM spectrum) cut to the cylindrical shape and kept in place by circular PVC hoses tightly stretched against the inner walls of the tank [9].

Two slightly different experimental setups were used for this work: one for collecting muons decaying inside the detector and the other for collecting muons traversing the detector vertically. The main difference between them is the way the trigger is done. In the first case the trigger is produced by the occurrence of two consecutive pulses at the tank PMT within a time window of 25.6 μs. The trigger for the second case is simply given by the time coincidence of the PMT pulses from two scintillation hodoscopes placed vertically, one above and the other below the tank. A single Hamamatsu 8" R1408 PMT looking downwards was located at the center of the tank with the PMT slightly immersed in the water. The PMT signal was transmitted via RG58 coaxial cable to a digital osciloscope, Tektronix TDS220, and to a commercial NIM [10] discriminator module. The dimensions of the scintillation counters were 20x40 cm^2 and the discrimination threshold of their PMT's was -30 mV. A 1" slab of steel was placed between the bottom of the tank and the lower hodoscope in order to harden the energy spectrum of the triggering muons. The high voltage of the PMT was 1.38 kV, providing a single-pulse rate of 800 Hz for the 1.86 m^2 WCD, i.e., 430 Hz/m^2. A custom-made CAMAC [11] TDC module was used to measure the time interval between consecutive pulses coming out of the discriminator. The CAMAC controller used was the LeCroy 8901; it was connected to a National Instruments GPIB port on a pentium PC running at 133 MHz. The DAQ program was written in LabView, which is a graphic programming package from National Instruments.

The time interval between consecutive pulses and the waveform of the PMT signal for a time interval of about 22 μs prior to and including the second pulse were written to hard disk for about 25 000 events. The second pulse occurring inside a time window of 25.6 μs after the first pulse is used to trigger the DAQ by means of a 10 bit counter fed by a 40 MHz clock in the muon module. The waveform was digitized using the digital scope with a sampling period of 10 ns. The integrated charge and the amplitude of the PMT pulses were obtained offline from the recorded waveform for every event. The analysis was carried out by using PAW [12].

RESULTS AND DISCUSSION

After using a number of cuts, we found out that the cut that most efficiently filters out the structure in the time distribution of pairs of consecutive pulses coming out of the PMT, leaving a clear exponential distribution with a decay constant compatible with the muon lifetime, and with the lowest value for the χ^2 over number of degrees of freedom for the fit, is the one that requires that the integrated charge of the second pulse be greater than the integrated charge of the first pulse.

By comparing the peak position for the charge distribution of the decay electron with the peak position for the charge distribution of muons crossing the WCD vertically which turned out to be 250 pC, we conclude that the average Čerenkov light radiated by the decay electron is 0.18 times the Čerenkov signal coming from muons crossing the WCD vertically. The rate of raw data events was measured by reducing the processing time of each event so as to render a DAQ system with no dead time; it turned out to be 50 Hz. Therefore the rate of events that pass the final selection cuts is about 4 Hz. By extrapolating this number to a full-size WCD of 10 m^2 we expect a rate of muon decay events useful for calibration and monitoring of about 20 Hz. The average Čerenkov signal produced by a stopping muon is 0.70 times that produced by a decay electron or equivalently 0.12 times that produced by an average vertical muon. Additional information can be found somewhere else [13,14].

CONCLUSIONS

We have demonstrated experimentally that the "muon-decay trigger", characterized by the cuts: $C_2 > C_1$ and time between two consecutive pulses $< 8\mu s$, selects stopping-muon events useful for remote calibration and monitoring of WCDs. The muon-decay trigger as described in this paper has already been incorporated to the trigger hierarchy of the front-end electronics for the Auger WCD's [15].

REFERENCES

1. Auger collaboration, "The Pierre Auger Project – Design Report", http://www-td-auger.fnal.gov:82, March 1997.
2. Greisen, K., Phys. Rev Letters 16, 748 (19965)
3. Zatsepin, G.T. and Kuz'min, V.A., JETP Letters 4, 78 (1966)
4. Lawrence, M.A., Reid, R.J.O. and Watson, A.A., J. Phys. G17, 733 (1991).
5. Bird, D.J. et al., Astrophysics J. 424, 491 (1994).
6. Aguilar-Benitez, M. et al, Review of Particle Properties, Phys. Rev. D50, (1994).
7. Kutter, T. et al., Auger Project Technical Note GAP-97-025, (1997)
8. Prike, C., Auger Project Technical Note GAP-97-015, (1997).
9. Alcaraz, F. et al., Auger Project Technical Note GAP-97-050, (1997).
10. Leo, W.R., Techniques for Nuclear and Particle Physics Experiments, Springer Verlag, Berlin, Heidelberg (1987). (1989).
11. CAMAC Updated Specifications, ANSI/IEEE SH-08482 (1982).
12. Brun, R. et al, PAW, Physics Analysis Workstation, Cern Computer Center Program Librery, Q121, (1989).
13. Villasenor, L. et al., Proc. of the 25th. Intl. Cosmic Ray Conference (Durban), Vol. 7, 197 (1997).
14. Villaseñor, L., Proc. of the SOMI XII, Congress on Instrumentation (San Luis Potosi), 400 (1997).
15. Nitz, D., Proc. of the 25th. Intl. Cosmic Ray Conference (Durban), Vol. 5, 293 (1997).

List of Participants

ALTARELLI, Guido	CERN, Genève
ALVAREZ, Erica	UNAM, México
ARAIZA, Moises	BUAP, Puebla
ARANDA, Jorge	FCFM-UMSNH, Morelia
ASTORGA, Francisco	IFyM-UMSNH, Morelia
AVILA, Manuel	FC-UAEM, Cuernavaca
AYALA, Alejandro	ICN-UNAM, México
BARBERIS, Pablo	UNAM, México
BENITEZ, Fernando	UNAM, México
BESPROSVANY, Jaime	IF-UNAM, México
BORUNDA, Mónica	UNAM, México
BOUZAS, Antonio	CINVESTAV, Mérida
BRAMON, Albert	Universitat Autònoma de Barcelona
CÁZAREZ, Federico	UNAM, México
CÁZAREZ, Jesús	UAZ, Zacatecas
CERÓN, Victoria	CINVESTAV, México
CIFUENTES, Edgar	Universidad de San Carlos, Guatemala
CORONA, Gustavo	UMSNH, Morelia
CUAUTLE, Eleazar	CINVESTAV, México
DE COSS, Maritza	CINVESTAV, Mérida
DELÉPINE, David	U.C. Louvain-la-Neuve, Belgique
DÍAZ, Lorenzo	IF-BUAP, Puebla
DILCHER, Sebastián	UNAM, México
D'OLIVO, Juan Carlos	ICN-UNAM, México
DOMÍNGUEZ, Galileo	USLP, San Luis Potosí
ESPINOZA, Catalina	UNAM, México
FÉLIX, Julián	IF-UG, León
FUENTES, Ivette	UNAM, México
GARCÍA, Augusto	CINVESTAV, México
GARCÍA, Fulgencio	UMSNH, Morelia
GARCÍA, Gerardo	IF-UG, León
GARCÍA, José Luis	CINVESTAV, México
GERMÁN, Gabriel	IF-UNAM, Cuernavaca

Goiz, Klara	CINVESTAV, México
Gómez, César	IMFF-CSIC, Madrid
González, Mauricio	UMSNH, Morelia
González Sprinberg, Gabriel	Universidad de la República, Uruguay
Gutiérrez, Alejandro	BUAP, Puebla
Gutiérrez, Benjamín	UNAM, México
Gupta, Virendra	CINVESTAV, Mérida
Hernández, Jaime	BUAP, Puebla
Hernández, Javier	CINVESTAV, México
Herrera, Gerardo	CINVESTAV, México
Herrera, León	UMSNH, Morelia
Hojvat, Carlos	Fermilab, Illinois
Ibáñez, Araceli	BUAP, Puebla
Jerónimo, Gilberto	UMSNH, Morelia
Jerónimo, Yasser	CETIS, México
Kielanowski, Piotr	CINVESTAV, México
Kirchuk, Ernesto	ICN-UNAM, México
Kreisler, Michael	Livermore National Laboratory, California
Larios, Francisco	Michigan State Univ./CINVESTAV, México
Linares, Edgar	CINVESTAV, México
Linares, Román	UNAM, México
López Castro, Gabriel	CINVESTAV, México
López, Mauricio	Centro Nacional de Metrología, México
Lucio, José Luis	IF-UG, León
Macorra, Axel de la	IF-UNAM, México
Madriz, José	UMSNH, Morelia
Magaña, Leonel	CINVESTAV, México
Martínez, Jesús	IPN, México
Martínez, Juan	UNAM, México
Marzari, Alberta	INFN, Torino
Matos, Tonatiuh	IFyM-UMSNH/CINVESTAV, México
Medina, Martín	UMSNH, Morelia
Medina, Pablo	CINVESTAV, México
Mondragón, Myriam	IF-UNAM, México
Mora, Gerardo	CINVESTAV, México
Morelos, Antonio	IF-UASLP, San Luis Potosí
Moreno, Gerardo	IF-UG, León
Moreno, Matías	IF-UNAM, México
Morfín, Jorge	Fermilab, Illinois
Mota, Gregorio	UNAM, México

Muñoz, Laura	IF-UG, León
Murguía, Gabriela	UNAM, México
Nellen, Lukas	IF-UNAM, México
Oso, Alfredo del	UNAM, México
Padilla, Elizabeth	UNAM, México
Pepe-Altarelli, Mónica	INFN, Frascati
Pérez, Bolivia	BUAP, Puebla
Pereyra, Orlando	UNESP, Sao Paulo
Quevedo, Fernando	IF-UNAM, México
Quirós, Mariano	CSIC, Madrid
Quiroz, Norma	CINVESTAV, México
Ramírez, Carlos	Univ. Industrial de Santander, Colombia
Ramírez, Cupatitzio	FCFM-BUAP, Puebla
Raya, Alfredo	UMSNH, Morelia
Ríos, Maribel	UMSNH, Morelia
Riquer, Verónica	UNAM, México
Ritto, Pavel	CINVESTAV, Mérida
Rodríguez, Ezequiel	UNAM, México
Román, Sergio	BUAP, Puebla
Rosado, Alfonso	IF-BUAP, Puebla
Ruiz-Altaba, Martí	IF-UNAM, México
Salazar, Humberto	BUAP, Puebla
Santiago, José	UNAM, México
Segovia, José	UNAM, México
Sheaf, Marleigh	CINVESTAV, México
Sosa, Modesto	IF-UG, León
Stenson, Kevin	Fermilab, Illinois
Stephens, Chirstopher	ICN-UNAM, México
Tavares, Gilberto	CINVESTAV, México
Toledo, Genaro	CINVESTAV, México
Torres, Manuel	IF-UNAM, México
Toscano, Jesús	CINVESTAV, México
Urrutia, Luis	ICN-UNAM, México
Vaandering, Eric	Fermilab, Illinois
Vargas, Marciano	BUAP, Puebla
Velasco, Liliana	UNAM, México
Vergara, José David	ICN-UNAM, México
Villanueva, Román	UG, León
Villanueva, Víctor	IF-UG, León
Villaseñor, Luis	IFyM-UMSNH, Morelia

WALDO, Reynaldo	UMSNH, Morelia
WEBER, Axel	ICN-UNAM, México
ZAS, Enrique	Universidad de Santiago de Compostela
ZAVALA, Erica	UNAM, México
ZEPEDA, Arnulfo	CINVESTAV, México
ZWIEBACH, Bartón	MIT, Massachussetts

Author Index

A

Alessandro, B., 5
Altarelli, G., 93
Avila, M. A., 297
Avilez, C., 80, 320
Ayala, A., 228

B

Borunda, M., 301

C

Cabo, A., 348
Christian, D. C., 80, 320
Church, M. D., 80, 320
Criscuolo, A., 305

D

de Coss, M., 309
Delépine, D., 312
Dilcher, S., 365
D'Olivo, J. C., 316

F

Félix, J., 80, 320
Forbush, M., 80, 320

G

Germán, G., 325
Gómez, C., 243
González-Sprinberg, G. A., 332
Gottschalk, E. E., 80, 320
Govaerts, J., 373
Gutiérrez, A., 336
Gutierrez, G., 80, 320

H

Hartouni, E. P., 80, 320
Hernández, R., 243
Hess, P. O., 365
Holmes, S. D., 320
Huerta, R., 309
Huson, F. R., 80, 320

J

Jensen, D. A., 320
Jiang, Y., 325

K

Knapp, B. C., 80, 320
Kobayashi, T., 344
Kreisler, M. N., 80, 320
Kubo, J., 344

L

López Vieyra, J. C., 365
Lucio M., J. L., 348, 373

M

Marzari Chiesa, A., 5
Mondragón, A., 361
Mondragón, M., 344
Montero, J. C., 353
Moreno, G., 80, 320

N

Nieves, J. F., 316

P

Pepe Altarelli, M., 42
Pleitez, V., 353

Q

Quirós, M., 143
Quiroz, N., 279

R

Ramírez, C., 357
Ravínez, O., 353
Rodríguez-Jáuregui, E., 361
Rosado, A., 336
Rosenbaum, M., 305
Ruiz-Altaba, M., 301

S

Salazar, H., 64
Stephens, C. R., 365
Stern, B. J., 80, 320

U

Uribe, J., 80, 320

V

Velasco-Sevilla, L. E., 369
Vergara, J. D., 305
Villanueva, V. M., 348, 373
Villaseñor, L., 378

W

Wang, M. H. L. S., 80, 320
Weber, A., 365
Wehmann, A., 80, 320
White, J. T., 80, 320
Wiencke, L. R., 80, 320

Z

Zas, E., 185
Zoupanos, G., 344
Zwiebach, B., 279